GROUNDWATER SYSTEMS PLANNING AND MANAGEMENT

Robert Willis

Department of Environmental Resources Engineering
Humboldt State University

William W-G. Yeh

Department of Civil Engineering
University of California, Los Angeles

Prentice-Hall, Inc.
Englewood Cliffs, New Jersey 07632

Library of Congress Cataloging-in-Publication Data

Willis, Robert (date)
Groundwater systems planning and management.

Includes bibliographies and index.
1. Water, Underground—Management—Mathematical models. 2. Water, Underground—Quality—Mathematical models. 3. Groundwater flow—Mathematical models.
I. Yeh, William W-G. II. Title.
TD403.W55 1987 333.91'0415 86-16893
ISBN 0-13-365651-9

Cover design: Lundgren Graphics, Inc.

Manufacturing buyer: Rhett Conklin

Printed in the United States of America

10 9 8 7 6 5 4 3 2 1

ISBN 0-13-365651-9 025

PRENTICE-HALL INTERNATIONAL (UK) LIMITED, *London*
PRENTICE-HALL OF AUSTRALIA PTY. LIMITED, *Sydney*
PRENTICE-HALL CANADA INC., *Toronto*
PRENTICE-HALL HISPANOAMERICANA, S.A., *Mexico*
PRENTICE-HALL OF INDIA PRIVATE LIMITED, *New Delhi*
PRENTICE-HALL OF JAPAN, INC., *Tokyo*
PRENTICE-HALL OF SOUTHEAST ASIA PTE. LTD., *Singapore*
EDITORA PRENTICE-HALL DO BRASIL, LTDA., *Rio de Janeiro*

To Peggy, Jennie, and all the children

CONTENTS

3. GROUNDWATER QUALITY—THE MASS TRANSPORT PROBLEM

4. NUMERICAL METHODS IN GROUNDWATER MANAGEMENT

5. OPTIMIZATION METHODS FOR GROUNDWATER MANAGEMENT

6. GROUNDWATER SUPPLY MANAGEMENT MODELS

7. GROUNDWATER QUALITY MANAGEMENT MODELS

8. THE INVERSE PROBLEM IN GROUNDWATER SYSTEMS

PREFACE

The management of groundwater resources involves the allocation of groundwater supplies and water quality to competing water demands and uses. The resource allocation problem is characterized by conflicting objectives and complex hydrologic, environmental, and economic constraints. The development of mathematical simulation models in the early 1970s provided groundwater planners with quantitative techniques for analyzing alternative groundwater pumping or recharge schedules and identifying the probable environmental impacts associated with subsurface waste disposal. Costs and benefits could, in principle, be developed for each planning, design, or management alternative and optimal control schedules could be developed for the groundwater system.

Although simulation models provide the resource planner with important tools for managing the groundwater system, the predictive models do not identify the optimal groundwater development, design, or operational policies for an aquifer system. Instead, the simulation models provide only localized information regarding the response of the groundwater system to pumping and/or artificial recharge. In contrast, groundwater optimization models can identify the optimal groundwater planning or design alternatives in the context of the system's objectives and constraints.

The development and application of groundwater optimization models for the control of groundwater hydraulics and water quality, and the inverse problem of parameter identification, are the primary emphasis of this book. The text is designed to introduce readers to the methods and approaches of systems analysis techniques applied to groundwater management problems. Our objective has been to incorporate in each chapter the current developments in groundwater optimization modeling rather than providing a detailed review or survey of the available literature. The modeling techniques discussed for the most part have been used in field applications in Taiwan, China, and Chile. We recognize that the groundwater optimization models are only a part of the solution to regional water resources planning problems. Generally, these problems require an interdisciplinary team of economists, social scientists, hydrologists, and water resources planners. And, as a result, the models, methods, and techniques presented here have to be considered in the more general framework of integrated basinwide water resources planning.

Groundwater optimization models are predicated on numerical models of the aquifer system. The imbedding, response, or transfer equation approach is emphasized in the development of the management models for groundwater planning. Because these techniques, and the underlying equations describing flow and mass transport in confined and unconfined aquifer systems, are pivotal in the development of the management models, Chapters 2, 3, and 4 summarize the governing equations of the groundwater system and survey current numerical methods. Chapter 5 presents some of the more common optimization methods and algorithms that have been used for the solution of groundwater management models. The methods include, for example, linear and quadratic programming, dynamic programming, penalty function methods, and an overview of multi-objective optimization and projected Lagrangian methods. The book requires an introductory background in calculus, matrix algebra, hydrology, and operations research.

The problems associated with the regional management of groundwater supplies are presented in Chapter 6. The chapter discusses the general groundwater allocation problems, optimal groundwater development and operation, the capacity expansion problem, and the conjunctive management of groundwater and surface water supplies. Numerous example problems demonstrate the application of optimization modeling for the solution of these problems.

Groundwater quality management is discussed in Chapter 7. Several numerical methods are presented for the solution of the general mass transport problem in porous media. The techniques include the Galerkin finite-element method, finite-difference techniques, the method of characteristics, and the multiple cell balance method. The problems of numerical dispersion and oscillations associated with the solution of the mass transport problems are also presented. The chapter concludes with the development and application of a conjunctive water supply and water quality management model.

In Chapter 8, parameter identification methods are classified under the error criterion used in the formulation of the inverse problem. The problem of ill-posedness in connection with the inverse problem is addressed. Typical inverse solution techniques are outlined. Several example problems illustrate the application of the Gauss-Newton algorithm for the identification of the transmissivity distribution in regional groundwater systems. The techniques used to compute the sensitivity coefficients are discussed. The problems of parameter uncertainty and optimum parameter dimension are addressed, as are statistical methods that can be used to characterize the parameter uncertainty.

The material contained in the text may be taught in several ways to meet the interests and/or needs of the reader or instructor. The authors have taught one-quarter courses in groundwater systems at Humboldt State University and UCLA, respectively. The course normally includes Chapters 1, 2, 4, and 6. A two-quarter or one-semester course would include the chapters on inverse problem of parameter identification and groundwater quality modeling.

Many individuals have contributed to the preparation of the book. The initial impetus for the text was provided by Pete Loucks who wanted to include some groundwater material in his book on water resource systems planning while the senior author was at Cornell University. The authors would also like to thank Barbara Smith of Humboldt State University and Debby Haines of UCLA who typed, word processed, and edited many drafts of the book. We also thank our colleagues and students at other universities for their contributions to the text. Special thanks are due to Miguel Mariño, Wen-Sen Chu, Tom Maddock, III, Brad Finney, LaDon Jones, Philip Liu, John Dracup, James McCarthy, Tracy Nishikawa, and Chuching Wang for their review and criticisms of earlier versions of the text. The remaining errors are, of course, our responsibility. We welcome any comments that will improve future editions of the book.

Robert Willis
William W-G. Yeh

1. GROUNDWATER RESOURCES

1.1 INTRODUCTION

In the hydrologic cycle, groundwater occurs whenever surface water occupies and saturates the pores or interstices of the rocks and soils beneath the earth's surface. The geologic formations that are capable of storing and transmitting the subsurface water are known as *aquifers* when the groundwater can be removed economically and used as a source of water supply (Todd, 1980). A geographic area, watershed, or drainage basin may be underlain by a series of aquifers, semipermeable formations (aquitards), or aquicludes, which may contain, but are incapable of transmitting water. Each formation has varying transmissive, storage, and water quality properties that affect the response of the basin to pumping and artificial recharge. The formations are collectively referred to as a *groundwater reservoir* or *groundwater system* (see Figure 1.1).

Groundwater reservoirs are found in a variety of geologic formations. Among the most common are unconsolidated or partially consolidated sand and gravel deposits, sandstone formations, carbonate (limestone) formations, or faulted (or fractured) crystalline or metamorphic rocks. In the United States alone, the water resource contained in the void space, joints, or fractures of these formations is twenty times greater than the available surface water sources (U.S. Water Resources Council, 1973).

Groundwater systems are traditionally developed as sources of domestic, industrial, or agricultural water supply. The generally good quality of the water and its accessibility in many regions of the world, where surface water is nonexistent or extremely costly to develop, have been important factors in stimulating the development of this relatively low-cost, reliable water resource.

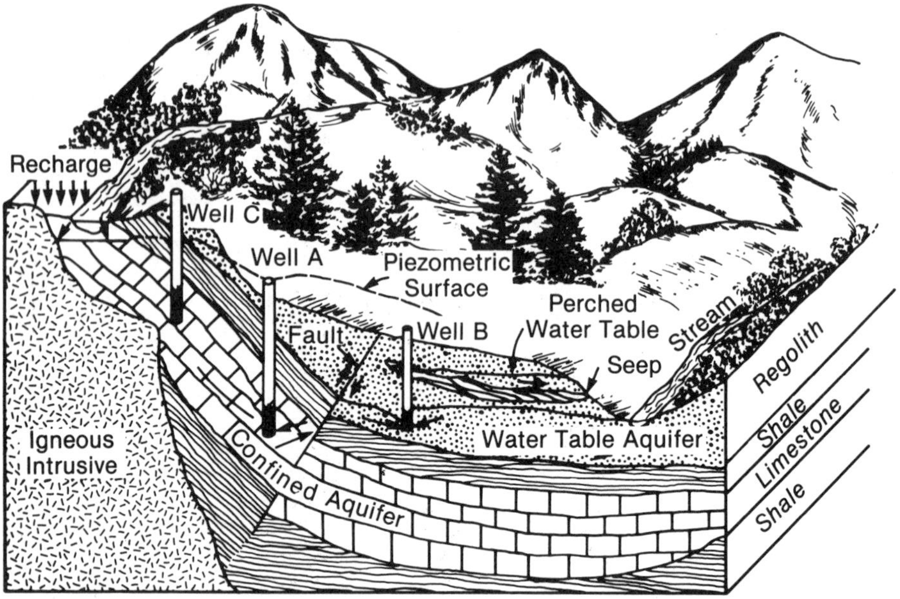

FIGURE 1.1 Groundwater system (Adapted from D.B. McWhorter and D.K. Sunada, *Groundwater Hydrology and Hydraulics*, 1977, Figure 1-4, with permission from Water Resources Publications, Littleton, Colorado 80161)

Groundwater systems can also provide temporary or long-term storage and treatment of wastewater. The injection of oil field brines and radioactive substances in deep saline formations can, under controlled conditions, isolate these materials from potable surface and groundwater supplies (see Figure 1.2). The natural filtering capacity of the porous media, adsorption and ion exchange processes, dilution, and chemical and biochemical reaction can also reduce nutrients, carbonaceous oxygen demand, and certain metals and bacteriological pathogens (EPA, 1973). This water quality resource, which is analogous to the assimilative capacity of surface waters, is utilized in basin spreading, overland flow, and spray irrigation waste treatment systems (see Figure 1.3).

An aquifer system can also represent an energy resource. In regions of the world with exceptionally high geothermal gradients, as for example in Wairakee, New Zealand, and the Geysers in northern California, steam and hot water extracted from the reservoirs may be used for power production (see Figure 1.4). Shallow aquifers are also used for temporary heat storage. When, for example, energy production exceeds demand, the surplus energy can be used to heat water which is then injected into the aquifer for short-term storage. Heat losses of less than 20 percent have been reported in preliminary field experiments (Meyer, 1976).

Groundwater reservoirs that underlie a stream, lake, or river may also be hydraulically coupled to the surface water system. For example, in the northwestern

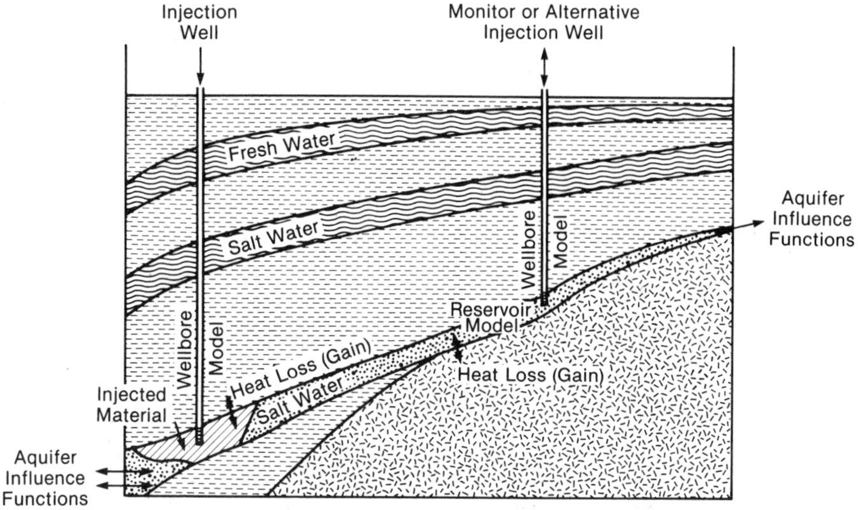

FIGURE 1.2 Deep well injection (Intercomp, Inc., 1976)

United States, groundwater usually provides the baseflow that sustains surface water flow during summer periods. Conjunctive groundwater and surface water management exploits the hydraulic and water quality interaction between the resources. During periods of low streamflow, groundwater may be used to satisfy municipal, industrial, or domestic water demands. In periods of high flow, surface water can be diverted to wells, spreading basins, or stream channels to recharge groundwater basins.

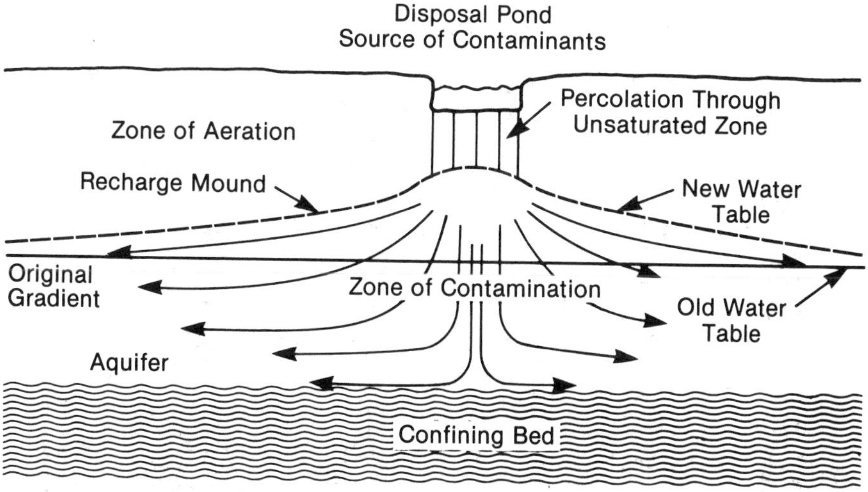

FIGURE 1.3 Spreading basin for wastewater disposal

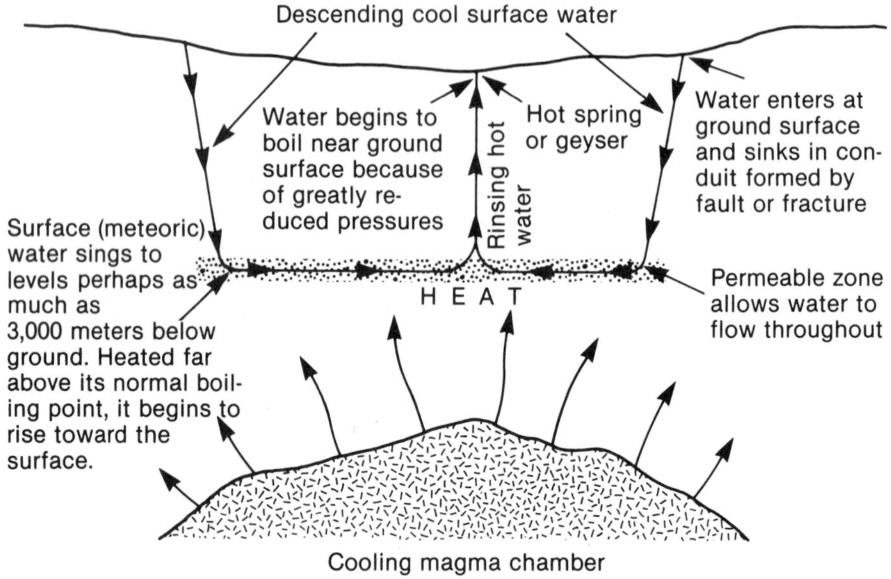
Descending cool surface water
Water begins to boil near ground surface because of greatly reduced pressures
Rinsing hot water
Hot spring or geyser
Water enters at ground surface and sinks in conduit formed by fault or fracture
Surface (meteoric) water sings to levels perhaps as much as 3,000 meters below ground. Heated far above its normal boiling point, it begins to rise toward the surface.
HEAT
Permeable zone allows water to flow throughout
Cooling magma chamber

FIGURE 1.4 A hydrothermal system (Adapted from D.K. Todd, *Groundwater Hydrology*. New York: Wiley, 1980, Figure 2.17. Copyright © 1980 by John Wiley & Sons, Inc. Reprinted with permission)

Groundwater management is broadly concerned with the evaluation of the environmental, hydrologic, and economic impacts and trade-offs associated with the development and allocation of groundwater supply and quality to competing water uses or demands. The analysis of these management and planning problems is predicated on a system's representation of the underlying physical, chemical, and hydraulic transport processes occurring within the groundwater basin.

The groundwater system is defined by:

1. The set of controlled and partially controlled inputs to the system. For example, subsurface inflows, natural recharge, precipitation, and replenishment from irrigation return flows, streams, and artificial recharge practices are major inputs to the aquifer system.
2. The system outputs, which include subsurface outflows, discharges to surface waters, naturally occurring springs, and evapotranspiration losses.
3. The parameters of the groundwater system. The parameters define the flow, quality, and thermal properties of the aquifer system, e.g., the storativity, transmissivities, and dispersion parameters.
4. The control or decision variables. These decisions detail the pumping, injection, and artificial recharge schedules of the ground and surface water system.

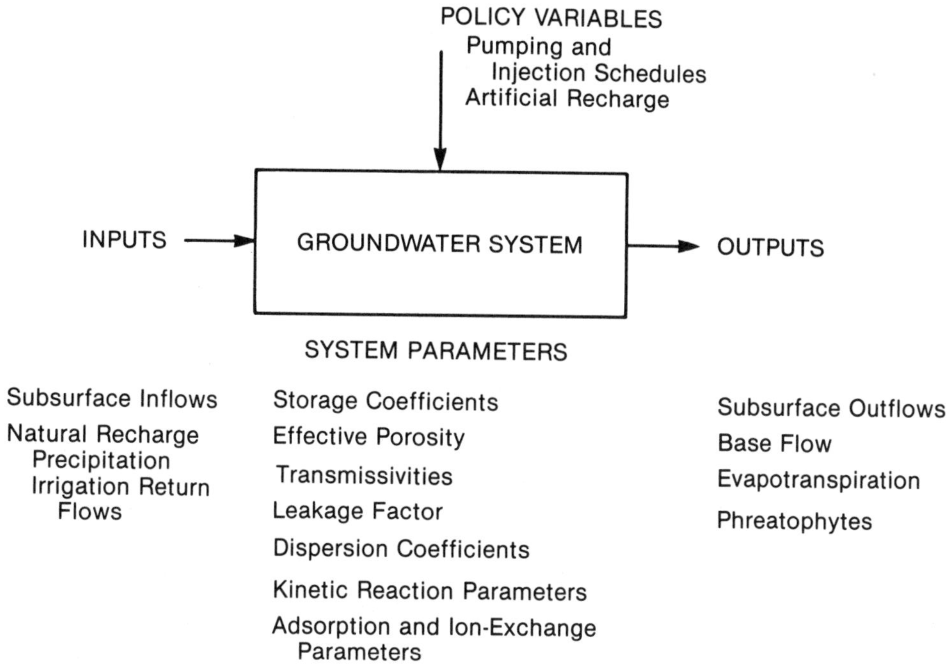

FIGURE 1.5 The groundwater system

5. The state variables that characterize the condition of the system, e.g., the hydraulic head, pressure, or temperature distribution or the concentrations of all constituents in the groundwater system.

An example groundwater system is shown in Figure 1.5.

1.2 GROUNDWATER SYSTEMS—AN OVERVIEW

Groundwater systems can be classified by considering the properties of the parameters, state, and decision variables. Typically, groundwater systems are *distributed parameter* systems. That is, the parameters vary spatially in the basin. Hydraulic parameters that vary spatially are also said to be nonhomogeneous. In contrast, an aquifer where the parameters do not vary spatially (but may vary temporally) is a *lumped parameter* system.

The groundwater system is also a dynamic and possibly nonlinear system. The transient response of the groundwater system may be a linear or a nonlinear function of the state and decision variables of the problem. The prediction of regional unconfined flow is, for example, a nonlinear, time-varying problem.

The inputs, boundary conditions, initial conditions, and, possibly, the parameters of the groundwater system may also be considered as random variables. The state variables of the system are then also random. If, for example, the annual groundwater recharge and/or demands for groundwater are random events, then water levels in wells will exhibit stochastic fluctuations.

In the analysis of groundwater systems it is also useful to consider three classes of management problems: the instrument problem, the parameter estimation or inverse problem, and the prediction problem. In the instrument or detection problem, the set of inputs to the system is unknown. The identification of the recharge or leakage in semiconfined or unconfined aquifers, from the response properties of the aquifer system, is an example of the instrument problem.

The inverse problem is concerned with the estimation of the system parameters and, possibly, unknown boundary and initial conditions from field observations of the state variables of the system. Assuming that all inputs and outputs are known with some degree of reliability, the inverse problem is the identification of the unknown parameters embedded in the partial differential equations that characterize the flow or mass transport occurring in the groundwater system. This is an extremely important problem in groundwater management and, in Chapter 8, algorithms will be presented for the solution of the inverse problem in regional groundwater systems.

The prediction problem is the third systems problem that occurs in groundwater management. The prediction of the response or dynamic behavior of the system is obtained from the analytical or numerical solution of the ordinary or partial differential equations characterizing the groundwater system. The prediction problem assumes, however, that all the system parameters and inputs are known throughout the aquifer system. The predictive model may be used in the simulation or the optimization of the groundwater system.

1.3 MODEL FORMULATION AND DEVELOPMENT

Simulation and optimization models of groundwater hydraulics or quality are mathematical representations of the groundwater system. The development and application of these models—the model building process—consists of several interrelated stages. The process is summarized in Figure 1.6.

Groundwater models are based on the available hydrologic and geohydrologic information. And, in this first phase of the modeling process, well logs, pumping test data, groundwater contour maps, water levels, precipitation, streamflow, and recharge information are compiled and statistically analyzed to determine the probable aquifer parameters, recharge conditions, initial and boundary conditions, and current groundwater extraction patterns. On the basis of the statistical reliability of this information, a choice has to be made regarding the most appropriate and realistic type of mathematical model. The model choice problem, the second phase of the process, is characterized by a hierarchy of mathematical models. For example,

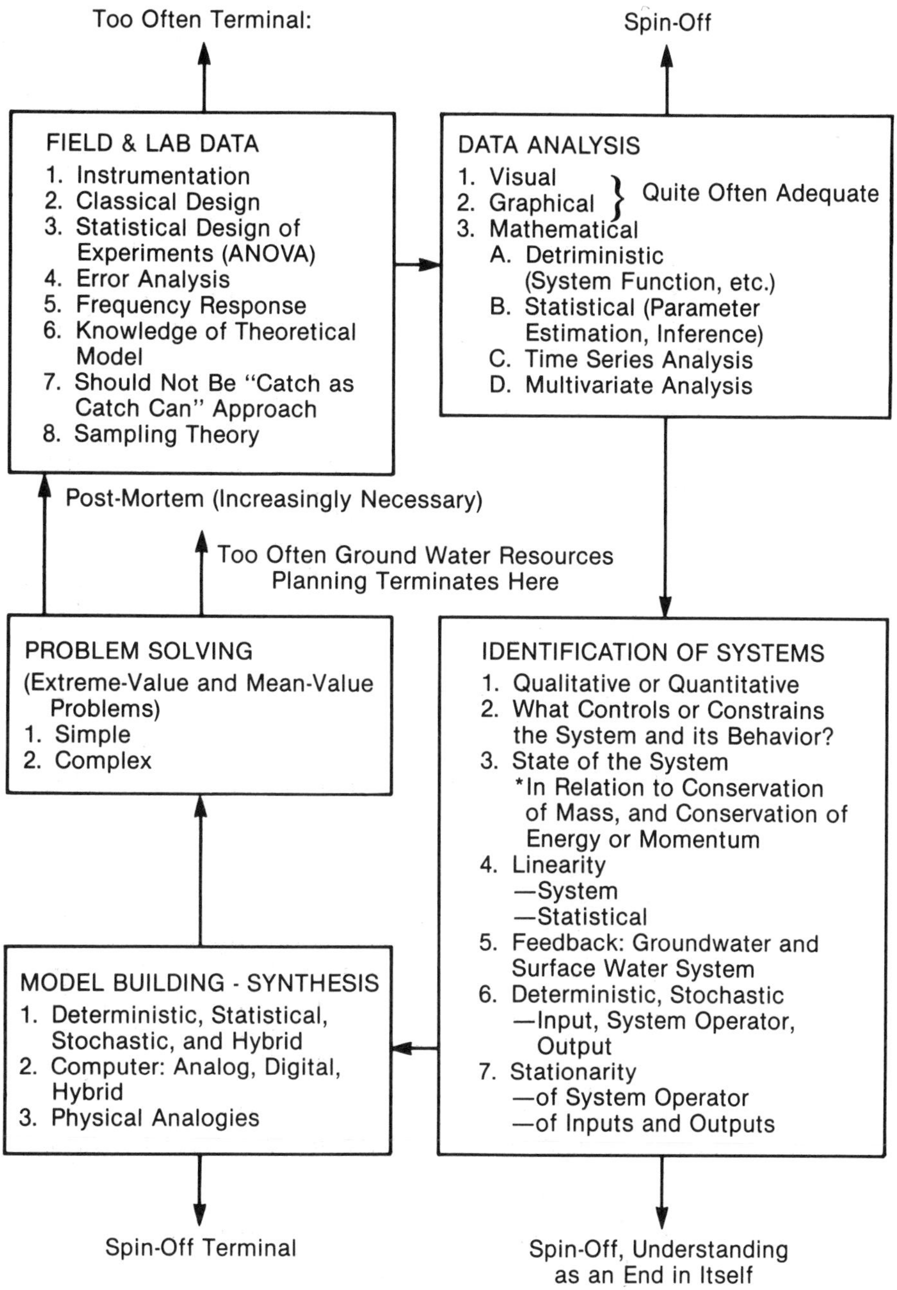

FIGURE 1.6 The modeling process (Adapted from C.C. Kisiel and L. Duckstein, "Operations Research Study of Water Resources—Part II: Case Study of the Tucson Basin (Arizona)." *The Water Resources Bulletin*, 6(6): 857–867, 1970. Reprinted with permission from the Water Resources Association)

in the case of limited or imprecise information regarding the underlying physical processes occurring in the aquifer system, a statistical or time series approach may be warranted. The analysis of trends, periodicities, and the correlation structure of the historical records may further define the mechanisms responsible for groundwater movement and water quality variations. From a planning perspective, however, mathematical models of the groundwater flow or mass transport will consist of:

1. Algebraic, differential, or integral equations that characterize the flow or transport processes. These equations relate the state variables of the system (i.e., hydraulic head or mass concentration), the policy or decision variables that provide control over the state variables (pumping or injection rates), the parameters that define the flow, and quality and properties of the system (storage coefficients, transmissivities, and dispersion coefficients).
2. Boundary conditions that define the inputs to the system such as recharge (precipitation and subsurface inflows) and outputs such as discharges to surface waters or evapotranspiration.
3. Initial conditions that portray the state of the system prior to development or operation of the aquifer system.

The third phase of the model-building process is the calibration and validation of the mathematical model. Model calibration compares the model's response or predictions with the historical groundwater data. Error criteria based on the differences between predicted and historical groundwater levels can be used to judge the predictive capability of the model. If the predictions differ significantly from the historical groundwater conditions, then the parameters, boundary, and initial conditions may be systematically varied to improve the model's performance. However, the possibility exists that, although the model's performance may be acceptable, the parameter values may be physically unrealistic. In this event, the underlying assumptions of the model would have to be re-examined to determine their appropriateness in the context of the validation results. This feedback element of the methodology typifies the iterative nature of the model-building process.

Differing sets of baseline hydrologic and geohydrologic data should be available for the calibration and validation of the mathematical model. Using the parameters and boundary conditions determined from the calibration of the model, the validation process predicts groundwater levels or water quality concentrations with a new set of surface and groundwater hydrologic data. Again, the model's performance is evaluated on the basis of how well, in a statistical sense, the model predictions match the observed historical conditions.

Following the calibration-validation process, the mathematical model can be used for the prediction (simulation) and/or optimization of the groundwater resources of the aquifer system. In a simulation approach, a set of planning, design, or operational policies can be analyzed by examining the simulated response of the aquifer system to the proposed management alternatives. Costs and benefits can be determined and sensitive hydrologic or environmental areas of the groundwater systems can be identified. From this information, the "best" management plan can be found. It should be emphasized, however, that the policy is optimal only in relation to the other simulated policies. Furthermore, simulation analysis provides only localized or limited information regarding the response properties of the system, and the possible hydrologic and environmental trade-offs of the groundwater system.

In contrast, optimization models identify the optimal planning, design, or operational policies within the context of the system's quantifiable objectives and hydrologic, economic, and environmental constraints. As a byproduct of the optimization process, the trade-offs (shadow prices) are also determined. Multi-objective programming techniques can also be used to delineate the set of noninferior solutions of the management problem. However, the same equations that would normally be used in a simulation approach can be used directly in an optimization analysis. The embedding, transfer, or response equation approach allows optimization models to capture all of the information normally associated with a simulation analysis.

It is this later approach, the response equation methodology, that is the common element of all the groundwater planning and management models presented in this text. Specifically, the planning models that will be discussed are:

1. The optimal dynamic management (development and utilization) of groundwater supplies.
2. The conjunctive management of ground and surface water supplies.
3. Regional groundwater quantity and quality management.
4. The parameter identification problem in regional groundwater basins.

The later chapters of the book detail the applications of these techniques to groundwater supply and quality management problems. Although the introductory chapters develop the underlying hydraulic and mass transport equations of the groundwater system, the reader is assumed to have an introductory background in groundwater hydrology (see, for example, Todd, 1980; Freeze & Cherry, 1979, or McWhorter & Sunada, 1977) and operations research, as presented by Haith (1982).

1.4 REFERENCES

Environmental Protection Agency. *Wastewater Treatment and Reuse by Land Application—Vol. II*. EPA-660/2-73-0066, August 1973.

Freeze, R.A., & Cherry, J.A., *Groundwater*. Englewood Cliffs NJ: Prentice-Hall, 1979.

Haith, D.A. *Environmental Systems Optimization*. New York: John Wiley and Sons, 1982.

Intercomp Resources Development and Engineering, Inc. "A Model for Calculating Effects of Liquid Waste Disposal in Deep Saline Aquifers, Parts I and II." *WRI-76/056*. Reston VA: United States Geological Survey, 1976.

Kisiel, C.C., & Duckstein, L. "Operations Research Study of Water Resources—Part II: Case Study of the Tucson Basin (Arizona)." *Water Resources Bulletin*, 6(6): 857–867, 1970.

McWhorter, D.B., & Sunada, D.K. *Groundwater Hydrology and Hydraulics*. Fort Collins CO: Water Resources Publications, 1977.

Meyer, C.F. "Status Report on Heat Storage Wells," *Water Resources Bulletin*, 12(2): 237–252; 1976.

Todd, D.K. *Groundwater Hydrology*. New York: John Wiley and Sons, 1980.

U.S. Water Resources Council. *Essentials of Groundwater Hydrology Pertinent to Water Resources Planning*. Bulletin No. 16. Washington DC, 1973.

2. GROUNDWATER FLOW EQUATIONS

2.1 INTRODUCTION

The mathematical description of groundwater flow in confined and unconfined aquifers is predicated on the fundamental physical processes of conservation of mass, energy, and momentum. The objective of this chapter is to develop from these principles the underlying equations describing groundwater flow in regional aquifer systems. We will assume, however, an isothermal groundwater system; enthalpy or temperature variations will be neglected in the analysis.

We begin by considering a momentum balance for the aquifer system. Rather than developing the momentum equations directly from the Navier-Stokes equations, we approach the problem empirically via Darcy's law. The interested reader is referred to Bear (1972) for a detailed development of the momentum equations.

2.2 DARCY'S LAW

Groundwater flows from states of higher energy levels to states of lower energy. The energy state at any point within an aquifer can be represented by the hydraulic head, $h(x, y, z, t)$ which is a function of the pressure and potential energy per unit weight of the fluid (assuming isothermal conditions in the aquifer). The kinetic energy of the water in the aquifer is neglected because of the small magnitude of groundwater velocities.

For a compressible fluid, the hydraulic head h (L) is defined as:

$$h(x, y, z, t) = \int_{p_0}^{p} \frac{dp}{\gamma(p(x, y, z, t))} + z \tag{2.1}$$

where $p(x, y, z, t)$ is the fluid pressure, γ is the specific weight of the fluid (and is assumed to be a function of the fluid pressure), z is the elevation above an arbitrary datum (a surrogate measure of potential energy), and p_0 is a reference pressure, usually taken to be atmospheric. For an incompressible fluid, Equation 2.1 becomes:

$$h(x, y, z, t) = \frac{p(x, y, z, t)}{\gamma} + z + p_0 \tag{2.2}$$

where p_0 is the atmospheric (gauge) pressure ($p_0 = 0$).

Darcy's law, which is an empirical law, states that the specific discharge (volume flow rate per unit cross-sectional area normal to the direction of flow) is proportional to the gradient of the hydraulic head

$$\mathbf{q} = -K\nabla h \tag{2.3a}$$

or

$$Q = -KA\nabla h \tag{2.3b}$$

where $\mathbf{q}$ is the bulk or seepage velocity vector with components, q_x, q_y, q_z (in the x, y, z coordinate directions), Q is the flow rate, A is the cross-sectional area, and ∇h is the gradient vector of the head,

$$\nabla h = \left(\frac{\partial h}{\partial x}\mathbf{i} + \frac{\partial h}{\partial y}\mathbf{j} + \frac{\partial h}{\partial z}\mathbf{k}\right)$$

($\mathbf{i}$, $\mathbf{j}$, and $\mathbf{k}$ are unit vectors in the x, y, z coordinate directions, respectively). The negative sign indicates that the direction of flow is from regions of higher head to lower head. The constant of proportionality is the hydraulic conductivity of the aquifer. The hydraulic conductivity K represents the rate of flow of water through a unit cross-sectional area perpendicular to the direction of flow under a unit hydraulic gradient. K has the dimensions of velocity and, in a sense, represents the flow rate at which water traverses the aquifer.

The hydraulic conductivity is a function of both fluid and medium properties. As can be shown by dimensional analysis using the basic units of length (L), mass (M), and time (T), the hydraulic conductivity can be expressed as (see, for example, DeWiest, 1965):

$$K = \frac{Cd^2\gamma}{\mu} = \frac{k\gamma}{\mu} \tag{2.4}$$

where d (L) is the average pore size or mean grain diameter of the granular material within the aquifer, μ $(ML^{-1}T^{-1})$ is the dynamic viscosity, γ $(ML^{-2}T^{-2})$ is the specific weight of the water, and C is a constant or shape factor which accounts for the effects of stratification and packing. k is referred to as the *intrinsic permeability* and is dependent solely on the medium properties ($k = Cd^2$).

The porous media is said to be homogeneous if the hydraulic conductivity is independent of the position within the aquifer. If not, the aquifer is inhomogeneous, i.e., $K = K(x, y, z)$. The isotropy or anisotropy of the aquifer system reflects the directional variability of the hydraulic conductivity. If the hydraulic conductivity in the x, y, and z directions is different, the system is anisotropic. If the hydraulic conductivity is the same in any coordinate direction, the aquifer is isotropic. The conditions of inhomogeneity and anisotropy are common occurrences in the soils and geologic formations of aquifers.

Because the specific discharge may not be colinear with the gradient of the hydraulic head, nor have equal specific discharge components in the x, y, and z directions, the hydraulic conductivity may be represented as a second-order tensor quantity (Eagleson, 1970). Darcy's law can then be generalized as:

$$\begin{bmatrix} q_x \\ q_y \\ q_z \end{bmatrix} = - \begin{bmatrix} K_{xx} & K_{xy} & K_{xz} \\ K_{yx} & K_{yy} & K_{yz} \\ K_{zx} & K_{zy} & K_{zz} \end{bmatrix} \begin{bmatrix} \dfrac{\partial h}{\partial x} \\ \dfrac{\partial h}{\partial y} \\ \dfrac{\partial h}{\partial z} \end{bmatrix} \tag{2.5a}$$

or

$$\mathbf{q} = -K\nabla h \tag{2.5b}$$

This linear seepage law assumes, however, that the inertial force (the product of the mass and acceleration) is negligible in comparsion with the viscous force. The Reynolds number of the groundwater flow domain is the ratio of the inertial and viscous forces:

$$N_{Re} = qd/\mu \tag{2.6}$$

Darcy's law is valid as long as the Reynolds number does not exceed a value between 1 and 10, ($N_{Re} \leq 1 - 10$), corresponding to a laminar flow regime.

Example Problem 2.1

Determine the discharge of the confined aquifer system shown in Figure 2.1.

From Darcy's law, the flow, Q, is

$$Q = -KA\frac{dh}{dx} \quad \text{and} \quad \frac{dh}{dx} = -\frac{Q}{KA}$$

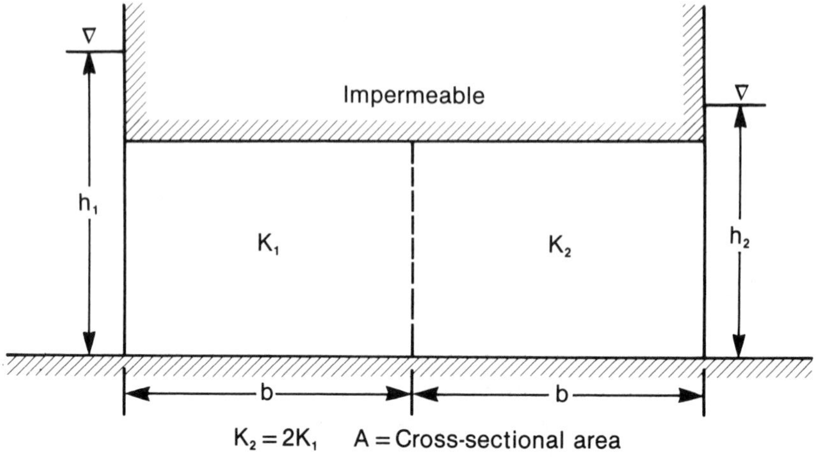

FIGURE 2.1 Example aquifer

Integrating the equation for each region of the aquifer produces

$$\int_{h_1}^{h_2} dh = -\frac{Q}{A}\left[\frac{1}{K_1}\int_0^b dx + \frac{1}{2K_1}\int_b^{2b} dx\right]$$

and

$$h_2 - h_1 = -\frac{Q}{AK_1}[b + b/2]$$

or

$$Q = \frac{2AK_1(h_1 - h_2)}{3b} \tag{2.7}$$

Example Problem 2.2

For the stratified aquifer system shown in Figure 2.2, determine the total discharge per unit width of the aquifer.

The total discharge can be expressed using Darcy's law as the sum of the discharges through each layer, or

$$Q = -b_1K_1\left[\frac{h_2 - h_1}{L}\right] - b_2K_2\left[\frac{h_2 - h_1}{L}\right]$$

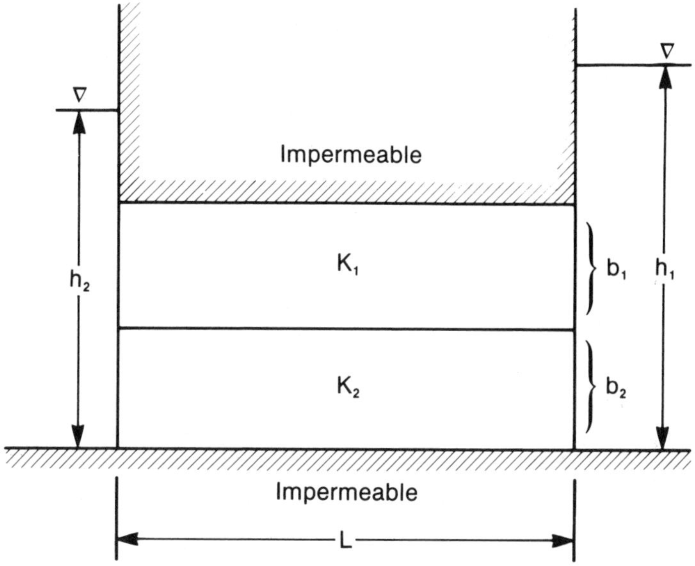

FIGURE 2.2 Flow in a stratified aquifer

The total discharge can also be expressed as:

$$Q = -\bar{b}\,\bar{K}\left[\frac{h_2 - h_1}{L}\right]$$

where $\bar{b} = b_1 + b_2$ and $\bar{K}$ is the average hydraulic conductivity of the aquifer system in the horizontal direction. Equating the expressions for the discharge, we have

$$\bar{K} = \frac{K_1 b_1 + K_2 b_2}{b_1 + b_2} \tag{2.8a}$$

In a similar manner, for n layers, the average hydraulic conductivity can be expressed as:

$$\bar{K} = \frac{\sum_{i=1}^{n} b_i K_i}{\sum_{i=1}^{n} b_i} \tag{2.8b}$$

2.3 THE CONTINUITY EQUATION

The continuity equation for the groundwater system can be developed by considering a nondeforming elemental volume located within the aquifer (Figure 2.3). The principle of mass conservation requires

$$\left[\begin{matrix}\text{mass inflow} \\ \text{rate}\end{matrix} - \begin{matrix}\text{mass outflow} \\ \text{rate}\end{matrix} = \begin{matrix}\text{rate of change with time in} \\ \text{mass storage}\end{matrix}\right] \tag{2.9}$$

The mass flow rate through an elemental area is $\rho q_n A$ where q_n is the component of the mean seepage fluid velocity normal to the area, A. Assuming that the volume is completely saturated with a single-phase, homogeneous fluid of density ρ, and the mean seepage velocity of the fluid and density are given at point P, the centroid of the control volume, then a Taylor series expansion may be used to obtain the net mass inflow to the elemental volume. For example, the mass inflow in the x direction, I_x (MT^{-1}) is

$$I_x = \rho q_x \, \Delta y \, \Delta z - \frac{\partial}{\partial x}(\rho q_x) \frac{1}{2} \, \Delta x \, \Delta y \, \Delta z$$

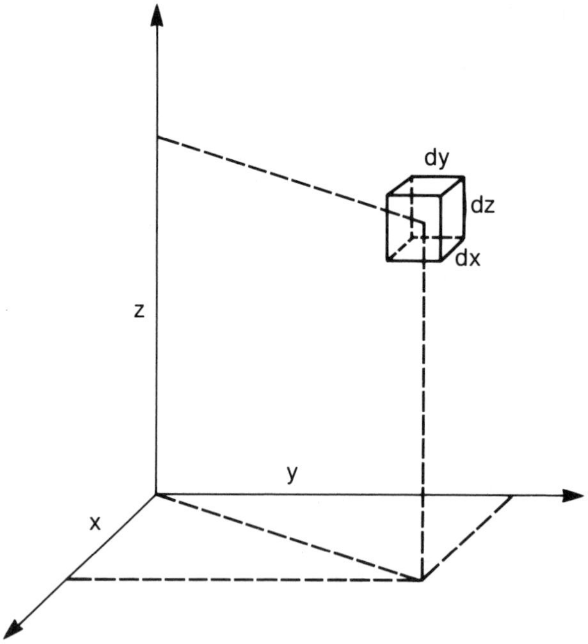

FIGURE 2.3 Small elemental rectangular parallel-piped (Adapted from I. Remson, G.M. Hornberger, and F.J. Moltz, *Numerical Methods in Subsurface Hydrology*. New York: Wiley, 1970. Copyright © 1970 by John Wiley & Sons, Inc. Reprinted with permission)

where q_x is the component of the fluid velocity in the x direction.

Simiarly, the mass outflow in the x direction, O_x is

$$O_x = \rho q_x \, \Delta y \, \Delta z + \frac{\partial}{\partial x} (\rho q_x) \frac{1}{2} \Delta x \, \Delta y \, \Delta z$$

Summing the contributions to the inflow and outflow in the x, y, z directions, the net mass inflow is

$$-\left[\frac{\partial}{\partial x} (\rho q_x) + \frac{\partial}{\partial y}(\rho q_y) + \frac{\partial}{\partial z}(\rho q_z)\right] \Delta x \, \Delta y \, \Delta z \qquad (2.10)$$

or in vector notation,

$$(-\nabla \cdot \rho q) \, \Delta x \, \Delta y \, \Delta z$$

where (∇) is the differential operator,

$$\nabla = \frac{\partial}{\partial x}\mathbf{i} + \frac{\partial}{\partial y}\mathbf{j} + \frac{\partial}{\partial z}\mathbf{k},$$

and **i**, **j**, and **k** are unit vectors in the x, y, z coordinate directions, respectively.

The net increase in the mass flow rate is equal to the rate of change of the mass stored within the elemental volume. If the effective porosity of the aquifer is n (L^0), then the mass M of fluid contained in the volume is $n\rho\Delta x \, \Delta y \, \Delta z$. For a nondeforming control volume, we have

$$\frac{\partial M}{\partial t} = \frac{\partial}{\partial t}(\rho n)\Delta x \, \Delta y \, \Delta z \qquad (2.11)$$

The continuity equation for the groundwater system is then,

$$-\left[\frac{\partial}{\partial x}(\rho q_x) + \frac{\partial}{\partial y}(\rho q_y) + \frac{\partial}{\partial z}(\rho q_z)\right] = \frac{\partial}{\partial t}(\rho n) \qquad (2.12a)$$

or

$$-\nabla \cdot \rho \mathbf{q} = \frac{\partial(\rho n)}{\partial t} \qquad (2.12b)$$

Equation 2.12 relates the mass fluxes and the temporal change in mass storage for any location in the groundwater system.

Example Problem 2.3

The hydraulic conductivity of an aquifer can be expressed as:

$$K = K_0(1 + e^{ax})$$

where K_0 and a are system parameters. Assuming steady-state conditions, develop, using the continuity equation, the governing equation for flow in the aquifer. Express the equation in terms of the hydraulic head.

For steady-state conditions and two-dimensional incompressible flow, the continuity equation may be expressed as

$$\frac{\partial q_x}{\partial x} + \frac{\partial q_y}{\partial y} = 0 \tag{2.13}$$

Introducing Darcy's law, the continuity equation becomes

$$\frac{\partial}{\partial x}\left(K \frac{\partial h}{\partial x}\right) + \frac{\partial}{\partial y}\left(K \frac{\partial h}{\partial y}\right) = 0$$

or,

$$\frac{\partial}{\partial x}\left[K_0 (1 + e^{ax}) \frac{\partial h}{\partial x}\right] + \frac{\partial}{\partial y}\left[K_0(1 + e^{ax}) \frac{\partial h}{\partial y}\right] = 0$$

Finally,

$$(1 + e^{ax}) \frac{\partial^2 h}{\partial x^2} + ae^{ax} \frac{\partial h}{\partial x} + (1 + e^{ax}) \frac{\partial^2 h}{\partial y^2} = 0 \tag{2.14}$$

The governing equation is a second-order linear partial differential equation.

2.4 *PARTIALLY SATURATED FLOW*

Many groundwater problems involve prediction of the moisture variation or water quality changes within the zone of aeration or soil system. The problems occur, for example, in the design of irrigation and drainage systems and in simulation of the infiltration process, a critical element in rainfall-run-off modeling.

The development of a mathematical model characterizing the moisture variation in the soil system is again based on the continuity Equation 2.12 or

$$-\nabla \cdot \rho \mathbf{q} = \frac{\partial}{\partial t}(\rho\theta)$$

where θ is the volumetric water content or the volume of water relative to the total volume of the porous media. The water or moisture content is related to the porosity (n) and the degree of saturation (s) through the relation,

$$\theta = ns$$

The *degree of saturation* is defined as the fractional volume of water in the total void volume. Assuming that Darcy's law is applicable for partially saturated flow, then

$$\mathbf{q} = -\mathbf{K}\nabla\phi$$

where now ϕ is the total potential in the zone of aeration and $\mathbf{K}$, the hydraulic conductivity, is a function of the volumetric water content. The relation between the hydraulic conductivity (permeability) and the moisture content for a light cohesive soil is shown in Figure 2.4(a). Assuming (1) an isothermal system and (2) the properties of the porous media are not affected by biological and chemical reaction, the total potential consists of the gravitational potential z and the capillary potential ψ. The *capillary potential* is the capillary suction head, which is a measure of capillary forces that hold water in the unsaturated zone. Defining z as positive upward, the total potential can be expressed as

$$\phi = \psi + z \tag{2.15}$$

Assuming incompressible flow, the continuity equation may be expressed as

$$\nabla \cdot K\nabla(\psi + z) = \frac{\partial\theta}{\partial t} \tag{2.16a}$$

$$\nabla \cdot K\nabla\psi + \frac{\partial K}{\partial z} = \frac{\partial\theta}{\partial t} \tag{2.16b}$$

In Cartesian coordinates, the equation is

$$\frac{\partial\theta}{\partial t} = \frac{\partial}{\partial x}\left(K\frac{\partial\psi}{\partial x}\right) + \frac{\partial}{\partial y}\left(K\frac{\partial\psi}{\partial y}\right) + \frac{\partial}{\partial z}\left(K\frac{\partial\psi}{\partial z}\right) + \frac{\partial K}{\partial z} \tag{2.17}$$

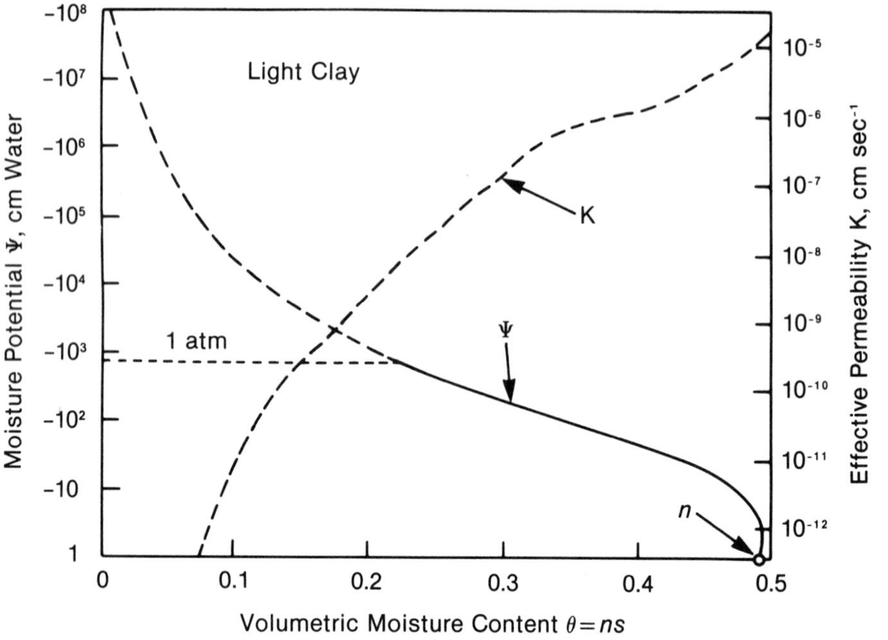
Light Clay
Moisture Potential Ψ, cm Water
Effective Permeability K, cm sec⁻¹
1 atm
K
Ψ
n
Volumetric Moisture Content θ = ns

(a)

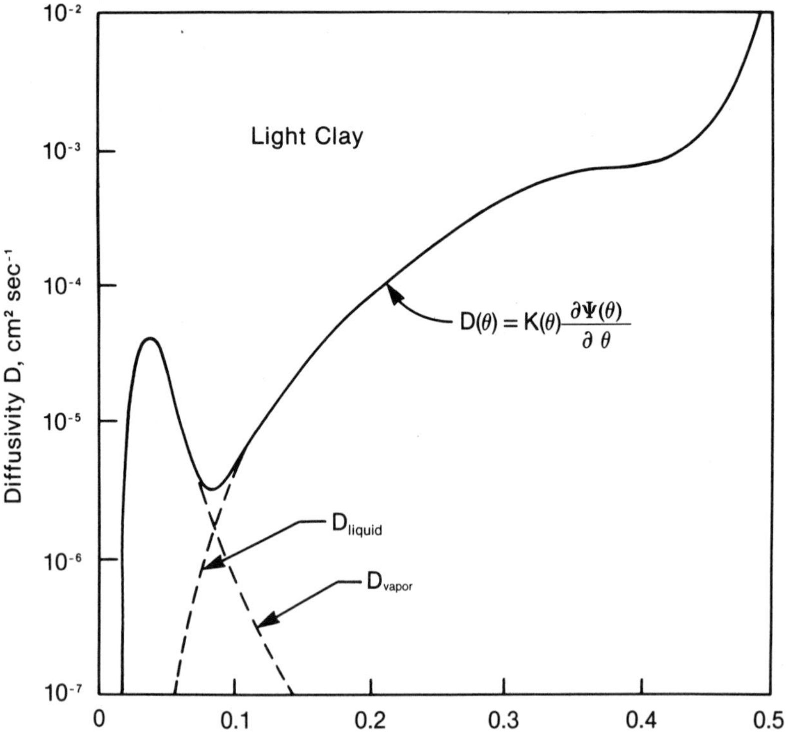
Light Clay
Diffusivity D, cm² sec⁻¹
D(θ) = K(θ) ∂Ψ(θ)/∂θ
D liquid
D vapor
Volumetric Moisture Content θ = ns

(b)

Assuming ψ and K are single-valued functions of θ, Equation 2.17 can be expressed as

$$\frac{\partial\theta}{\partial t} = \frac{\partial}{\partial x}\left(K\frac{\partial\psi}{\partial\theta}\frac{\partial\theta}{\partial x}\right) + \frac{\partial}{\partial y}\left(K\frac{\partial\psi}{\partial\theta}\frac{\partial\theta}{\partial y}\right) + \frac{\partial}{\partial z}\left(K\frac{\partial\psi}{\partial\theta}\frac{\partial\theta}{\partial z}\right) + \frac{\partial K}{\partial z} \qquad (2.18a)$$

and

$$\frac{\partial\theta}{\partial t} = \frac{\partial}{\partial x}\left[D(\theta)\frac{\partial\theta}{\partial x}\right] + \frac{\partial}{\partial y}\left[D(\theta)\frac{\partial\theta}{\partial y}\right] + \frac{\partial}{\partial z}\left[D(\theta)\frac{\partial\theta}{\partial z}\right] + \frac{\partial K(\theta)}{\partial z} \qquad (2.18b)$$

where we have introduced the soil moisture diffusivity $D(\theta) = K\frac{\partial\psi}{\partial\theta}$ in the continuity equation. (A typical diffusivity function is given in Figure 2.4(b).) Equation 2.18 is known as the Buckingham-Richards equation.

For the special case of vertical infiltration into a soil, Equation 2.18 may be expressed as

$$\frac{\partial\theta}{\partial t} = \frac{\partial}{\partial z}\left[D(\theta)\frac{\partial\theta}{\partial z}\right] + \frac{\partial K(\theta)}{\partial z} \qquad (2.19)$$

This diffusion equation is a nonlinear function of the volumetric water content.

Although we have assumed that the capillary potential and hydraulic conductivity are single-valued functions of the moisture content, the relationship between the soil tension and moisture content is generally not unique. Rather, the relationship between the capillary potential and the permeability depends on whether the soil is in a drying or wetting cycle. These hysteresis effects are illustrated in Figures 2.5a and 2.5b which show typical relations between the capillary potential and the effective permeability. In the figures, p_b is the babbling pressure, which is the first appearance of gas when dewatering a saturated sample, and k_{eff} (L^2) is the effective permeability, or the permeability when the medium is occupied by more than one fluid phase.

2.5 BOUNDARY AND INITIAL CONDITIONS—PARTIALLY UNSATURATED FLOW

The boundary conditions of the partially saturated flow problem prescribe how the state variable, the volumetric water content, changes in space and time on the boundaries of the soil system. Excluding hysteresis effects, the boundary conditions

FIGURE 2.4 (a) Moisture potentials and permeabilities for a cohesive soil. (b) Diffusivity for a cohesive soil. (Adapted from P.S. Eagelson, *Dynamic Hydrology*. New York: McGraw-Hill, 1970. Copyright © 1970 by McGraw-Hill Book Company. Reprinted with permission)

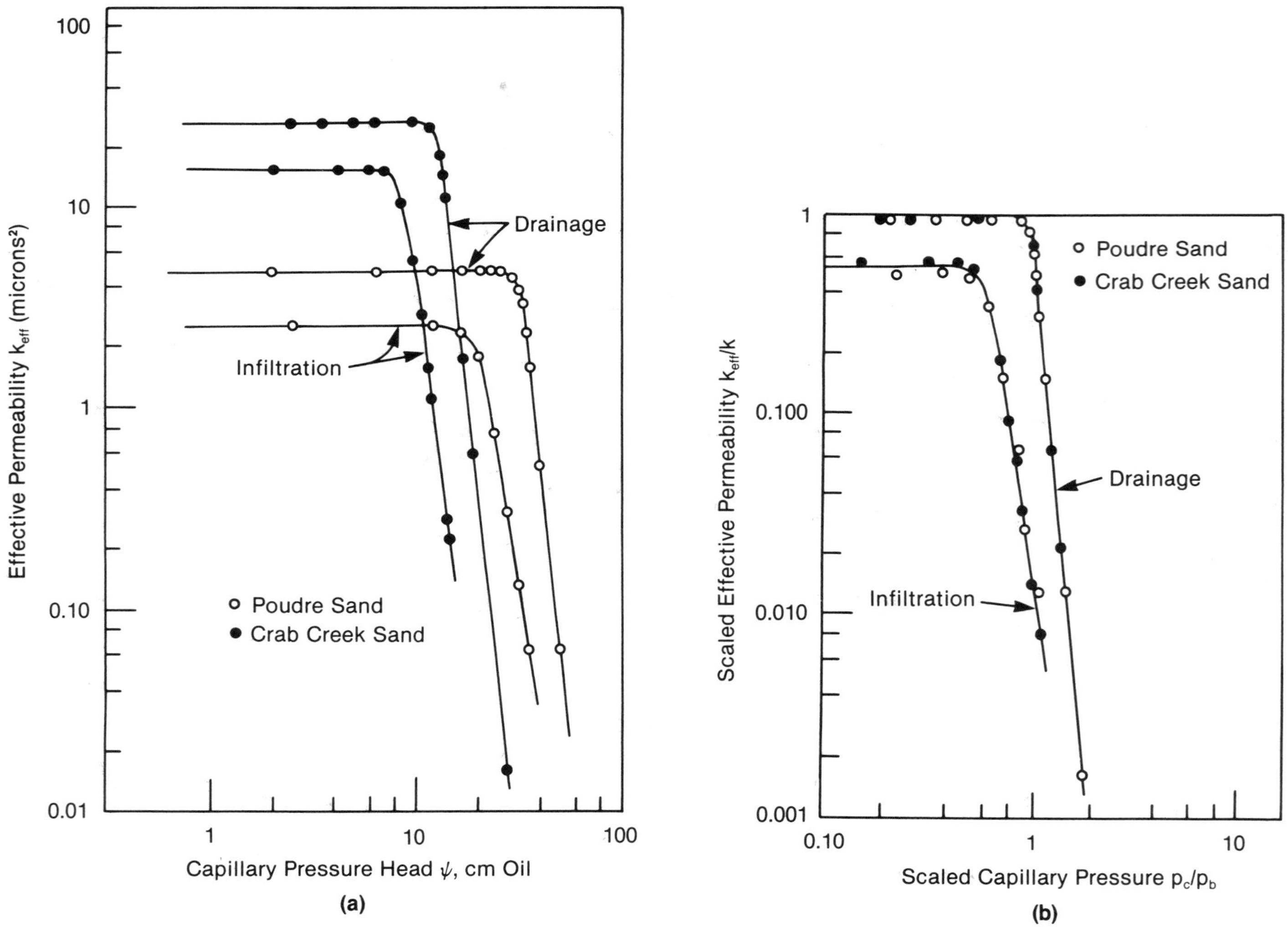

FIGURE 2.5 **Hysteresis effects** (Adapted from G.L. Corey and A.T. Corey, "Simulation Model for Drainage of Soils." *ASCE J. Irr. Drainage Div.*, IR3, 1967. Reprinted with permission from the American Society of Civil Engineers)

for the unsaturated flow problem are described by Dirichlet or Neumann boundary conditions.

Dirichlet boundary conditions occur when the water content is prescribed at all points on the system's boundaries. This occurs, for example, when there is ponded water on the soil surface. Defining Γ_1 as the soil system boundary, the boundary condition can be expressed as

$$\theta(x, y, z, t) = \theta_0(t), \ (x, y, z) \epsilon \ \Gamma_1 \tag{2.20}$$

where θ_0 is a known function of time.

Neumann boundary conditions are used to specify a known water flux at the boundary of the soil system. If rainfall or irrigation reaches the ground surface at the rate $R(x, t)$ then, the boundary condition is given as

$$\mathbf{R} \cdot \mathbf{n} = \mathbf{q} \cdot \mathbf{n} \tag{2.21a}$$

where $\mathbf{n}$ is a unit vector normal to the boundary and $\mathbf{R} = -R\,\mathbf{k}$ ($\mathbf{k}$ is a unit vector in the positive z direction). The velocity, $\mathbf{q}$, is given by Darcy's law, or

$$\mathbf{q} = -\mathbf{K}\,\nabla\,(\psi + z) = -\mathbf{K}\,\nabla\psi - \mathbf{Kk} \tag{2.21b}$$

The boundary condition can then be expressed as,

$$\mathbf{R} \cdot \mathbf{n} = -[\mathbf{K}\nabla\psi + \mathbf{Kk}] \cdot \mathbf{n}$$

A flux or Neumann boundary condition can also occur when a semipermeable layer is formed at the ground surface. If it can be assumed that the semipermeable layer is fully saturated, the boundary condition can be expressed as (Bear, 1979)

$$\mathbf{q} \cdot \mathbf{n} = \frac{\phi_o - \phi}{m_a/K_a} \tag{2.22}$$

where ϕ is the potential in the unsaturated system and ϕ_o is the potential in the external domain. m_a is the thickness of the semiconfining layer and K_a is its hydraulic conductivity.

In addition to the boundary conditions, the initial water content at $t = 0$ has to be known throughout the solution domain. These initial conditions can be expressed as

$$\theta(x, y, z, 0) = g(x, y, z) \tag{2.23}$$

where g defines the volumetric water content at all points in the interior of the soil system.

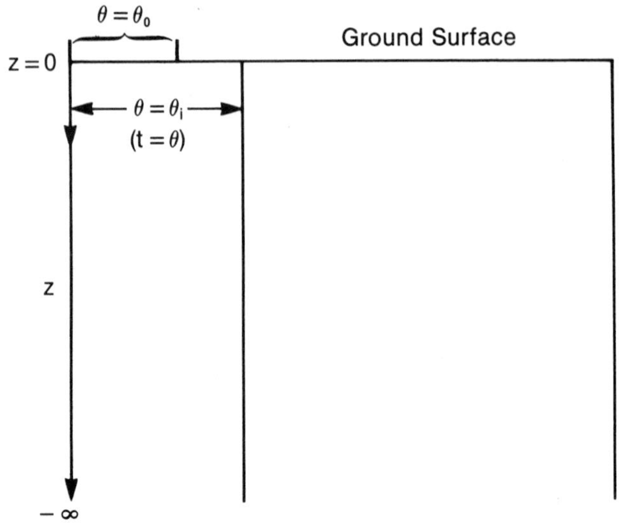

FIGURE 2.6 Example soil system

Example Problem 2.4

Determine the infiltration velocity (the velocity at $z = 0$, the ground surface) for the soil system shown in Figure 2.6. Assume a constant soil moisture diffusivity, D, and the following initial and boundary conditions,

$$\theta = \theta_i, \; z \leq 0, \; t = 0$$

$$\theta = \theta_0, \; z = 0, \; t > 0$$

$$\theta \text{ finite}, \; z \rightarrow -\infty, \; t > 0$$

The governing equation of the problem is, from Equation 2.19,

$$\frac{\partial \theta}{\partial t} = D \frac{\partial^2 \theta}{\partial z^2}$$

We have neglected the gravity term $\left(\frac{\partial K}{\partial z}\right)$ because we are examining the moisture distribution during the early stages of infiltration.

The solution of the diffusion equation is (Eagleson, 1970)

$$\frac{\theta - \theta_0}{\theta_i - \theta_0} = \text{erf}\left[\frac{|z|}{2(Dt)^{1/2}}\right] \tag{2.24}$$

where erf denotes the error function,

$$\operatorname{erf} \alpha = \left(\frac{4}{\pi}\right)^{1/2} \int_0^{\alpha} e^{-\xi^2}\, d\xi \tag{2.25}$$

The velocity at the ground surface, $z = 0$, is given by Darcy's law, or

$$q_z\bigg|_{z=0} = -D\frac{\partial \theta}{\partial z}\bigg|_{z=0} = (\theta_i - \theta_0)\left(\frac{D}{\pi t}\right)^{1/2} \tag{2.26}$$

or, after restoring the gravity term,

$$q_z|_{z=0} = (\theta_i - \theta_0)\left(\frac{D}{\pi t}\right)^{1/2} + K_z \tag{2.27}$$

Alternatively, we also observe that q_z is a solution of a diffusion equation since,

$$\frac{\partial q_z}{\partial t} = -D\frac{\partial}{\partial t}\left(\frac{\partial \theta}{\partial z}\right) = -D\frac{\partial}{\partial z}\left(\frac{\partial \theta}{\partial t}\right)$$

and

$$\frac{\partial q_z}{\partial t} = -D\frac{\partial}{\partial z}\left[D\frac{\partial^2 \theta}{\partial z^2}\right]$$

Since, however, $\dfrac{\partial q_z}{\partial z} = -D\dfrac{\partial^2 \theta}{\partial z^2}$, we have $\dfrac{\partial q_z}{\partial t} = D\dfrac{\partial^2 q_z}{\partial z^2}$ which is the diffusion equation.

A solution to this equation can be expressed as (Carslaw & Jaeger, 1959)

$$q_z = (A \cos mz + B \sin mz)e^{-Dm^2 t} + \text{constant}$$

where m^2 is the separation constant. Assuming the boundary conditions are

$$q_z|_{z=0} = f_\infty^*,\ t = \infty$$

$$q_z|_{z=0} = f_0^*,\ t = 0$$

then the infiltration velocity, q_z, is given as

$$q_z|_{z=0} = f_\infty^* + (f_0^* - f_\infty^*)e^{-Dm^2 t} \tag{2.28}$$

The equation is known as Horton's equation.

Example Problem 2.5

Determine the steady-state moisture distribution in the soil system shown in Figure 2.7. Assume the soil is uniform and the boundary conditions

$$q_z|_{z=0} \cdot \mathbf{k} = -W\mathbf{k}$$

$$\theta(-L) = n$$

where n is the porosity, **k** is a unit normal in the z direction (positive upward), and L is the approximate distance to the water table.

The hydraulics of the problem are described by the steady-state Buckingham-Richards equation,

$$\frac{d}{dz}\left[D\frac{\partial\theta}{\partial z} + K_z\right] = 0, \quad z \leqslant 0 \tag{2.29}$$

Directly integrating the equation produces

$$D\frac{d\theta}{dz} + K_z = C, \quad z \leqslant 0$$

The constant of integration may be evaluated using the flux boundary condition at the ground surface,

$$D\frac{d\theta}{dz} + K_z = W$$

At the ground surface, the velocity is W, the steady infiltration rate. The differential equation describing the steady-state moisture distribution is then

$$\frac{d\theta}{dz} = \frac{W - K_z}{D}, \quad \theta(-L) = n, \quad z \leqslant 0 \tag{2.30}$$

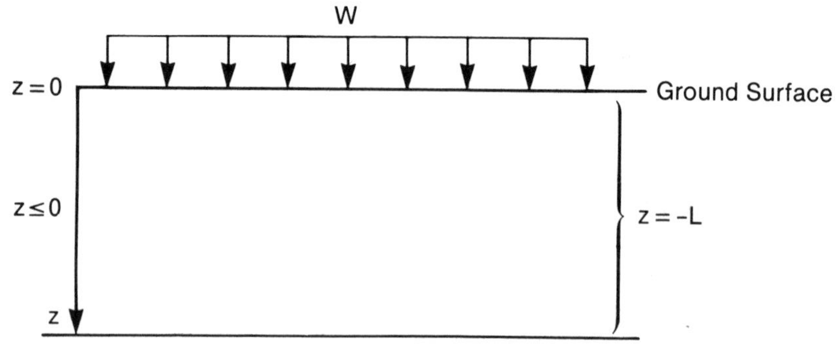

FIGURE 2.7 Soil system

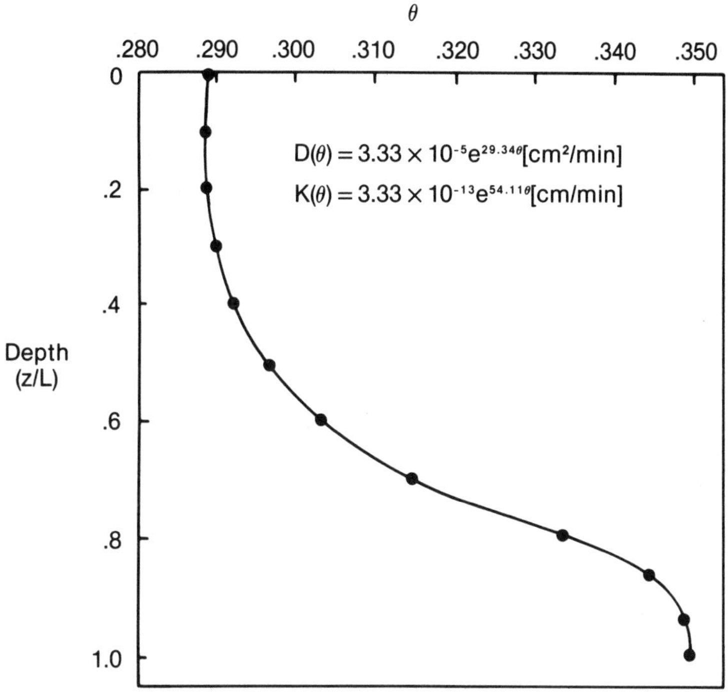

FIGURE 2.8 Steady-state distribution of the volumetric water content

For a uniform conductivity and diffusivity, the moisture distribution is given as

$$\theta = \left[\frac{W - K_z}{D}\right](z + L) + n, \quad z \leq 0 \tag{2.31a}$$

If, however, the conductivity and diffusivity are functions of θ, as soil data indicate, then the solution can be expressed as

$$\int_{\theta}^{n} \left\{\frac{D(\theta)}{W - K_z(\theta)}\right\} d\theta = \int_{0}^{-L} dz \tag{2.31b}$$

Typically, conductivity and diffusivity functions for many soils can be represented by equations of the form

$$K(\theta) = a_1 e^{b_1\theta}, \; D(\theta) = a_2 e^{b_2\theta}$$

where a_i, b_i are soil parameters.

A typical steady-state moisture profile is shown in Figure 2.8.

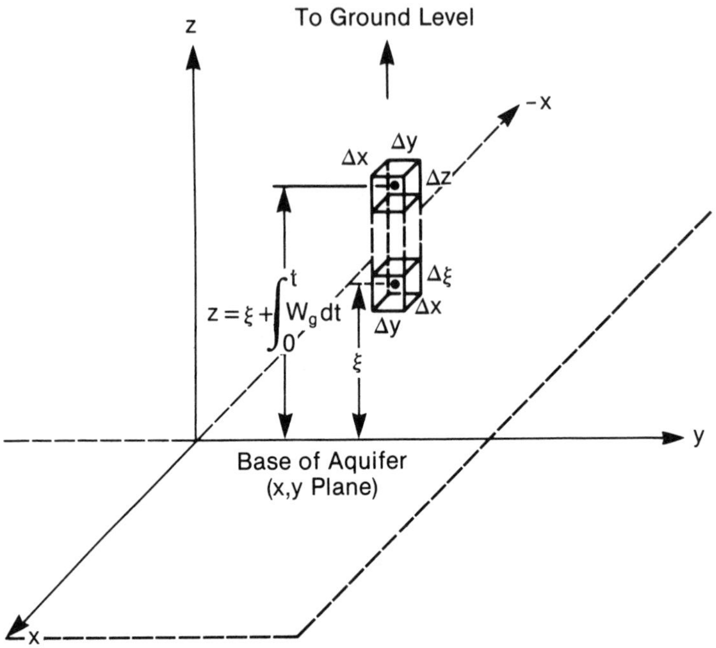

FIGURE 2.9 **Coordinate system showing an elemental aquifer volume $\Delta x \Delta y \Delta z$ which deforms and moves with the aquifer always to contain the same grains. Δz is viewed as the coordinate of a particular grain; t is time; $W_g = dz/dt$ is the grain velocity; and $z = \xi$ at $t = 0$.** (Adapted from I. Remson, G.M. Hornberger, and F.J. Moltz, *Numerical Methods in Subsurface Hydrology*. New York: Wiley, 1970, p. 38. Copyright © 1970 by John Wiley & Sons, Inc. Reprinted with permission)

2.6 CONSERVATION OF MASS IN A DEFORMING POROUS MEDIUM

In a deforming porous medium, the temporal changes in porosity and density are caused by variations in the fluid pressure. When pumping, injection, drainage, and external loadings alter the effective stress within the aquifer, the solid matrix (grains) of the aquifer deforms with time. Assuming the deformation occurs only in the vertical z direction, we can define an elemental volume, V', which moves and deforms with the solid grains (Figure 2.9). The grain velocity (W_g), which must be accounted for in all temporal derivatives of the aquifer's properties, is the rate of change of the vertical coordinate z of V', or

$$z = \xi + \int_0^t W_g dt$$

and

$$W_g = \frac{dz}{dt}$$

ξ is the initial location of the elemental volume at $t = 0$.

Following Cooper (1966) and Remson et al. (1971), we assume that the mean velocity of the fluid ($\mathbf{q}$) may be expressed as the sum of a Darcy velocity ($\mathbf{q}_D$) relative to the moving grains and $W_g\mathbf{k}$, the grain velocity or,

$$\mathbf{q} = \mathbf{q}_D + nW_g\mathbf{k} \tag{2.32}$$

where $\mathbf{k}$ is a unit vector in the z direction, and n is the porosity.

Substituting Equation 2.32 into the continuity Equation 2.12b, and simplifying the derivatives yields

$$-\nabla \cdot \rho(\mathbf{q}_D + nW_g\mathbf{k}) = \frac{\partial}{\partial t}(\rho n) \tag{2.33a}$$

or

$$-\nabla \cdot \rho\mathbf{q}_D - nW_g\frac{\partial \rho}{\partial z} - \rho W_g\frac{\partial n}{\partial z} - \rho n\frac{\partial W_g}{\partial z} = n\frac{\partial \rho}{\partial t} + \rho\frac{\partial n}{\partial t} \tag{2.33b}$$

Equation 2.33 can be written in terms of the substantial or material derivatives of n and ρ, i.e.,

$$\frac{dn}{dt} = \frac{\partial n}{\partial t} + \frac{dz}{dt}\frac{\partial n}{\partial z} = \frac{\partial n}{\partial t} + W_g \cdot \frac{\partial n}{\partial z} \tag{2.34a}$$

$$\frac{d\rho}{dt} = \frac{\partial \rho}{\partial t} + \frac{dz}{dt}\frac{\partial \rho}{\partial z} = \frac{\partial \rho}{\partial t} + W_g \cdot \frac{\partial \rho}{\partial z} \tag{2.34b}$$

or

$$-\nabla \cdot \rho\mathbf{q}_D = n\frac{d\rho}{dt} + \rho\frac{dn}{dt} + \rho n\frac{\partial W_g}{\partial z} \tag{2.35}$$

The temporal variation in the fluid density is dependent upon the compressibility of the fluid, β ($LM^{-1}T^2$), or

$$\beta = \frac{-dV_W/V_W}{dp} \tag{2.36}$$

where V_W is the fluid volume. From the conservation of the fluid mass, m, in the elemental volume, the compressibility may be written

$$\beta = \frac{d(m/\rho)/(m/\rho)}{dp} = \frac{1}{\rho}\frac{\partial \rho}{\partial p}$$

and

$$\beta \rho dp = d\rho$$

The derivative $\frac{dp}{dt}$ can then be expressed as

$$\frac{d\rho}{dt} = \beta\rho \frac{dp}{dt} \tag{2.37}$$

The porosity variation may be expressed in terms of the grain velocity, W_g. If we consider the mass conservation of the solid grains as they move through a nondeforming elemental volume, V^*, we have from the continuity equation,

$$\nabla \cdot \rho_s (1 - n) W_g \mathbf{k} = \frac{\partial}{\partial t} (\rho_s (1 - n)) \tag{2.38}$$

where ρ_s is the density of the solid grains. Assuming that the individual solid grains are rigid during deformation (ρ_s = constant), and again assuming vertical deformation, Equation 2.38 simplifies to

$$(1 - n) \frac{\partial W_g}{\partial z} - W_g \cdot \frac{\partial n}{\partial z} = \frac{\partial n}{\partial t}$$

or using the substantial derivative, Equation 2.34, the above expression becomes,

$$(1 - n) \frac{\partial W_g}{\partial z} = \frac{dn}{dt} \tag{2.39}$$

The spatial variation in the grain velocity is related to the pressure changes within the aquifer and the vertical compressibility of the granular skeleton. The compressibility $\alpha(LM^{-1}T^2)$ expresses the variation in the z coordinate of the deforming volume, V', or

$$\alpha = - \frac{d(\Delta z)/\Delta z}{d\sigma_z} \tag{2.40}$$

where σ_z is the effective intergranular stress $(ML^{-1}T^{-2})$ in the z direction. The temporal variations in the intergranular stress may be expressed as

$$\alpha \frac{d\sigma_z}{dt} = -\frac{1}{\Delta z}\frac{d(\Delta z)}{dt} \tag{2.41}$$

At any depth in the aquifer, the intergranular stress and the pore pressure combine to support the total weight of the overburden, neglecting any arching effects, or

$$p + \sigma_z = \text{constant}$$

and

$$\frac{d\sigma_z}{dt} = -\frac{dp}{dt}$$

Equation 2.41 may then be written in terms of the pore pressure

$$\alpha \frac{dp}{dt} = \frac{1}{\Delta z}\frac{d(\Delta z)}{dt} \tag{2.42}$$

The right-hand side of Equation 2.42 is the rate of the change of volume per unit volume, or the divergence of the grain velocity. The dilatation of the porous media from the pressure variation is then,

$$\alpha \frac{dp}{dt} = \nabla \cdot W_g \mathbf{k} = \frac{\partial W_g}{\partial z} \tag{2.43}$$

The continuity equation may now be expressed in terms of the grain velocity and the pressure variation within the aquifer, or from Equations 2.37 and 2.39,

$$\nabla \cdot \rho \mathbf{q}_D = n\rho\beta \frac{dp}{dt} + \rho(1-n)\frac{\partial W_g}{\partial z} + \rho n \frac{\partial W_g}{\partial z} \tag{2.44}$$

and, with Equation 2.43,

$$-\nabla \cdot \rho \mathbf{q}_D = \rho(\alpha + n\beta)\frac{dp}{dt} \tag{2.45}$$

For many groundwater flow problems, the spatial variations in density are usually much smaller than the velocity variations. Neglecting these density terms in Equation 2.45, and expressing the pore pressure in terms of the hydraulic head via

Equation 2.1, the continuity equation is

$$-\nabla \cdot \mathbf{q}_D = \rho g(\alpha + n\beta)\left[\frac{dh}{dt} - W_g\right] \tag{2.46a}$$

$$-\nabla \cdot \mathbf{q}_D = S_s\left[\frac{dh}{dt} - W_g\right] \tag{2.46b}$$

S_s is the specific storage of the aquifer resulting from compression of the granular skeleton ($\rho g\alpha$) and expansion of the pore water ($\rho gn\beta$). Physically, the specific storage represents the volume of water released or taken into storage per unit volume of the aquifer per unit change in the hydraulic head provided that (1) water is instantaneously released from (or stored in) the void space and (2) the aquifer is compressible only in the vertical direction.

Assuming the grain velocity is small in comparison with the Darcy velocity and that flow is predominantly in the horizontal plane, i.e., a mildly sloping aquifer, $\frac{\partial h}{\partial z} << 1$, the continuity equation characterizing transient groundwater flow is (Jacob, 1950)

$$-\nabla \cdot \mathbf{q}_D = S_s\frac{\partial h}{\partial t} \tag{2.47}$$

2.7 GROUNDWATER FLOW EQUATION FOR A CONFINED OR LEAKY AQUIFER

In confined or artesian aquifer systems, the amount of water released from groundwater storage is dependent on the compressibility of the water and of the porous media, and not on the quantity of water contained in the pore space. Confined systems are bounded above and below by confining layers composed of impermeable rocks or clay-like materials. In contrast, leaky or semiconfined aquifers have partially permeable confining layers or aquitards that are capable of leakage and storage. These multi-aquifer systems are hydraulically interdependent as changes in the potential in one system, caused by groundwater pumping, or recharge, can induce flow to and from adjacent aquifers. To simplify our discussion of these systems, we assume that the leakage or flow between aquifers and aquitards occurs only in the vertical direction and that any storage effects in the aquitards are negligible. Figures 2.10 and 2.11 illustrate typical confined and semiconfined aquifer systems.

The governing equation characterizing two-dimensional horizontal flow in semiconfined aquifers can be obtained by averaging the continuity equation over the saturated thickness of the aquifer. Defining $b(x, y)$ as the thickness of the main

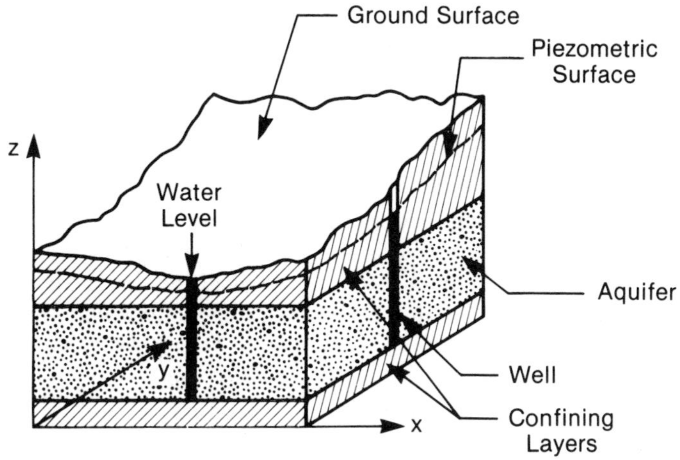

FIGURE 2.10 Confined aquifer system (Adapted from G.F. Pinder and W.G. Gray, *Finite Element Simulation in Surface and Subsurface Hydrology*. New York: Academic Press, 1977, p. 128. Reprinted with permission)

aquifer at any point (x, y) in the horizontal plane and assuming an impermeable, horizontal lower boundary, the vertically averaged continuity equation is

$$-\left\{\int_0^b \frac{\partial q_x}{\partial x} dz\right\} - \left\{\int_0^b \frac{\partial q_y}{\partial y} dz\right\} - \left\{\int_0^b \frac{\partial q_z}{\partial z} dz\right\} = \left\{\int_0^b \frac{\partial}{\partial t} S_s h dz\right\} \tag{2.48}$$

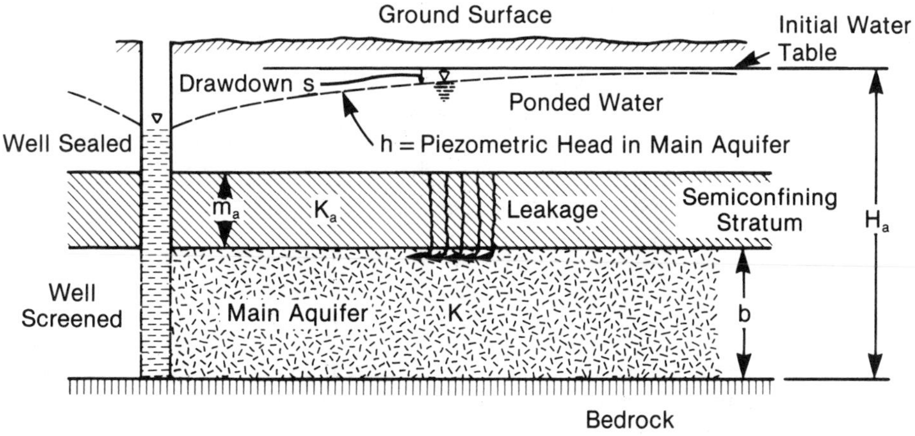

FIGURE 2.11 Leakage into aquifer (Adapted from R.J.M. DeWiest, *Geohydrology*. New York: Wiley, 1965, p. 272. Copyright © 1965 by John Wiley & Sons, Inc. Reprinted with permission)

Equation 2.48 may be simplified using Leibnitz's rule for differentiation of an integral (Pipes, 1958), i.e.,

$$\frac{d}{dx}\left\{\int_{u(x)}^{v(x)} g(x, y)dy\right\} = g(v, y)\frac{dv}{dx} - g(x, u)\frac{du}{dx} + \int_{u(x)}^{v(x)} \frac{\partial g}{\partial x}dy \tag{2.49}$$

and Darcy's law, if the dependence of K on the compressibility of the fluid is neglected. Assuming average or effective values for the conductivity and the specific storage over the thickness of the aquifer, and that the coordinate system is colinear with the principal directions of the conductivity field, the flow equation is

$$\frac{\partial}{\partial x}\left(T_{xx}\frac{\partial \bar{h}}{\partial x}\right) + \frac{\partial}{\partial y}\left(T_{yy}\frac{\partial \bar{h}}{\partial y}\right) + q_z(b) = S_s b\frac{\partial \bar{h}}{\partial t} \tag{2.50}$$

where $\bar{h}$ is the average value of the hydraulic head,

$$\bar{h} = \frac{1}{b}\int_0^b h dz \tag{2.51}$$

and T_{xx} and T_{yy} are the components of the transmissivity, the product of the conductivity and the aquifer's thickness, in the x and y coordinate directions. Although the use of the transmissivity parameter should be restricted to groundwater flow systems where vertical flow is negligible or can be ignored, it is the most commonly used parameter for measuring an aquifer's ability to transmit water.

The right-hand side of Equation 2.50 may be expressed in terms of the storage coefficient, S, or the volume of water released or taken into storage per unit cross-sectional area per unit change in the hydraulic head. The storage coefficient is simply the specific storage integrated over the thickness of the aquifer.

The final term, $q_z(b)$ in Equation 2.50 is the vertical recharge or discharge from the aquifer, e.g., leakage or evapotranspiration. In confined aquifers, the normal velocity at the boundary is zero. However, if water is transmitted through an upper semiconfining layer, the vertical leakage from the overlying aquitard may be expressed using Darcy's law

$$q_z(b) = \frac{K_a(H_a - \bar{h})}{m_a} \tag{2.52}$$

where K_a, H_a, m_a are the vertical hydraulic conductivity of the semiconfining stratum, the external head in the overlying aquifer, and the thickness of the semiconfining layer. The flow model assumes that all flow through the semiconfining layer is vertical (Jacob, 1946). The assumption is valid when K_a is at least two orders of magnitude smaller than the conductivity in the aquifer (Cooley, et al., 1972).

Storage in the confining layer is also neglected. Aquitard storage is considered in models presented by Hantush (1960), Neuman and Witherspoon (1969), Bredehoeft and Pinder (1970), and Pinder and Gray (1977).

The effect of pumping and injection wells within the groundwater system may also be simulated by representing the wells as point sources or point sinks provided that the wells fully penetrate the thickness of the aquifer. For example, if the index set Ω defines the location of all the pumping and injection wells within the system, then the point sources/sinks can be expressed as:

$$\pm \sum_{w\in\Omega} Q_w \delta(x - x_w, y - y_w) \tag{2.53}$$

where $-Q_w$ is the discharge ($+Q_w$ recharge) from the wth pumping (injection) well located at x_w, y_w and $\delta(x - x_w, y - y_w)$ is the Dirac delta function, where

$$\delta(x - x_w, y - y_w) = \begin{cases} = 1, \text{if} x = x_w, y = y_w \\ = 0, \text{if} x \neq x_w, y \neq w_w \end{cases} \tag{2.54}$$

As originally presented by Hantush (1949, 1959), the equation characterizing flow in a semiconfined or leaky aquifer system can be expressed as

$$\frac{\partial}{\partial x}\left(T_{xx}\frac{\partial h}{\partial x}\right) + \frac{\partial}{\partial y}\left(T_{yy}\frac{\partial h}{\partial y}\right) + \frac{K_a(H_a - h)}{m_a} \pm \sum_{w\in\Omega} Q_w \delta(x - x_w, y - y_w) = S\frac{\partial h}{\partial t} \tag{2.55}$$

For convenience, we have dropped the overbar notation.

Example Problem 2.6

An extensive confined aquifer is to be used as a source of municipal water supply. The aquifer is composed of alluvial sand and gravel. Well log data indicate that the aquifer is homogeneous and isotropic. Determine the steady-state response equation of the groundwater system relating the piezometric head and the groundwater pumping rate.

The hydraulic response equation may be easily developed from the confined flow equation (Equation 2.55). If we consider a fully penetrating well, as shown in Figure 2.12, located in the center of the groundwater basin, the flow equation can be expressed in radial coordinates as

$$\frac{1}{r}\frac{d}{dr}\left(r\frac{dh}{dr}\right) = 0 \tag{2.56}$$

The model is a second-order ordinary differential equation. Integrating the equation

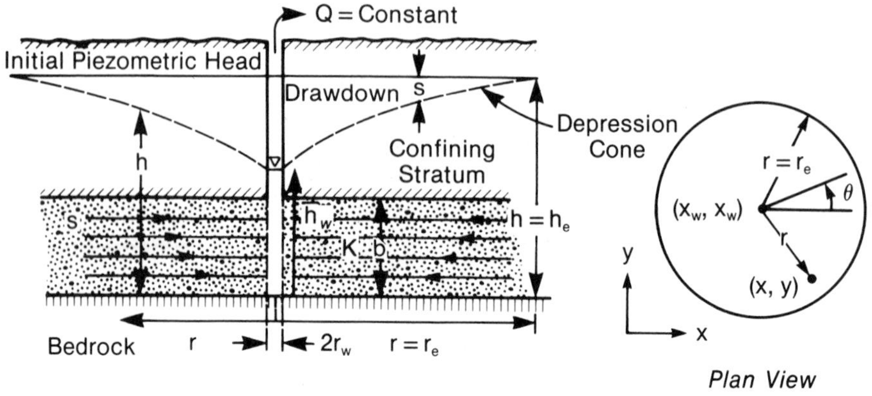

FIGURE 2.12 Radial flow to a well completely penetrating a confined aquifer (Adapted from R.J.M. DeWiest, *Geohydrology*. New York: Wiley, 1965, p. 242. Copyright © 1965 by John Wiley & Sons, Inc. Reprinted with permission)

we have

$$r\frac{dh}{dr} = \text{constant}$$

The constant of integration may be evaluated from Darcy's law, or

$$r\frac{dh}{dr} = \frac{Q}{2\pi T} \tag{2.57}$$

In this and the following example problems, we have incorporated the minus sign of the Darcy boundary condition in Q, i.e., a pumping well is now defined as $Q \geq 0$.

The final response equation relating the head, pumping rate, Q, and the radial distance from the well site is

$$h = \frac{Q}{2\pi T}\ln r + C$$

The equation is known as the *equilibrium* or *Theim equation* (Theim, 1906).

If there are M pumping wells in the system, the total head in the aquifer system may be found from superposition of the individual well responses, or

$$h(x, y) = \sum_{w} \frac{Q_w}{4\pi T}\ln\left\{(x - x_w)^2 + (y - y_w)^2\right\} + C \tag{2.58}$$

where x_w and y_w are the coordinates of the pumping $(+Q)$ or injection $(-Q)$ well. The constant of integration may be evaluated using a prescribed boundary condition.

Example Problem 2.7

An extensive homogeneous and isotropic leaky aquifer of thickness b is overlain by a bed of glacial till (see Figure 2.11). The thickness of the semiconfining stratum is m_a and the hydraulic conductivity is K_a. The head in the aquifer overlying the aquitard is H_a and is assumed to be unaffected by pumping in the aquifer. Determine the steady-state response equation for a fully penetrating well in the aquifer system.

Again, assuming that the pumping well is located in the center of the aquifer system, the flow equation in radial coordinates is

$$\frac{T}{r}\frac{d}{dr}\left(r\frac{dh}{dr}\right) + \frac{K_a(H_a - h)}{m_a} = 0 \tag{2.59}$$

Further assuming that the well diameter is infinitesimally small, the boundary condition at the well is given by Darcy's law, or

$$\lim_{r\to 0} r\frac{dh}{dr} = \frac{Q}{2\pi T}$$

Defining the new dependent variable, y as

$$y = H_a - h$$

the differential equation may be written as

$$\frac{d^2y}{dr^2} + \frac{1}{r}\frac{dy}{dr} - \frac{K_a}{m_a}y = 0$$

Introducing the new independent variable, ρ,

$$\rho = r\left(\frac{K_a}{m_a T}\right)^{1/2}$$

the differential equation becomes

$$\frac{d^2y}{d\rho^2} + \frac{1}{\rho}\frac{dy}{d\rho} - y = 0 \tag{2.60}$$

which is a modified form of Bessel's equation. The general solution of the equation is

$$y = AI_0(\rho) + BK_0(\rho) \tag{2.61}$$

where A and B are constants of integration and I_0 and K_0 are modified Bessel functions of the first and second kind of order zero. Requiring y to be finite as $\rho \to \infty$, A is 0 and the solution may be expressed as

$$y = BK_0(\rho)$$

Using Darcy's law to evaluate the constant of integration and recovering the original variables, the steady-state response equation is

$$h = \frac{Q}{2\pi T} K_0 \left\{ r\left(\frac{K_a}{m_a T}\right)^{1/2} \right\} \tag{2.62}$$

For a system of pumping and injection wells, the solutions again may be superimposed to obtain the total hydraulic response of the aquifer system. The solutions are valid for finite diameter wells as shown by Mariño and Luthin (1982).

Example Problem 2.8

An extensive, homogeneous, isotropic confined aquifer is to be developed as a source of agricultural water supply. Determine the dynamic response equation of the aquifer system relating the drawdown and the pumping rate, Q. Assume that the aquifer is of infinite areal extent.

Since groundwater flow is essential radial, the confined flow equation may be expressed in radial coordinates as

$$\frac{\partial^2 s}{\partial r^2} + \frac{1}{r}\frac{\partial s}{\partial r} = \frac{S}{T}\frac{\partial s}{\partial t} \tag{2.63}$$

where s is the drawdown ($s = h_o - h$) and h_o is the initial piezometric head. The initial and boundary conditions are

$$s(r, 0) = 0 \tag{2.64a}$$

$$s(\infty, t) = 0 \tag{2.64b}$$

$$\lim_{r \to 0} r\frac{\partial s}{\partial r} = -\frac{Q}{2\pi T} \tag{2.64c}$$

We have also assumed that the pumping well can be represented as a line sink.

The confined flow equation, a second-order, linear diffusion equation, can be transformed into an ordinary differential equation using a similarity or Boltzmann transformation. Introducing the Boltzmann variable, u,

$$u = \frac{r^2 S}{4Tt} \tag{2.65}$$

and using the chain rule, the flow equation in radial coordinates may be expressed as

$$\frac{d^2 s}{du^2} + (1 + 1/u)\frac{ds}{du} = 0 \tag{2.66}$$

The transformed initial and boundary conditions are

$$s(\infty) = 0 \tag{2.67a}$$

$$\lim_{u \to 0} u\frac{ds}{du} = -\frac{Q}{4\pi T} \tag{2.67b}$$

Using the substitution, $p = \frac{ds}{du}$, a first integration of the equation yields

$$u\frac{ds}{du} = C_1 e^{-u}$$

where C_1 is a constant of integration. Using the second boundary condition, the equation simplifies to

$$\frac{ds}{du} = \frac{Q}{4\pi T}\frac{e^{-u}}{u}$$

Performing the final integration and invoking the first boundary condition, the drawdown equation is (Theis, 1935)

$$s = \frac{Q}{4\pi T}\int_u^{\infty}\frac{e^{-x}}{x}\,dx \tag{2.68a}$$

or

$$s = \frac{Q}{4\pi T}W(u) \tag{2.68b}$$

where $W(u)$ is the well function. The well function has the infinite series representation (DeWiest, 1965)

$$W(u) = -0.5772 - \ln u + u - \frac{u^2}{2 \cdot 2!} + \frac{u^3}{3 \cdot 3!} - \frac{u^4}{4 \cdot 4!} + \quad (2.69)$$

The Theis model can also be generalized to incorporate a variable groundwater pumping schedule. For example, assume the aquifer is pumped at a variable extraction rate for three planning periods (see Figure 2.13). Because the response of the aquifer to a step change in pumping or recharge is linear, the drawdown for any period may be found from superposition of the well responses, or

$$s = \frac{Q_1}{4\pi T} W\left(\frac{r^2S}{4Tt}\right), \quad 0 \leq t \leq t_1 \quad (2.70a)$$

$$s = \frac{Q_1}{4\pi T} W\left(\frac{r^2S}{4Tt}\right) + \left(\frac{Q_2 - Q_1}{4\pi T}\right) W\left(\frac{r^2S}{4T(t - t_1)}\right), \quad t_1 \leq t \leq t_2 \quad (2.70b)$$

$$s = \frac{Q_1}{4\pi T} W\left(\frac{r^2S}{4Tt}\right) + \left(\frac{Q_2 - Q_1}{4\pi T}\right) W\left(\frac{r^2S}{4T(t - t_1)}\right) + \left(\frac{Q_3 - Q_2}{4\pi T}\right) W\left(\frac{r^2S}{4T(t - t_2)}\right), \quad t \geq t_2 \quad (2.70c)$$

where Q_k is the extraction rate for planning period k. The response equations are linear functions of the extraction rates.

Example Problem 2.9 (after Hantush, 1962)

Determine the hydraulic head distribution in the confined aquifer shown in Figure 2.14. Assume the aquifer is homogeneous with isotropic and steady-flow conditions.

For the one-dimensional aquifer system, the governing equation is given by

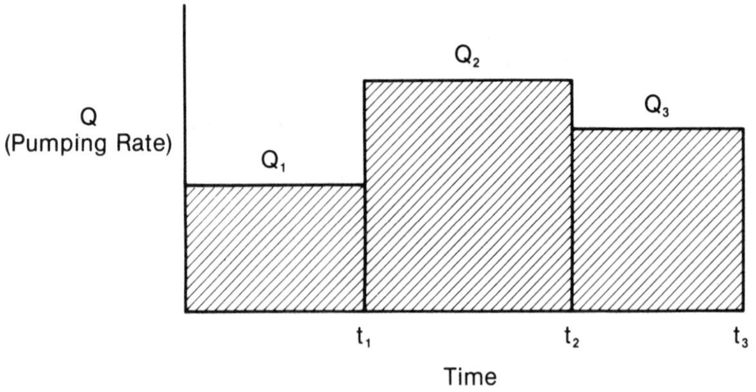

FIGURE 2.13 Example pumping schedule

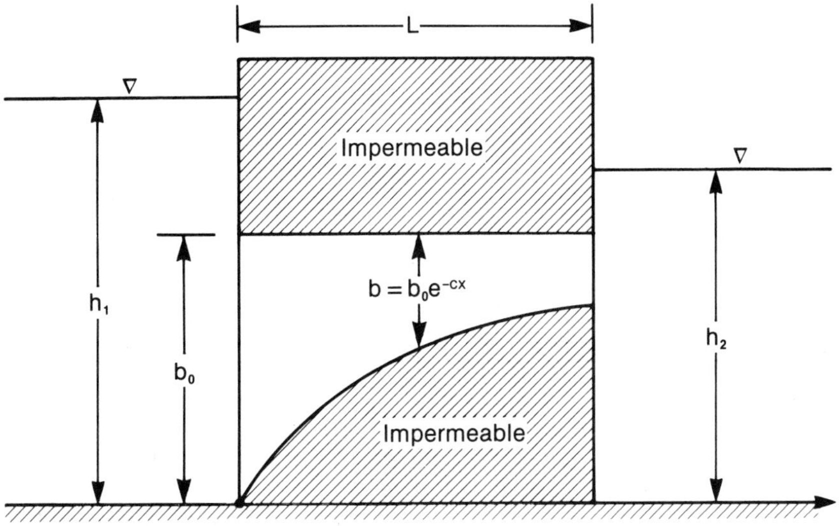

FIGURE 2.14 Example aquifer

Equation 2.55, or

$$\frac{d}{dx}\left(Kb\frac{dh}{dx}\right) = 0 \tag{2.71}$$

with the boundary conditions,

$$h(0) = h_1,\ h(L) = h_2 \tag{2.72a}$$

and

$$b = b_0 e^{-cx} \tag{2.72b}$$

Simplifying the equation,

$$Kb\,\frac{d^2h}{dx^2} + K\,\frac{db}{dx}\frac{dh}{dx} = 0$$

or

$$\frac{d^2h}{dx^2} - c\,\frac{dh}{dx} = 0$$

The solution of this second-order, linear ordinary differential equation is

$$h = C_1\, e^{cx} + C_2$$

where C_1 and C_2 are constants of integration. Using the boundary conditions to evaluate these constants, the steady-state head distribution is described by the equation

$$h = h_1 - \left[\frac{(h_1 - h_2)(e^{cx} - 1)}{(e^{cL} - 1)}\right] \tag{2.73}$$

2.8 GROUNDWATER FLOW EQUATION FOR AN UNCONFINED AQUIFER

In contrast to confined aquifers, unconfined groundwater systems have a free surface boundary, a boundary at atmospheric pressure. Water released from storage occurs because of gravity drainage as the water level in the aquifer responds to pumping, drainage, or natural or artificial recharge. Typical unconfined aquifer systems are shown in Figures 2.15 and 2.16.

Mathematical models of unconfined groundwater flow are again based on the continuity equation for the groundwater system. Assuming that the compressibility of the medium and fluid are relatively insignificant (2 percent or less), the continuity equation for the groundwater system is

$$-\nabla \cdot \mathbf{q}_D \pm S^* = 0$$

where, again, S^* represents sources or sinks in the system.

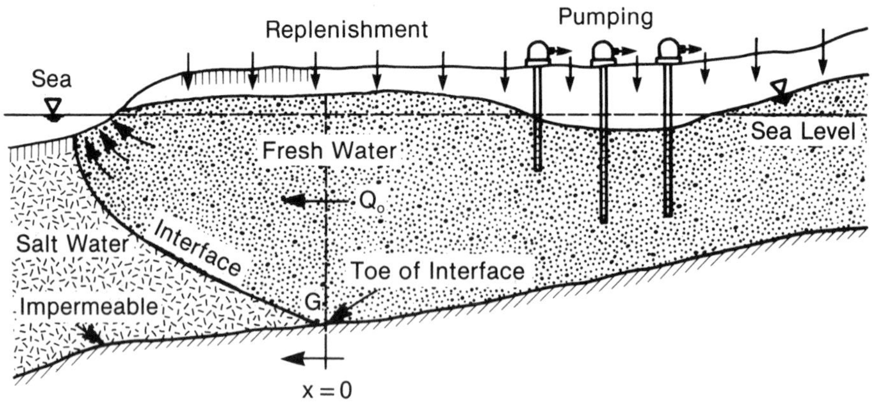

FIGURE 2.15 A typical cross section of a coastal aquifer with pumping (Adapted from J. Bear, *Hydraulics of Groundwater*. New York: McGraw-Hill, 1979, p. 38. Copyright © 1979 by McGraw-Hill Book Company. Reprinted with permission)

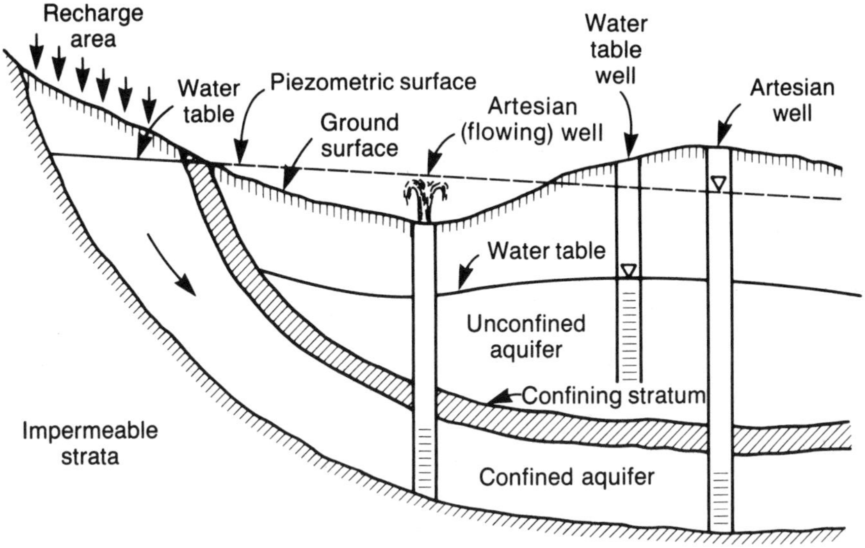

FIGURE 2.16 Schematic cross section illustrating unconfined and confined aquifers (Adapted from D.K. Todd, *Groundwater Hydrology*. New York: Wiley, 1980, p. 42. Copyright © 1980 by John Wiley & Sons, Inc. Reprinted with permission)

Introducing Darcy's law, we have

$$\nabla \cdot \mathbf{K}\nabla h \pm S^* = 0 \tag{2.74a}$$

If the aquifer system is homogeneous and isotropic, or anisotropic, Poisson's equation describes unconfined groundwater flow, or

$$\nabla^2 h \pm \frac{S^*}{K} = 0 \tag{2.74b}$$

We also note that, in the absence of pumping or injection, the flow problem is described by Laplace's equation. Neither of these diagnostic equations for h explicitly involve time. They are, however, valid for both steady and unsteady flow conditions.

The dynamic variation in the hydraulic head, h, is caused by changes in the elevation of the free surface. The free surface will rise or fall as pumping, drainage, or recharge takes place. The free surface boundary condition can be developed by considering Figure 2.17. If we neglect the capillary fringe, all points on the free surface $F(x,y,z,t)$, are at atmospheric pressure ($p = 0$). The potential or head (h) on the free surface is given as

$$h(x,y,z,t) = p/\gamma + z = z \quad \text{on} \quad F$$

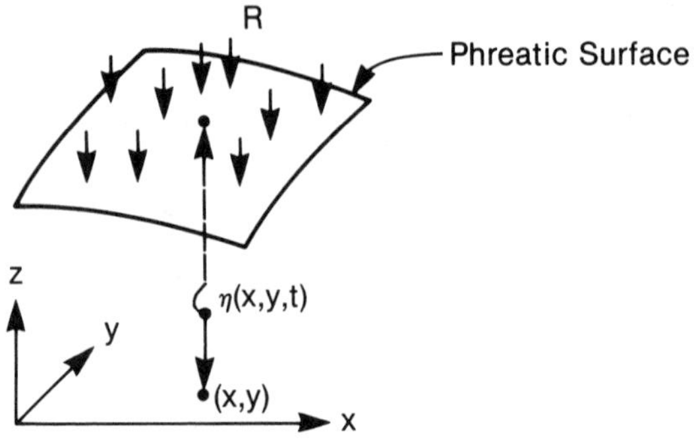

FIGURE 2.17 Example-free surface (Adapted from J. Bear, *Dynamics of Fluids in Porous Media*. New York: Elsevier Science Publishing Co., Inc., 1972, p. 423)

The equation of the free surface is then

$$F(x,y,z,t) = h(x,y,z,t) - z = 0 \tag{2.75}$$

Defining **R** ($\mathbf{R} = -R\mathbf{k}$) as the rate of accretion or recharge (**k** is the unit vector in the positive z direction), then the kinematic boundary condition that must be satisfied on the free surface is

$$\frac{dF}{dt} = \frac{\partial F}{\partial t} + \frac{1}{n}(\mathbf{q} - \mathbf{R}) \cdot \nabla F = 0 \tag{2.76a}$$

where n is the porosity.

Using Darcy's law, the kinematic boundary condition can be expressed as

$$n\frac{\partial F}{\partial t} = (\mathbf{K}\nabla h + \mathbf{R}) \cdot \nabla F \tag{2.76b}$$

Introducing the equation of the free surface (Eq. 2.75), the boundary condition becomes

$$n\frac{\partial F}{\partial t} = \left[K_{xx}\frac{\partial h}{\partial x}\,\mathbf{i},\ K_{yy}\frac{\partial h}{\partial y}\,\mathbf{j},\ \left(K_{zz}\frac{\partial h}{\partial z} - R\right)\mathbf{k}\right]\ \left[\frac{\partial h}{\partial x}\,\mathbf{i},\ \frac{\partial h}{\partial y}\,\mathbf{j},\ \left(\frac{\partial h}{\partial z} - 1\right)\mathbf{k}\right]$$

and

$$n\frac{\partial h}{\partial t} = \left[K_{xx}\left(\frac{\partial h}{\partial x}\right)^2 + K_{yy}\left(\frac{\partial h}{\partial y}\right)^2 + K_{zz}\left(\frac{\partial h}{\partial z}\right)^2 - \frac{\partial h}{\partial z}(K_{zz} + R) + R\right] \tag{2.77}$$

The unconfined flow problem is then described by a Laplace or Poisson equation and a nonlinear boundary condition. From a computational perspective, the solution of the problem is difficult because of the unknown location of the free surface. Furthermore, the nonlinearity of the free surface boundary condition is likely to be so severe that conventional finite difference and finite element numerical techniques are also likely to succeed only in the most elementary flow models.

Recognizing the numerical problems associated with the solution of the free surface problem, an obvious approach to the solution of the unconfined flow problem is to approximate phreatic flow using the confined aquifer flow equation. And, for many groundwater problems, this is indeed practical as long as the drawdowns occurring within the aquifer are small in relation to the initial saturated thickness. As a result, the transmissivity and storage coefficients, which are functions of the saturated thickness, will not vary appreciably and the vertical velocities will remain small in relation to the horizontal velocities within the aquifer. The drawdowns are not likely to be small, however, in the vicinity of any pumping or injection wells and correction terms must be added to the equations to predict drawdowns near the wells (see, for example, Jacob, 1950).

The unconfined aquifer problem is commonly analyzed using the Dupuit theory of unconfined flow, provided the system cannot be approximated by a linear flow model. The Dupuit assumptions essentially prescribe uniform and horizontal flow within any vertical cross section and a mildly sloping aquifer, i.e., a hydrostatic pressure distribution (Remson, et al., 1971). With the Dupuit assumptions and the assumptions used in the development of the confined flow equation, an approximate equation for two-dimensional unconfined flow can be obtained by integrating the continuity equation over the saturated thickness of the aquifer. However, in this case, the saturated thickness is a function of space and time. Replacing b in Equation 2.48 by the height of the free surface, $h = z = \eta\,(x,y,t)$, we have

$$-\left\{\int_0^{\eta}\frac{\partial q_x}{\partial x}dz\right\} - \left\{\int_0^{\eta}\frac{\partial q_y}{\partial y}dz\right\} - \left\{\int_0^{\eta}\frac{\partial q_z}{\partial z}dz\right\} = \left\{\int_0^{\eta}\frac{\partial}{\partial t}S_s h\,dz\right\} \tag{2.78}$$

Once again assuming effective values of the conductivity and storage coefficient and alignment of the coordinate system with the principal axes of the conductivity tensor, Equation 2.78 becomes

$$-\left\{\frac{\partial}{\partial x}(\eta\bar{q}_x) + \frac{\partial}{\partial y}(\eta\bar{q}_y)\right\} + \left\{q_x\frac{\partial\eta}{\partial x} + q_y\frac{\partial\eta}{\partial y} - q_z(\eta)\right\}\bigg|_F = \frac{\partial}{\partial t}(S_s\eta\bar{h}) - S_s\eta\frac{\partial\eta}{\partial t} \tag{2.79}$$

where $\bar{h}$, $\bar{q}_x$, and $\bar{q}_y$ are again vertical averaged values. Assuming,

$$\bar{h} \simeq \eta = h$$

which is analogous to the Dupuit assumptions (Cooley, 1974), the free-surface, kinematic boundary condition is

$$q_z(\eta) = S_y \frac{\partial h}{\partial t} - R + q_x \frac{\partial \eta}{\partial x} + q_y \frac{\partial \eta}{\partial y} \quad \text{on} \quad F \tag{2.80}$$

where S_y is the specific yield of the aquifer and R is the net recharge (LT^{-1}). The specific yield (or effective porosity) differs from the storage coefficient in that it is the quantity of water that can be drained out of a saturated volume of porous media. The magnitude of the specified yield is on the order of the porosity.

Combining Equations 2.79 and 2.80, introducing Darcy's law, and dropping the overbar notation we have

$$\frac{\partial}{\partial x}\left[K_{xx} h \frac{\partial h}{\partial x}\right] + \frac{\partial}{\partial y}\left[K_{yy} h \frac{\partial h}{\partial y}\right] + R = [S_y + S_s h] \frac{\partial h}{\partial t} \tag{2.81}$$

The specific storage resulting from compressibility of the aquifer and the water is usually much smaller than the specific yield for unconfined aquifers (Walton, 1970). Equation 2.81 may then be rewritten as

$$\frac{\partial}{\partial x}\left[K_{xx} h \frac{\partial h}{\partial x}\right] + \frac{\partial}{\partial y}\left[K_{yy} h \frac{\partial h}{\partial y}\right] + R = S_y \frac{\partial h}{\partial t} \tag{2.82a}$$

or

$$\nabla \cdot \mathbf{K} \nabla h^2 + 2R = 2S_y \frac{\partial h}{\partial t} \tag{2.82b}$$

which is the Boussinesq equation, a nonlinear, second-order partial differential equation. Pumping and injection wells may also be incorporated via Equation 2.53, in the recharge terms of the equation, provided the well can be represented as point sources or sinks.

Two methods of linearizing the Boussinesq problem are often used in groundwater hydrology. The first method is based on the assumption that the depth of flow varies slightly within the groundwater system. The assumption is analogous to the use of the confined flow equation to approximate unconfined flow (Polubarinova-Kochina, 1962). The head may then be expressed as

$$h = \bar{h} + \hat{h}$$

where $\bar{h}$ is the average depth of flow and $\hat{h}$ is the deviation of the head from $\bar{h}$. If it is assumed that $\hat{h} \ll \bar{h}$, the Boussinesq equation becomes,

$$\frac{\partial}{\partial x}\left[K_{xx} \bar{h} \frac{\partial h}{\partial x}\right] + \frac{\partial}{\partial y}\left[K_{yy} \bar{h} \frac{\partial h}{\partial y}\right] + R = S_y \frac{\partial h}{\partial t} \tag{2.83a}$$

or

$$\nabla \cdot \mathbf{T} \nabla h + R = S_y \frac{\partial h}{\partial t} \tag{2.83b}$$

where $\mathbf{T} = \mathbf{K}\bar{h}$. The equation has the same form as the confined flow equation (Eq. 2.55).

The second method of linearization is based on the variation of the temporal derivative. Rewriting the derivative as

$$S_y \frac{\partial h}{\partial t} = \frac{S_y}{2h} \frac{\partial h^2}{\partial t}$$

and assuming that $\bar{S} = S_y/2h$ is approximately constant and equal to $S_y/2\bar{h}$, the Boussinesq equation is linear in h^2. Unconfined flow is then described by the equation,

$$\frac{\partial}{\partial x}\left[K_{xx}\frac{\partial h^2}{\partial x}\right] + \frac{\partial}{\partial y}\left[K_{yy}\frac{\partial h^2}{\partial y}\right] + 2R = 2\bar{S}\frac{\partial h^2}{\partial t} \tag{2.84a}$$

or

$$\nabla \cdot \mathbf{K} \nabla h^2 + 2R = 2\bar{S}\frac{\partial h^2}{\partial t} \tag{2.84b}$$

The initial and boundary conditions are also required to be linear in h^2. This second linearized approximation predicts more accurately the water levels in unconfined groundwater systems provided $[h(x, y, t) - h(x, y, 0)] \leq .5h(x, y, 0)$ and $R \leq 0.2K$ (Brutsaert & Ibrahim, 1966; Mariño, 1964).

Example Problem 2.10

For the unconfined aquifer shown in Figure 2.18, determine the steady-state head distribution. Assume the aquifer is homogeneous and isotropic and has a uniform recharge rate, R (L/T).

The hydraulics of the aquifer system are described by the Boussinesq equation

$$\frac{d}{dx}\left[Kh\frac{dh}{dx}\right] + R = 0 \tag{2.85}$$

and the boundary conditions

$$h(-L/2) = h_0 \tag{2.86a}$$

$$\left.\frac{dh}{dx}\right|_{x=0} = 0 \tag{2.86b}$$

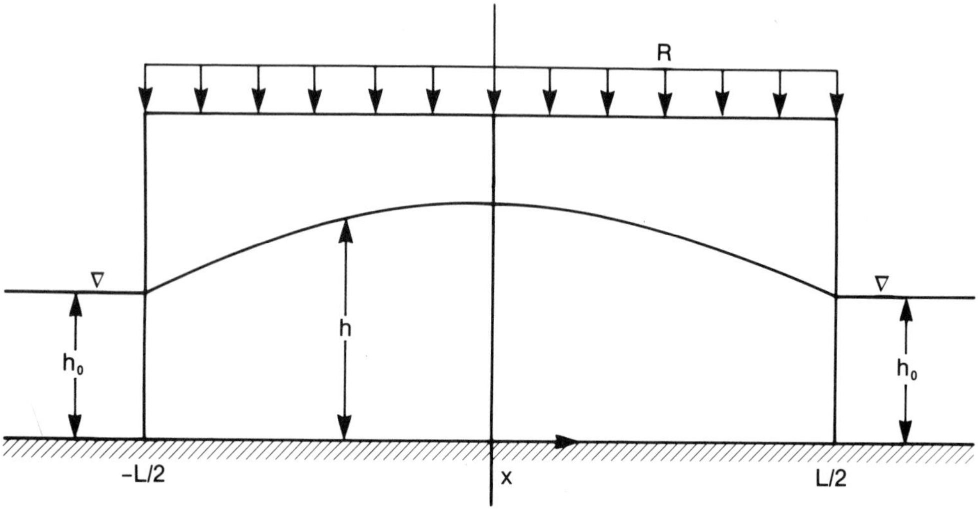

FIGURE 2.18 Unconfined flow example

Direct integration of the equation produces

$$h^2 = -\frac{Rx^2}{K} + C_1 x + C_2$$

where C_1 and C_2 are the constants of integration. Evaluating these constants from the known boundary condition information, the steady-state head distribution is described by the ellipse equation, or

$$h^2 = h_0^2 - \frac{R}{K}\left\{x^2 - \left(\frac{L}{2}\right)^2\right\} \tag{2.87}$$

Example Problem 2.11

Artificial recharge of groundwater can often take place in spreading basins or stream channels. For the recharge system shown in Figure 2.19, determine the steady-state head distribution. Assume the aquifer is homogeneous and isotropic and that for $x > x_e$, (the effective radius) $h = d$, the initial head in the system.

The hydraulics of the aquifer system are described by the Boussinesq equation, or

$$\frac{d}{dx}\left[Kh\frac{dh}{dx}\right] + f(x) = 0 \tag{2.88}$$

where $f(x)$ is the source term that is defined as

$$f(x) = \begin{cases} R & |x| \leq L \\ 0 & |x| > L \end{cases} \tag{2.89a}$$

The boundary conditions of the problem require that

$$h(x_e) = d, \quad \left.\frac{dh}{dx}\right|_{x=0} = 0 \tag{2.89b}$$

and that there be continuity of the head and the velocity at the boundary, $x = L$, or

$$h|_{L^-} = h|_{L^+}, \quad \left.\frac{dh}{dx}\right|_{L^-} = \left.\frac{dh}{dx}\right|_{L^+}$$

For the flow region, $|x| \leq L$, the governing equation is

$$\frac{d}{dx}\left[h\frac{dh}{dx}\right] + \frac{R}{K} = 0$$

Direct integration of the equation produces

$$h^2 = -\frac{R}{K}x^2 + C_2$$

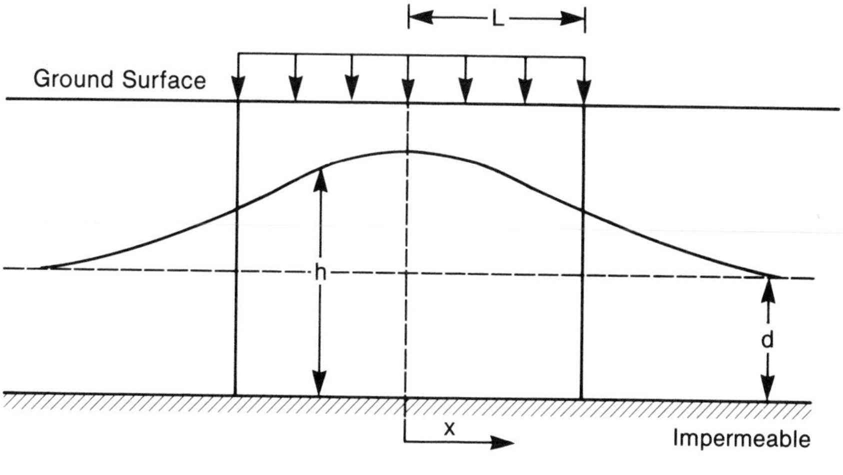

FIGURE 2.19 Unconfined aquifer system

where C_2 is a constant of integration and we have used the gradient condition at $x = 0$ to evaluate the first constant of integration, $C_1 = 0$.

For the flow domain, $|x| > L$, the governing equation is simply,

$$\frac{d}{dx}\left[h\frac{dh}{dx}\right] = 0$$

The solution is given as

$$h^2 = C_3x + C_4$$

where C_3 and C_4 are constants of integration. We evaluate these constants from the given boundary conditions. For example, continuity of h and dh/dx at $x = L$ requires that

$$2h\frac{dh}{dx}\bigg|_{L^-} = -\frac{2RL}{K} = 2h\frac{dh}{dx}\bigg|_{L^+} = C_3$$

and

$$C_3 = -\frac{2RL}{K}$$

At the effective distance, x_e, we also have

$$h(x_e) = d \Rightarrow d^2 = \frac{2(-RL)}{K}x_e + C_4$$

or

$$C_4 = d^2 + \frac{2RLx_e}{K}$$

Also at $x = L$,

$$h^2 = \frac{-RL^2}{K} + C_2 = \frac{2(-RL)}{K}L + d^2 + \frac{2RL}{K}x_e$$

or,

$$C_2 = \frac{-RL^2}{K} + d^2 + \frac{2RL}{K}x_e$$

The head distribution in the aquifer system is then given by the equations

$$h^2 = -\frac{R}{K}x^2 - \frac{RL^2}{K} + d^2 + \frac{2RLx_e}{K}, \quad |x| \leq L$$

$$h^2 = -\frac{2RL}{K}x + d^2 + \frac{2RLx_e}{K}, \quad L < |x| \leq x_e \tag{2.90}$$

$$h = d, \quad |x| \geq x_e$$

Example Problem 2.12

The extensive, unconfined aquifer shown in Figure 2.20 is to be developed as a source of water supply. Assuming that the aquifer is homogeneous and isotropic, develop the steady-state response equation relating the velocities (q_r, q_θ) and the groundwater extraction rate(s), Q.

In radial coordinates, the Boussinesq equation may be expressed as

$$\frac{1}{r}\frac{d}{dr}\left[rh\frac{dh}{dr}\right] = 0 \tag{2.91}$$

with the Darcy boundary condition,

$$\lim_{r \to 0}\left[rh\frac{dh}{dr}\right] = \frac{Q}{2\pi K} \tag{2.92}$$

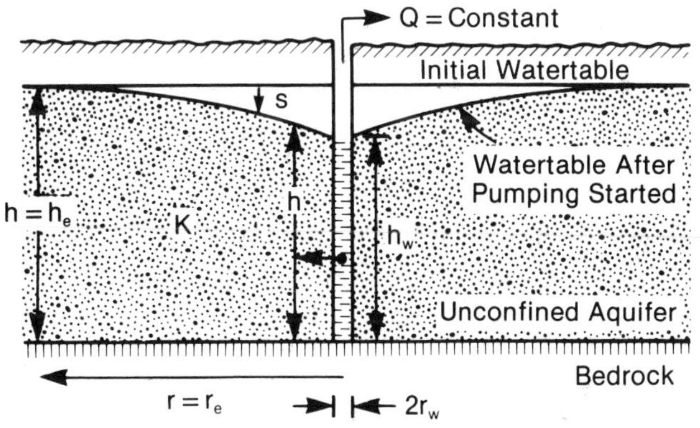

FIGURE 2.20 Radial flow in a water table aquifer (Adapted from R.J.M. DeWiest, *Geohydrology*. New York: Wiley, 1965, p. 242. Copyright © 1965 by John Wiley & Sons, Inc. Reprinted with permission)

The solution of the differential equation is

$$h^2 = \frac{Q}{\pi K} \ln r + C$$

or

$$h^2 = \frac{Q}{2\pi K} \ln \{(x - x_w)^2 + (y - y_w)^2\} + C \tag{2.93a}$$

where C is a constant of integration and (x_w, y_w) are the coordinates of the pumping well.

For m pumping or injection $(-Q)$ wells in the basin, the system response equation is found from superposition of the individual well responses, or

$$h^2 = \sum_w \frac{Q_i}{2\pi K} \ln\{(x - x_w)^2 + (y - y_w)^2\} + C \tag{2.93b}$$

The velocity field is given by Darcy's law, or

$$q_r = -K\frac{\partial h}{\partial r}, \quad q_\theta = \frac{K}{r}\frac{\partial h}{\partial \theta}$$

Evaluating the partial derivatives, the velocities are

$$q_r = -\frac{K}{\pi}\sum_w Q_i\left\{\frac{(r\cos\theta - x_w)\cos\theta + (r\sin\theta - y_w)\sin\theta}{(r\cos\theta - x_w)^2 + (r\sin\theta - y_w)^2}\right\}/(2h) \tag{2.94a}$$

$$q_\theta = \frac{K}{\pi}\sum_w Q_i\left\{\frac{(r\cos\theta - x_w)(-r\sin\theta) + (r\sin\theta - y_w)(r\cos\theta)}{(r\cos\theta - x_w)^2 + (r\sin\theta - y_w)^2}\right\}/(2rh) \tag{2.94b}$$

The velocities, in this case, are nonlinear functions of the pumping or injection rates.

2.9 BOUNDARY AND INITIAL CONDITIONS OF THE AQUIFER SYSTEM

No two aquifers are alike in size, shape, complexity, or composition. Groundwater basins may consist of sand and gravel deposits or are lensed at random with silt and clay. Other basins may be located in mountain valleys and contain sedimentary materials from meandering streams and rivers or consist of materials formed by chemical or faulting processes (consolidated sandstone or fissured limestone). What groundwater systems do have in common is that they are a part of the hydrologic cycle. The boundary conditions of the groundwater system represent the aquifer's hydraulic interaction with other portions of the hydrologic cycle. Recharge can occur, for example, from precipitation, streamflow, or leakage from semiconfining formations. In contrast, evapotranspiration and baseflow are examples of the discharge of groundwater to the surface water environment. In reality, these hydrologic boundaries are not sharply defined. There may not be a distinct or sudden change in underground conditions to distinguish between an aquifer and the confining region, or an aquifer and partially saturated porous media. Because of these problems, a conceptual or model aquifer with distinct boundaries is often used to approximate the actual diffuse boundaries of the groundwater basin.

The conceptual model of the aquifer system is based on the assumption of a two-dimensional flow regime. All boundary effects are then assumed to be uniformly distributed over the vertical thickness of the aquifer. The head or a function of the head must be prescribed on all the boundaries of the aquifer system to complete the mathematical characterization of the problem. The most common type of boundary conditions are Dirichlet, Neumann, or Cauchy.

1. *Dirichlet conditions* occur when a portion of the boundary is at a prescribed head level. For example, if an aquifer is adjacent to a stream or lake then

$$h(x, y, z, t) = h^*(t), \ (x, y, z) \in \Gamma_1 \tag{2.95}$$

where $h^*(t)$ is the boundary potential, Γ_1, is the boundary curve, and t is time.

2. *Neumann conditions* occur when a portion of the boundary has specified flow transversing it normal to the boundary. If $g^*(x, y, z, t)$ is such a flow and n is the normal to the bounding curve Γ_2, then the Neumann condition states

$$\mathbf{q} \cdot \mathbf{n} = -\mathbf{K}\nabla h \cdot \mathbf{n} = g^*(x, y, z, t), \ (x, y, z) \in \Gamma_2 \tag{2.96}$$

3. The *Cauchy* or mixed boundary condition occurs when both the potential and its normal derivatives are specified on the boundary Γ_3 or

$$\delta_1 T_{\alpha\beta} \frac{\partial h}{\partial x_\beta} d_\alpha + \delta_2 h + \delta_3 = 0 \text{ on } \Gamma_3, \ \alpha, \beta = x, y \tag{2.97}$$

where d_α are the direction cosines associated with the boundary of the system and δ_1, δ_2, δ_3 are given functions of space and time. This boundary condition approximates the induced infiltration in a coupled aquifer and stream system (Prickett & Lonnquist, 1971).

Prior to solving the groundwater flow equation, the initial state or condition of the aquifer system must be specified. Typically, the initial conditions can be expressed as

$$h(x, y, z, 0) = g(x, y, z) \tag{2.98}$$

where $g(x, y, z)$ defines the head at every point within the aquifer system.

2.10 CLOSING COMMENTS

Groundwater flow equations have been developed for confined and unconfined groundwater systems. Because of the mathematical complexity of these linear or possibly nonlinear equations, numerical techniques such as finite difference and finite element methods are often used to approximate and predict flow variations in regional aquifer systems. More importantly, these techniques can be used to describe how groundwater management alternatives affect the hydraulic and water quality response of the aquifer system. However, before considering the development of these response or transfer equations, the generalized groundwater quality problem will be presented in Chapter 3.

2.11 REFERENCES

Bear, J. *Dynamics of Fluids in Porous Media.* New York: Elsevier, 1972.

Bear, J. *Hydraulics of Groundwater.* New York: McGraw-Hill, 1979.

Bredehoeft, J. D., & Pinder, G. F. "Digital Analysis of Areal Flow in Multiaquifer Groundwater Systems: A Quasi-Three Dimensional Model." *Water Resources Research,* 6(3): 885–888, 1970.

Brutsaert, W., & Ibrahim, H. A. "On the First and Second Linearization of the Boussinesq Equation." *J. Amer. Society of Geophysics,* 11: 549–554, 1966.

Carslaw, H. S., & Jaeger, J. C. *Conduction of Heat in Solids.* London: Oxford University Press, 1959.

Cooley, R. L., Harsh, J. F., & Lewis, D. C. *Principles of Groundwater Hydrology,* Vol. 10, *Hydrologic Engineering Methods for Water Resources Development.* Davis CA: Hydrologic Engineering Center, 1972.

Cooley, R. L. *Finite-Element Solutions for the Equations of Groundwater Flow,* Technical Report Series H-W, No. 18, Desert Research Institute, University of Nevada, Reno, 1974.

Cooper, H. H. "The Equation of Groundwater Flow in Fixed and Deforming Coordinates," *Journal of Geophysical Research,* 71(20): 4785–4790, 1966.

Corey, G. L., & Corey, A. T. "Simulation for Drainage of Soils." ASCE, *J. Irr. Drainage Div.,* IR3, pp. 3–23, 1967.

DeWiest, R. J. M. *Geohydrology.* New York: John Wiley and Sons, 1965.

Eagleson, P. S. *Dynamic Hydrology*. New York: McGraw-Hill, 1970.

Hantush, M. S. *Plane Potential Flow of Groundwater with Linear Leakage*, Ph.D. dissertation, University of Utah, 1949.

Hantush, M. S. "Nonsteady Flow to Flowing Wells in Leaky Aquifer." *J. of Geophysical Research*, 64:1043–1052, 1959.

Hantush, M. S. "Modification of the Theory of Leaky Aquifers," *J. of Geophysical Research*, 65:3713–3725, 1960.

Hantush, M. S. "Flow of Groundwater in Sands of Nonuniform Thickness, 2., Approximate Theory." *J. of Geophysical Research*, 67(2):711–720, 1962.

Jacob, C. E. "Radial Flow in a Leaky Artesian Aquifer." *Transanctions of the American Geophysical Union*, 27:198–205, 1946.

Jacob, C. E. "Flow of Groundwater," in *Engineering Hydraulics*, H. Rouse, Ed. New York: John Wiley and Sons, 1950.

Mariño, M. A. "Growth and Decay of Groundwater Ridges in Response to Deep Percolation," Master's thesis, New Mexico Institute of Mining and Technology, Socorro NM, 1964.

Mariño, M. A., & Luthin, J. N. *Seepage and Groundwater*. In *Developments in Water Science*, vol. 13, New York: Elsevier, 1982.

Neuman, S. P., & Witherspoon, P. A. "Applicability of Current Theory of Flow in Leaky Aquifers." *Water Resources Research*, 5(4): 817–829, 1969.

Philip, J. R. "Evaporation and Moisture and Heat Fields in The Soil." *J. of Meteorology*, (414), 1957.

Pinder, G. F., & Gray, W. G. *Finite Element Simulation in Surface and Subsurface Hydrology*. New York: Academic Press, 1977.

Pipes, L. A. *Applied Mathematics for Engineers and Physicists*. New York: McGraw-Hill, 1958.

Polubarinova-Kochina, P. Ya. *Theory of Groundwater Movement*. Princeton NJ: Princeton University Press, 1962.

Prickett, T. A., & Lonnquist, C. O. "Selected Digital Computer Techniques for Groundwater Resource Evaluation." *Bulletin No. 55*, Illinois State Water Survey, Urbana, 1971.

Remson, I., Hornberger, G. M., & Molz, F. J. *Numerical Methods in Subsurface Hydrology*. New York: Wiley-Interscience, 1971.

Theim, G. *Hydrologische Methoden*. Leipzig: J. M. Gebhardt, 1906.

Theis, C. V. "The Relation Between Lowering of the Piezometric Surface and the Rate and Duration of Discharge of a Well Using Groundwater Storage." Trans. Amer. Geophysical Union, 16th Annual Meeting, pp. 519–524, 1935.

Todd, D. K. *Groundwater Hydrology*. New York: John Wiley and Sons, 1980.

3. GROUNDWATER QUALITY—THE MASS TRANSPORT PROBLEM

3.1 MASS TRANSPORT EQUATION

The management of an aquifer's assimilative waste capacity, or the prediction of contaminant transport in groundwater systems, is predicated upon the conservation of mass equation (Eq. 2.12). The state variables of the groundwater system are the concentrations of all constituents within the groundwater aquifer. If $c_s(ML^{-3})$ is the mass concentration of constituent s, and if the index set χ defines all the constituents in the system, then assuming isothermal conditions, the continuity equation may be written as

$$-\nabla \cdot c_s\mathbf{q}_s = \frac{\partial}{\partial t}(c_s n) \qquad s \in \chi \tag{3.1}$$

where $\mathbf{q}_s$ is now the mass average velocity of the sth constituent.

The mass average velocity of the overall solution, $\mathbf{q}$, can be related to the mass velocities of each constituent, $\mathbf{q}_s$, and the concentration of each constituent, or

$$\mathbf{q} = \frac{\sum_{s\in\chi} c_s\, q_s}{\sum_{s\in\chi} c_s} = \frac{\sum_{s\in\chi} c_s\, q_s}{c} \tag{3.2}$$

where c is the density or mass concentration of the solution.

The product of the concentration and the velocity of the constituent is the local rate at which mass passes through a unit area perpendicular to $\mathbf{q}$ (relative to a fixed coordinate system). The velocity is referred to as the *barycentric velocity* (see Figure 3.1).

In addition to external forces affecting fluid flow, the movement of constituents in the aquifer system is also caused by the concentration gradient of each constituent. It is possible to define the mass flux of each constituent s, $\mathbf{j}_s$, relative to the mass average velocity of the solution, as

$$\mathbf{j}_s = c_s(\mathbf{q}_s - \mathbf{q}) \tag{3.3}$$

The mass velocity, $\mathbf{q}_s$, of a constituent relative to the mass average velocity of the solution is thought to be described by Ficks's law of diffusion (Bear, 1972). The mass transport that occurs from the differences in the mass fluxes is proportional to the concentration gradient and acts in the direction to decrease the concentration gradient, or

$$\mathbf{j}_s = -c\,\mathbf{D}\nabla(c_s/c) \tag{3.4}$$

where $\mathbf{D}$ is the hydrodynamic dispersion tensor. The dispersion phenomenon is the result of the molecular diffusion and hydrodynamic mixing that occurs during groundwater flow. Variations in the magnitude and direction of the local fluid velocity and heterogeneity in the porosity initiate a spreading of the contaminant within the groundwater system. As time increases, the constituent occupies an ever-increasing volume of the porous media. On a large scale, however, spatial variations

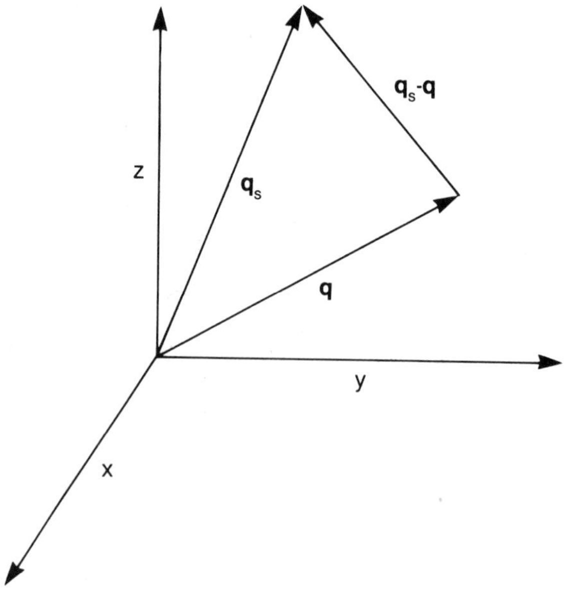

FIGURE 3.1 Barycentric velocity

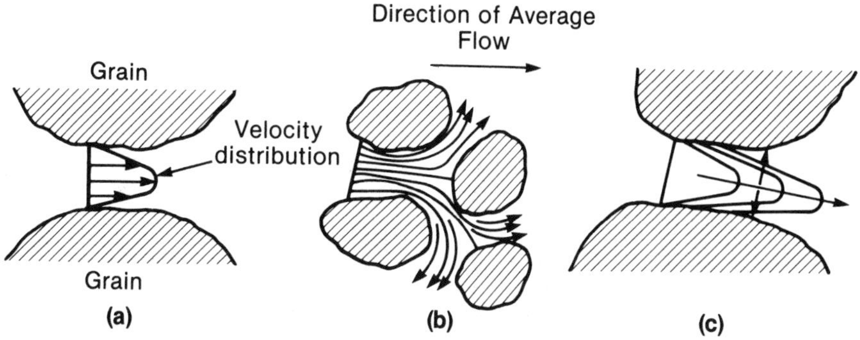

FIGURE 3.2 Spreading due to mechanical dispersion and molecular diffusion (Adapted from J. Bear, *Hydraulics of Groundwater*. New York: McGraw-Hill, 1979, p. 227. Copyright © 1979 by McGraw-Hill Book Company. Reprinted with permission)

in the hydraulic conductivity also contribute to the dispersion or mixing of the constituents within the aquifer system. Figures 3.2 and 3.3 illustrate the dispersion process.

Although the representation of the dispersion tensor does not appear to be unique for all field situations, the equations typically used to relate dispersion to

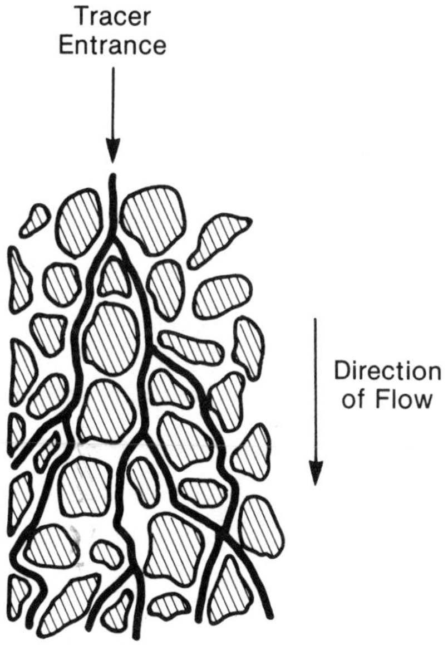

FIGURE 3.3 Lateral dispersion of a tracer originating from a point source in a porous medium (Adapted from D.K. Todd, *Groundwater Hydrology*. New York: Wiley, 1980, p. 95. Copyright © 1980 by John Wiley & Sons, Inc. Reprinted with permission)

the velocity field are (Scheidegger, 1961)

$$D_{\alpha\beta} = \alpha_{II}\delta_{\alpha\beta}q + (\alpha_I - \alpha_{II})q_\alpha q_\beta/q + D_d\tau, \quad \alpha,\beta = x,y,z \tag{3.5}$$

where α_I and α_{II} (L) are the longitudinal and transverse dispersivities of the aquifer; $\delta_{\alpha\beta}$ is the Kronecker delta ($\alpha = \beta$, $\delta = 1$; $\alpha \neq \beta$, $\delta = 0$); q is the magnitude of the velocity; and q_α, q_β are the velocities in the x, y, z coordinate direction. D_d is the molecular diffusion coefficient which is dependent on the kinetic properties of individual fluid molecules, and τ is the medium tortuosity. Equation 3.5 assumes, however, that the medium is isotropic with respect to the dispersivity.

Assuming that the average velocity of the solution is again composed of a Darcy velocity ($\mathbf{q}_D$) and the grain velocity ($\mathbf{q}_g$), then the mass flux of a constituent is (Bredehoeft & Pinder, 1973):

$$c_s\,\mathbf{q}_s = \mathbf{j}_s + c_s\,\mathbf{q} = \mathbf{j}_s + c_s(\mathbf{q}_D + \mathbf{q}_g) = -c\,\mathbf{D}\nabla(c_s/c) + c_s(\mathbf{q}_D + \mathbf{q}_g) \tag{3.6}$$

The continuity equation can then be expressed as

$$-\nabla\cdot\{-c\,\mathbf{D}\nabla\,(c_s/c)\} - \nabla\cdot c_s(\mathbf{q}_D + \mathbf{q}_g) = \frac{\partial}{\partial t}(nc_s) \tag{3.7}$$

Again assuming that deformation occurs only in the vertical direction, i.e., $\mathbf{q}_g = nW_g\mathbf{k}$, then

$$\nabla\cdot c\,\mathbf{D}\nabla(c_s/c) - \nabla\cdot c_s\mathbf{q}_D - \frac{\partial}{\partial z}(nc_sW_g) = \frac{\partial}{\partial t}(nc_s) \tag{3.8a}$$

and

$$\nabla\cdot c\mathbf{D}\nabla(c_s/c) - \nabla\cdot c_s\mathbf{q}_D - nc_s\frac{\partial W_g}{\partial z} - c_sW_g\frac{\partial n}{\partial z} - nW_g\frac{\partial c_s}{\partial z} = n\frac{\partial c_s}{\partial t} + c_s\frac{\partial n}{\partial t} \tag{3.8b}$$

Equation 3.8b may be simplified using the substantial derivatives introduced in Equation 2.34, again to account for the relative motion of the grains. Equation 3.8b then becomes

$$\nabla\cdot c\,\mathbf{D}\nabla(c_s/c) - \nabla\cdot c_s\mathbf{q}_D - nc_s\frac{\partial W_g}{\partial z} = n\frac{dc_s}{dt} + c_s\frac{dn}{dt} \tag{3.9}$$

As described in Chapter 2, the spatial variation in the grain velocity is related to the transient fluid pressure variations (p), or

$$\frac{\partial W_g}{\partial z} = \alpha\frac{dp}{dt} \tag{2.43}$$

where $\alpha(LM^{-1}T^{2})$ is the compressibility of the granular skeleton. Simplifying Equation (3.9)

$$\nabla \cdot c\mathbf{D}\nabla(c_s/c) - \nabla \cdot c_s\mathbf{q}_D = n\frac{dc_s}{dt} + c_s\frac{dn}{dt} + nc_s\alpha\frac{dp}{dt} \qquad (3.10)$$

Finally, assuming,

$$W_g\frac{\partial p}{\partial z} << \frac{\partial p}{\partial t}, \quad W_g\frac{\partial n}{\partial z} << \frac{\partial n}{\partial t} \quad \text{and} \quad W_g\frac{\partial c_s}{\partial z} << \frac{\partial c_s}{\partial t}$$

the general transport equation for a conservative, noninteracting constituent is,

$$\nabla \cdot c\,\mathbf{D}\nabla(c_s/c) - \nabla \cdot c_s\mathbf{q}_D = nc_s\alpha\frac{\partial p}{\partial t} + \frac{\partial}{\partial t}(nc_s), \quad s \in \chi \qquad (3.11)$$

Equation 3.11, which is a second-order parabolic equation, relates the differences in the dispersive and convective mass fluxes to the time rate of change of the concentration and pressure fields within the aquifer system.

3.2 GROUNDWATER QUALITY MODEL

The prediction or simulation of groundwater quality is predicated on the hydrodynamic and mass transport equations of the aquifer system. The dependent variables of the prediction problem are the mass concentrations, c_s, the solution concentration (c) or density (ρ), the fluid velocity, the pressure variation, and, possibly, the viscosity. The viscosity may vary as a result of changes in the pressure and the overall solution concentration. Mathematically, the prediction problem is characterized by the equations describing:

1. The motion or momentum balance for the system (Darcy's law) or

$$\mathbf{q} = -\frac{k\gamma}{\mu}\nabla\left(\frac{p}{\gamma} + z\right) \qquad (2.5)$$

where k is the intrinsic permeability and γ is the specific weight.

2. The continuity equation of the groundwater system, relating the divergence of the mass flux of the solution and the temporal variation in the fluid pressure,

$$-\nabla \cdot \rho\mathbf{q}_D = \rho(\alpha + n\beta)\frac{\partial p}{\partial t} \qquad (2.45)$$

3. The mass transport equation,

$$\nabla \cdot c\,\mathbf{D}\nabla(c_s/c) - \nabla \cdot c_s\mathbf{q}_D = nc_s\alpha\,\frac{\partial p}{\partial t} + \frac{\partial(nc_s)}{\partial t}, \quad s \in \chi \tag{3.11}$$

where χ is an index set defining all the constituents within the system.

4. The equation defining the solution density or concentration in the aquifer system,

$$\rho = c = \sum_{s\in\chi} c_s \tag{3.12a}$$

5. The equations of state of the system. These equations relate the solution density and viscosity to the fluid pressure, p, and the constituent concentrations,

$$\rho = f_1(c_s;\, p) \tag{3.12b}$$

$$\mu = f_2(c_s;\, p) \tag{3.12c}$$

The generalized flow and transport equations of the groundwater quality system cannot, in general, be solved by analytical methods because of the dependence of the transport problem on the velocity field, the solution density, and the pressure variations occurring in the aquifer. The nonlinearity of the momentum, transport, and continuity equations also complicates the numerical solution of the problem; therefore, iterative or linearization methods are often the only feasible techniques for solution of the problem.

In many groundwater problems, however, the system's equations may be simplified by examining the magnitude of the convective and dispersive components of the mass transport equation. The first of these assumptions concerns the dispersive transport term, $\nabla \cdot c\,\mathbf{D}\nabla(c_s/c)$. In most groundwater problems, molecular diffusion is negligible in comparison with hydrodynamic dispersion; therefore, the tortuosity terms of the equation may be neglected. Second, if it can be assumed that relatively dilute concentrations of the constituents are present in the aquifer system, then the overall solution density is constant. The dispersion terms are then independent of the solution density or,

$$\nabla \cdot c\,\mathbf{D}\,(c_s/c) \cong \nabla \cdot \mathbf{D}\,\nabla c_s$$

Third, if the time rate of change of the fluid pressure is small in comparison with the time rate of change of the mass concentration and porosity, the compressibility term in the transport equation can be neglected,

$$nc_s\alpha\,\frac{\partial p}{\partial t} \approx 0 \tag{3.13}$$

Finally, assuming negligible viscosity variation, a state equation (Eq. 3.12), can be eliminated from the groundwater system's equations.

The ramifications of these assumptions are that now, (1) the flow equation may be expressed in terms of the hydraulic head rather than the individual pressure and gravitational potential components, and (2) the flow and transport equation are *linear* partial differential equations. With these assumptions, the groundwater quality system is described by (1) the flow equation,

$$-\nabla \cdot \mathbf{q}_D = S_s \frac{\partial h}{\partial t} \tag{2.47}$$

(2) the mass transport equation,

$$\nabla \cdot \mathbf{D}\nabla c_s - \nabla \cdot c_s\mathbf{q}_D = \frac{\partial}{\partial t}(nc_s) , \quad s \in \chi \tag{3.14}$$

and (3) the momentum balance (Darcy's law) for the aquifer system,

$$\mathbf{q}_D = -\mathbf{K}\nabla h \tag{2.5}$$

3.3 VERTICALLY AVERAGED MASS TRANSPORT EQUATIONS

The majority of groundwater quality simulation models are based on two-dimensional flow and mass transport equations. These equations can be obtained by vertically averaging the transport equation (Eq. 3.14), over the thickness of the aquifer system. Assuming (1) a conservative, noninteracting constituent and (2) that the constituent concentration does not affect the overall solution density, the averaging process is described by the equation,

$$\int_0^m \frac{\partial}{\partial t}(c_s n)dz = -\int_0^m \frac{\partial}{\partial x}(c_s q_x)dz - \int_0^m \frac{\partial}{\partial y}(c_s q_y)dz - \int_0^m \frac{\partial}{\partial z}(c_s q_z)dz$$
$$+ \int_0^m \frac{\partial}{\partial \alpha} D_{\alpha\beta} \frac{\partial c_s}{\partial \beta} dz, \qquad \alpha,\beta = x, y, z \tag{3.15}$$

where m is the saturated thickness of the aquifer. Again using Leibnitz's rule and assuming average values of the parameters (n, $\mathbf{D}$) over the saturated thickness, yields

$$\frac{\partial}{\partial t}(\bar{n}\,\bar{c}_s m) = \nabla \cdot (\bar{c}_s \bar{\mathbf{q}} m) - \nabla \cdot m\bar{\mathbf{D}}\nabla \bar{c}_s + c_s q^*|_{z=m} \tag{3.16}$$

where q_z^* is the mass flux entering the aquifer and the overbar notation denotes a vertically averaged quantity, e.g. $\bar{c}_s = 1/m \int_0^m c_s dz$. Small deviations from the mean values have been neglected, i.e.,

$$\bar{c}_s\bar{\mathbf{q}} \cong \overline{c_s\mathbf{q}} \tag{3.17}$$

It is this form of the transport equation that will be used in the subsequent development of the groundwater quality models presented in Chapter 7.

The two-dimensional representation of the dispersion tensor is similarly defined as (Pinder & Gray, 1977)

$$D_{xx} = \bar{D}_L\bar{q}_x^2/\bar{q}^2 + \bar{D}_T\bar{q}_y^2/\bar{q}^2 + \bar{D}_d\bar{\tau} \tag{3.18a}$$

$$D_{yy} = \bar{D}_L\bar{q}_y^2/\bar{q}^2 + \bar{D}_T\bar{q}_x^2/\bar{q}^2 + \bar{D}_d\bar{\tau} \tag{3.18b}$$

$$D_{xy} = D_{yx} = (\bar{D}_L - \bar{D}_T)\bar{q}_x\bar{q}_y/\bar{q}^2 \tag{3.18c}$$

where $\bar{q}$ is the magnitude of the velocity and $\bar{D}_L = \bar{\alpha}_I\bar{q}$ and $\bar{D}_T = \bar{\alpha}_{II}\bar{q}$ and, $\bar{\alpha}_I$, $\bar{\alpha}_{II}$ are again the longitudinal and transverse dispersivities. $\bar{D}_d$ and $\bar{\tau}$ represent the averaged molecular diffusion and tortuosity parameters.

3.4 BOUNDARY AND INITIAL CONDITIONS

The complete mathematical description of the groundwater quality problem requires a set of unique initial and boundary conditions. As in the flow equations, the initial conditions of the mass transport problem are the concentration of all constituents present in the system at the beginning of the simulation or planning horizon, or

$$c_s(\mathbf{x},0) = g_s(\mathbf{x}), \quad s \in \chi \tag{3.19}$$

where g_s is the given initial concentration.

The boundary conditions prescribe the variation of the mass concentration on the boundaries of the groundwater system. These conditions express the conservation of the mass flux on both sides of the system's boundaries. If the superscript *a* denotes the aquifer system inside the boundary, and the superscript *e*, the domain external to the boundary, and **n** is a unit vector normal to the boundary, the general boundary condition is

$$\mathbf{J}^a \cdot \mathbf{n} = \mathbf{J}^e \cdot \mathbf{n} \tag{3.20}$$

where $\mathbf{J} \cdot \mathbf{n}$ is the vector dot product. The convective and dispersive mass flux is

$$\mathbf{J} = -\mathbf{D}\nabla(c_s) + c_s\mathbf{q}$$

Two boundary conditions commonly occur in groundwater quality problems. The first occurs when the external boundary is impervious to mass flow and diffusion. In this event, the flux boundary condition is,

$$\mathbf{J}^a \cdot \mathbf{n} = 0 \tag{3.21a}$$

The second occurs when the external boundary is a liquid continuum, e.g., a lake, reservoir, or the ocean. The boundary condition is then,

$$\mathbf{J}^a \cdot \mathbf{n} = q_n c_s, \quad s \in \chi \tag{3.21b}$$

where c_s is the known concentration in the external domain.

Example Problem 3.1

Determine the concentration distribution in the semi-infinite groundwater system shown in Figure 3.4. The boundary and initial conditions for the problem are

$$c(\infty, t) = 0,\ t \geqslant 0 \tag{3.22a}$$

$$c(0, t) = c_0,\ t > 0 \tag{3.22b}$$

$$c(z, 0) = 0 \tag{3.22c}$$

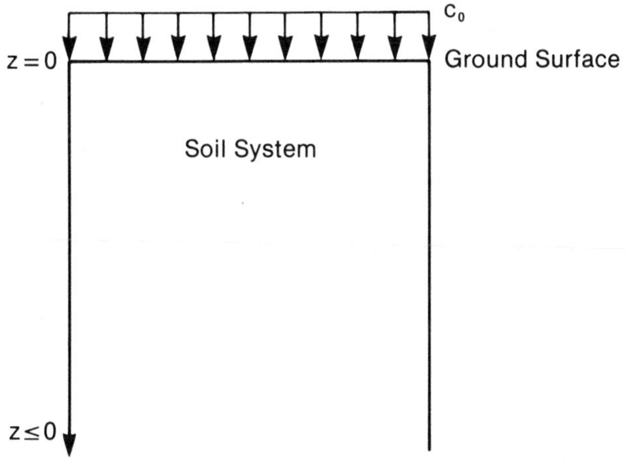

FIGURE 3.4 Semi-infinite groundwater system

The mass transport equation for the system can be expressed as

$$D\frac{\partial^2 c}{\partial z^2} - q_z\frac{\partial c}{\partial z} = \frac{\partial(nc)}{\partial t} \tag{3.23}$$

where we have assumed a constant velocity and dispersion field.

Taking the Laplace transform of the equation, i.e.,

$$\bar{c}(s) = \int_0^\infty e^{st}\, c(z,t)dt \tag{3.24}$$

the transformed equation becomes (with $\overline{D} = \overline{D}/n$ and $v_z = q_z/n$)

$$\overline{D}\frac{d^2\bar{c}}{dz^2} - v_z\frac{d\bar{c}}{dz} = s\bar{c}$$

with the conditions

$$\bar{c}(0) = c_0/s,\ \bar{c}(\infty) = 0$$

The solution of this linear, second-order, ordinary differential equation is

$$\bar{c} = \frac{c_0}{s}\exp\left\{\frac{[v_z - (v_z^2 + 4s\overline{D})^{1/2}]\, z}{2\overline{D}}\right\}$$

where exp is the exponential. After taking the inverse transform of the equation, the concentration distribution can be expressed as

$$c = \frac{c_0}{2}\left[\operatorname{erfc}\left\{\frac{z - v_z t}{2(\overline{D}t)^{1/2}}\right\} + \exp\left\{\frac{v_z z}{\overline{D}}\right\}\operatorname{erfc}\left\{\frac{z + v_z t}{2(\overline{D}t)^{1/2}}\right\}\right] \tag{3.25}$$

where erfc is the complementary error function.

An interesting generalization of the example problem occurs when there is a time-varying surface concentration, $c_0(t)$. If the boundary condition can be expressed as $c(0,t) = c_0e^{\gamma t}$, $t > 0$ then, introducing the transformation (Mariño, 1974a)

$$c(z,t) = c^*(z,t)\exp\left\{\frac{v_z z}{2\overline{D}} - \frac{v_z^2 t}{4\overline{D}}\right\}$$

the problem can be expressed as

$$\frac{\partial^2 c^*}{\partial z^2} = \frac{1}{\overline{D}} \frac{\partial c^*}{\partial t}$$

$$c^*(z, 0) = 0$$

$$c^*(0, t) = c_0\, e^{\eta t} \quad \text{and} \quad c^*(\infty, t) = 0$$

where $\eta = \frac{v_z^2}{4\overline{D}} + \gamma$. Again, taking the Laplace transform of the equation with respect to t yields

$$\frac{d^2\bar{c}^*}{dz^2} - \frac{s}{\overline{D}}\, c^* = 0$$

where

$$\bar{c}^*\,(0,\ s) = \frac{c_0}{s - \eta},\ \bar{c}^*\,(\infty,\ s) = 0$$

and the bar indicates the transformed function and s is the parameter of the transformation. The inverse Laplace transform of the equation is

$$c^*(z,\ t) = \frac{c_0 e^{\eta t}}{2}\left\{\exp\left[-z\left(\frac{\eta}{\overline{D}}\right)^{1/2}\right] \operatorname{erfc}\left[\frac{z}{2(\overline{D}t)^{1/2}} - (\eta t)^{1/2}\right] + \exp\left[z\left(\frac{\eta}{\overline{D}}\right)^{1/2}\right] \operatorname{erfc}\left[\frac{z}{2(\overline{D}t)^{1/2}} + (\eta t)^{1/2}\right]\right\}$$

where erfc is the complimentary error function. After recovering the original variables, the concentration distribution is described by the equation

$$c(z, t) = \frac{c_0 e^{\gamma t}}{2}\left\{\exp\left[\frac{z(v_z - \xi)}{2\overline{D}}\right] \operatorname{erfc}\left(\frac{z - \xi t}{2\sqrt{\overline{D}t}}\right) + \exp\left[\frac{z(v_z + \xi)}{2\overline{D}}\right] \operatorname{erfc}\left(\frac{z + \xi t}{2\sqrt{\overline{D}t}}\right)\right\} \tag{3.26}$$

where $\xi = (v_z^2 + 4\overline{D}\gamma)^{1/2}$. The concentration at any time and spatial location is again a linear function of the waste input concentration, c_0. Radioactive decay and adsorption can also be incoporated in the analytical solution as presented by Mariño (1974b).

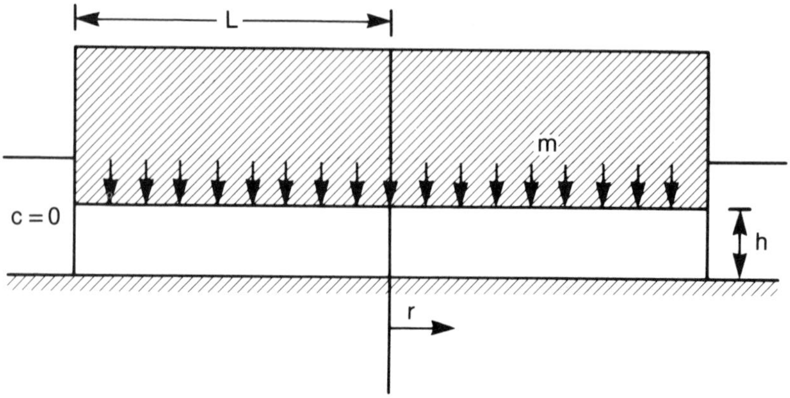

FIGURE 3.5 Groundwater contamination problem (Adapted from Wen-Hsiung Li, *Differential Equations of Hydraulic Transients, Dispersion, and Groundwater Flow*. Englewood Cliffs NJ: Prentice-Hall, 1972, p. 186. Copyright © 1972 by Prentice-Hall, Inc. Reprinted with permission)

Example Problem 3.2*

A contaminant is introduced to a circular aquifer at a constant rate, m (mass per unit time per unit horizontal area). The flow is negligible and dispersion takes place via molecular diffusion. The depth to length ratio of the aquifer (h/L) is also so small so as to eliminate any vertical variation in the concentration distribution. Determine the steady-state concentration. Assume at $r = L$, $c = 0$ (see Figure 3.5).

Neglecting convective transport and assuming that D is constant throughout the flow domain, the mass transport equation may be expressed in radial coordinates as

$$\frac{D}{r}\frac{d}{dr}\left(r\frac{dc}{dr}\right) + \frac{m}{n} = 0 \tag{3.27}$$

or, after simplifying,

$$\frac{d}{dr}\left[r\frac{dc}{dr}\right] = -\frac{mr}{Dn}$$

Directly integrating the equation produces

$$c = -\frac{mr^2}{4Dn} + C_1 \ln r + C_2$$

*From Wen-Hsiung Li, DIFFERENTIAL EQUATIONS OF HYDRAULIC TRANSIENTS, DISPERSION, AND GROUNDWATER FLOW, © 1972, p. 174. Reprinted by permission of Prentice-Hall, Inc., Englewood Cliffs, New Jersey.

where C_1 and C_2 are constants of integration. Requiring c to be finite everywhere in the domain implies that $C_1 = 0$. Furthermore, using the condition that $c(L) = 0$, the steady-state concentration distribution is given as

$$c(r) = \frac{m}{4Dn}(L^2 - r^2),\ 0 \le r \le L \tag{3.28}$$

Example Problem 3.3

A contaminant is introduced to the groundwater system shown in Figure 3.6. Assuming a constant velocity and dispersion field, determine the steady-state concentration distribution. The boundary and initial conditions are

$$c(x, 0, t) = \begin{cases} c_0,\ 0 \le |x| \le \varepsilon,\ t > 0 \\ 0,\ \varepsilon < |x| < \eta_o,\ t > 0 \end{cases} \tag{3.29a}$$

$$\left.\frac{\partial c}{\partial x}\right|_{x=0} = 0,\ t > 0, \tag{3.29b}$$

$$\left.\frac{\partial c}{\partial x}\right|_{x=\eta_0} = 0,\ t > 0 \tag{3.29c}$$

$$c(x, y, 0) = 0, 0 \le x < \eta_0, y > 0 \tag{3.29d}$$

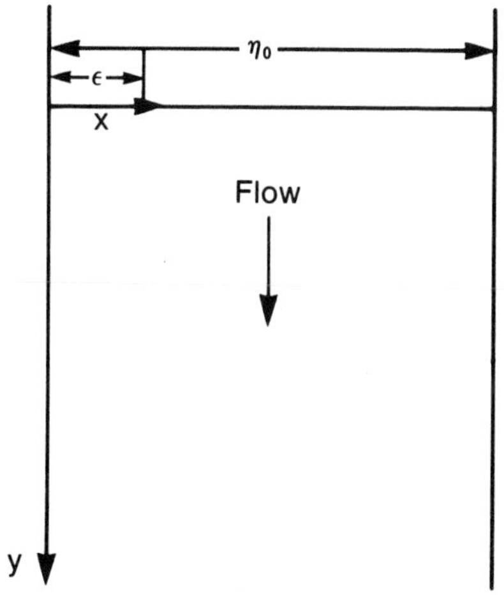

FIGURE 3.6 Two-dimensional dispersion example

The governing equation of the groundwater system is from Equation 3.14,

$$q_x \frac{\partial c}{\partial x} = D_{xx} \frac{\partial^2 c}{dx^2} + D_{yy} \frac{\partial^2 c}{\partial y^2} \qquad 0 \le x \le \eta_0,\ y > 0,\ t > 0 \qquad (3.30)$$

We begin the analysis by assuming, using separation of variables techniques, that the concentration, c, can be represented as the product of the functions,

$$c = X(x)Y(y)$$

Introducing this relation into the transport equation produces the two ordinary differential equations, or

$$\frac{d^2X}{dx^2} - \frac{\lambda^2}{D_{xx}} X = 0; \quad \frac{d^2Y}{dy^2} - \frac{q_x}{D_{yy}} \frac{dY}{dy} - \frac{\lambda^2}{D_{yy}} Y = 0$$

where λ^2 is the separation constant.

The solutions of these equations may be expressed as (Kamke, 1949)

$$X(x) = C_1 \cos \left\{ \frac{\lambda x}{(D_{xx})^{1/2}} \right\} + C_2 \frac{(D_{xx})^{1/2}}{\lambda} \sin \left\{ \frac{\lambda x}{(D_{xx})^{1/2}} \right\}$$

and

$$Y(y) = C_{3,n} \exp \left\{ 1/2 \left(\frac{q_x}{D_{yy}} + J_n \right) y \right\} + C_{4,n} \exp \left\{ 1/2 \left(\frac{q_x}{D_{yy}} - J_n \right) y \right\}$$

where C_1, C_2, $C_{3,n}$, and $C_{4,n}$ are constants of integration and

$$J_n = \left[\left(\frac{q_x}{D_{yy}} \right)^2 + \frac{4\eta^2\pi^2}{\eta_0^2} \left(\frac{D_{xx}}{D_{yy}} \right) \right]^{1/2}$$

Using the no-flow boundary condition yields

$$C_2 = 0$$

and

$$\lambda_n^2 = D_{xx} \left(\frac{n\pi}{\eta_0} \right)^2; \ n = 0, 1$$

The equation for X can then be written as

$$X_n(x) = C_n \cos\left(\frac{n\pi}{\eta_0} x\right)$$

Furthermore, boundedness of the solution requires that $C_{3,n} = 0$.

An infinite series solution can then be constructed

$$c(x, y) = \sum_{n=0}^{\infty} \left\{C_n \cos\left(\frac{\pi n x}{\eta_0}\right)\right\} \exp\left\{1/2\left(\frac{q_x}{D_{yy}} - J_n\right) y\right\}$$

The constants, C_n, can be evaluated using the boundary condition at the ground surface and the orthogonality of X_n or

$$c(x, 0) = \sum_{n=0}^{\infty} C_n \cos\left(\frac{n\pi}{\eta_0} x\right)$$

and

$$C_n = \frac{2}{\eta_0} \int_0^{\eta_0} c(x, 0) \cos\left(\frac{n\pi}{\eta_0} x\right) dx, \; n = 0, 1$$

The coefficients are then given as

$$C_0 = \frac{2\varepsilon C_0}{\eta_0}, \quad C_n = \frac{2C_0}{\pi n} \sin\frac{n\pi}{\eta_0}\varepsilon, \quad n = 1, 2 \tag{3.31a}$$

The steady-state concentration distribution can then be expressed as (Bruch & Street, 1967),

$$c(x, y) = \frac{C_0 \varepsilon}{\eta_0} + \sum_{n=1}^{\infty} \left[C_n \cos\frac{n\pi}{\eta_0} x\right] \exp\left\{1/2\left(\frac{q_x}{D_{yy}} - J_n\right) y\right\} \tag{3.31b}$$

3.5 *NONCONSERVATIVE PROCESSES*

Water quality constituents that undergo change or alteration in the aquifer system are nonconservative. Typical examples of these processes include coliform die-off, the transformation of ammonia to nitrate nitrogen via microbiological reactions, and various adsorption and ion-exchange processes. The mechanisms responsible for these transformations can also be incorporated in the source/sink term of the vertically averaged mass transport equation (Eq. 3.16).

If chemical or biochemical reaction can be represented by first-order kinetics, then the mass flux of the source/sink term of the transport equation may be expressed as

$$S_s^* = \pm k_s n c_s m \tag{3.32}$$

where k_s is the reaction coefficient (T^{-1}) for the constituent and m is the saturated thickness of the aquifer. A positive coefficient indicates the growth of the constituent within the system.

Another possible sink for constituents in the groundwater system is adsorption. *Adsorption* is the transfer of a constituent from the solution phase to the grains and interstices of the porous media. Adsorption is an extremely complicated process that is dependent on the pH of the system, the temperature, and aquifer materials. All these factors can be expressed in terms of an adsorption isotherm, which relates the concentration of a constituent in the adsorbed and solution phases. Several isotherms that are widely used in agricultural research to describe *equilibrium* adsorption are:

1. $A = Rc$, linear equilibrium adsorption.
2. $A = kc^n$, the Freundlich isotherm.
3. $A = \dfrac{kbc}{1 + kc}$, the Langmuir adsorption isotherm where k is the Langmuir equilibrium constant (L^3M^{-1}) and b (ML^3) is the maximum adsorptive capacity of the porous medium.

Phosphorous adsorption in soils has been represented by all three isotherms (Enfield & Bledsoe, 1975). The Freundlich isotherm has been used to simulate sulfate silver, atrazine, and phosphorus adsorption (Chao, et al., 1962; Cameron & Hute, 1977). Adsorption of boron and cadmium in certain types of soils can be described by the Langmuir adsorption isotherm (Tanji, 1970; John, 1972).

Nonequilibrium adsorption expresses the temporal change in the adsorbed constituents as a function of the solute concentration. Relations that have been widely used in the modeling of pesticides and herbicides are (Oddson, et al., 1970),

$$\frac{\partial A}{\partial t} = \alpha(kc - A) \tag{3.33a}$$

$$\frac{\partial A}{\partial t} = ac^b A^d \tag{3.33b}$$

where the parameters α, k, a, b, d are dependent on the soil conditions and the soil temperature.

Ion exchange is the simultaneous transfer of an ion between the solution and the solids of the porous media. A typical ion-exchange process can be expressed as

(Helfferich, 1962; Bolt, 1967)

$$b\bar{c}_1 + ac_2 \rightleftarrows a\bar{c}_2 + bc_1 \tag{3.34}$$

where 1 and 2 denote the chemical exchanging species with valences a and b. The adsorbed species is $\bar{c}$; the dissolved species is c. The ion selectivity coefficient ($\bar{k}_s$) relates the concentration of these species at equilibrium, or

$$\bar{k}_s = \frac{(\bar{c}_2)^a(c_1)^b}{(\bar{c}_1)^b(c_2)^a} \tag{3.35}$$

If it can be assumed that the concentration of the exchanging ion is small in relation to the concentration of the other ions, then the exchange process will not significantly affect the concentration of the ion in the solution or adsorbed phases. As a result, the adsorbed phase of the major ion is approximately equal to the cation exchange capacity (CE) in the solution phase, and the concentration of the major ion equals the total concentration c. Equation 3.35 becomes

$$\bar{k}_s = \frac{(\bar{c}_2)^a(c)^b}{(\mathrm{CE})^b(c_2)^a} \tag{3.36a}$$

and

$$k_d = \frac{\bar{c}_2}{c_2} = \left[\frac{\bar{k}_s(\mathrm{CE})^b}{(c_0)^b}\right]^{1/\alpha} \tag{3.36b}$$

where k_d is the ion-exchange distribution coefficient. For these conditions, a linear relationship exists between the adsorbed and solute species. The distribution coefficient is then the slope of the equilibrium ion-exchange isotherm.

These relations may be incorporated in the mass transport equation by modifying the derivative,

$$\frac{\partial}{\partial t}(nc_s m)$$

to reflect the change in mass in the adsorbed phase. If we assume that A_s is the concentration of constituent s in the adsorbed phase, i.e., the mass of the constituent per unit volume of the solids, then this term is simply,

$$\frac{\partial}{\partial t}\{nc_s m + (1 - n)A_s m\} \tag{3.37}$$

where A_s is again specified by the particular isotherm. In the case of a constituent that may also degrade in the adsorbed phase, e.g., ammonia nitrogen, the sink term is

$$\pm\, k_{a,s} m(1 - n)A_s \tag{3.38}$$

where $k_{a,s}$ is the reaction coefficient (T^{-1}) for the constituent s in the adsorbed phase.

Point sources can also occur when, for example, injection wells are used for the disposal of wastewaters. The mass flow rate per unit thickness of the aquifer discharged to the groundwater system may be expressed in terms of the mass concentration of the constituent in the source fluid, $c_s^*(ML^{-3})$, the rate of injection $Q_r^*(L^3T^{-1})$, the thickness of the aquifer, m, and the Dirac delta function, or

$$S_s^* = \sum_{r\in\psi} \frac{c_{sr}^* Q_r^*}{m}\, \delta(x - x_r,\, y - y_r) \tag{3.39}$$

where ψ is an index set defining the location of the injection wells (x_r, y_r) within the aquifer system and δ ( ) is the Dirac delta function.

The mass transport equation for a nonconservative constituent can then be expressed as

$$\nabla \cdot m\,\mathbf{D}\,\nabla c_s - \nabla \cdot m\, c_s \mathbf{q}_D \pm k_s n c_s\, m \pm k_{a,s}\, m\,(1 - n)A_s$$
$$+ \sum_{r\in\psi} \frac{c_{sr}^* Q_r^*}{m}\, \delta(x - x_r,\, y - y_r) = \frac{\partial}{\partial t}\,[m\,(nc_s + (1 - n)A_s)] \tag{3.40}$$

where $\mathbf{q}_D$ is Darcy's velocity and again A_s defines a particular adsorption isotherm. In the case of nonequilibrium adsorption, the mass transport equation is coupled with the time-dependent sorption equations (Eq. 3.33).

Example Problem 3.4

A nonconservative constituent is discharged into the soil system as shown in Figure 3.4. The constituent undergoes linear equilibrium adsorption and first-order decay in the soil system. Determine the concentration distribution in the soil. Assume that the initial and boundary conditions for the problem are

$$c(z, 0) = 0 \tag{3.41a}$$

$$c(0, t) = c_0, \qquad t > 0 \tag{3.41b}$$

$$c(\infty, t) = 0 \tag{3.41c}$$

For a constant velocity field and a uniform soil, the mass transport equation can be expressed as

$$\frac{\partial}{\partial t}(nc + (1-n)Rc) = D\frac{\partial^2 c}{\partial z^2} - q_z\frac{\partial c}{\partial z} - k_s\,nc - k_{as}(1-n)\,Rc \qquad (3.42)$$

where R is the distribution coefficient between the adsorbed and solution phases and k_s and k_{as} are the kinetic reaction coefficients in the solution and adsorbed phases. Simplifying the equations, we have

$$\frac{\partial c}{\partial t} = D^*\frac{\partial^2 c}{\partial z^2} - q^*\frac{\partial c}{\partial t} - k^* c$$

where now $D^* = [D/n(1 + eR)]$, $q_z^* = [q_z/n(1 + eR)]$ and $k^* = (k_s + k_{as}eR)/(1 + eR)$ where e is the reciprocal of the void ratio $[(1 - n)/n]$.

The solution of the model is similar to the solution presented in Example Problem 3.1 for a conservative constituent. Again, we first take the Laplace transform of the equation, or

$$\frac{d^2\bar{c}}{dz^2} - \frac{q_z^*}{D^*}\frac{d\bar{c}}{dz} - \frac{1}{D^*}(k^* + s)\bar{c} = 0$$

Using the boundary condition at the ground surface and the requirement that the transform should vanish at $z = \infty$, the solution of the differential equation is

$$\bar{c} = \frac{c_0}{s}e^{rz}$$

where

$$r = \frac{1}{2D^*}\left\{q_z^* - \sqrt{q_z^{*2} + 4D^*(k^* + s)}\right\}$$

The inverse transform of the solution can be obtained using the convolution theorem, or

$$L^{-1}(\bar{c}) = c_0\left\{\frac{q_z^* z}{2D^*}\right\}\int_0^t \frac{z}{\sqrt{4\pi D^*\eta^3}}\exp\left\{-\frac{\beta^2\eta}{4D^*} - \frac{x^2}{4D^*\eta}\right\}d\eta$$

where exp is the exponential, $\beta^2 = q_z^{*2} + 4D^*k^*$ and η is a variable of integration.

The concentration distribution can then be described by the equation (Cho, 1971)

$$c(z,t) = \frac{c_0}{2}\left[\exp\left\{\frac{z}{2D^*}\left(q_z^{*2} - \sqrt{q_z^{*2} + 4D^*k^*}\right)\right\}\operatorname{erfc}\left\{\frac{z - \sqrt{q_z^{*2} + 4D^*k^*t}}{\sqrt{4D^*t}}\right\}\right.$$
$$\left. + \exp\left\{\frac{z}{2D^*}\left(q_z^{*2} + \sqrt{q_z^{*2} + 4D^*k^*t}\right)\right\}\operatorname{erfc}\left\{\frac{z + \sqrt{q_z^{*2} + 4D^*k^*t}}{\sqrt{4D^*t}}\right\}\right] \quad (3.43)$$

Example Problem 3.5

A potential water quality problem in agricultural areas is nitrate contamination. The transport of NH_4^+ in soils is complicated by the fact that it is positively charged and participates in ion-exchange reactions, in contrast to the oxidized forms, NO_2^- and NO_3^-. Assuming first-order kinetics for the transformation process, develop the equation describing the steady-state distribution of NH_4^+ and NO_3^- in the soil system. Assume a constant velocity field and a homogeneous soil.

We first consider the microbiological transformation that occurs in the soil system. The first-order kinetic reactions, which are valid for low concentrations of the constituents, may be expressed as (McLaren, 1970)

$$S_1^* = -k_{s1}c_1 \quad (3.44a)$$

$$S_2^* = k_{s1}c_1 - k_{s2}c_2 \quad (3.44b)$$

$$S_3^* = k_{s2}c_2 - k_{s3}c_3 \quad (3.44c)$$

where the subscripts 1, 2, and 3 identify the ions, NH_4^+, NO_2^-, and NO_3^-, respectively; k_{si} is the rate coefficient for each constituent.

Ammonium also enters into ion-exchange reactions in the soil and is adsorbed on the negatively charged soil particle surfaces. This transfer is considered to be reversible in the sense that its movement in the soil is delayed temporarily until it is replaced by similar ions. This type of reversible sorption-desorption reaction can be described by a linear equilibrium isotherm that is valid for slowly moving dilute solutions (Lindstrom & Boersma, 1970), or

$$A_1 = R\,c_1 \quad (3.45)$$

R is again the distribution coefficient, which is assumed to be constant and independent of the solute concentration.

The mathematical model describing the nitrogen transformations can be expressed as

$$D^* \frac{d^2c_1}{dz^2} - v_z \frac{dc_1}{dz} - k_{s1}^* c_1 = 0 \qquad (3.46a)$$

$$D^* \frac{d^2c_2}{dz^2} - v_z \frac{dc_2}{dz} - k_{s2}c_2 + k_{s1}c_1 = 0 \qquad (3.46b)$$

$$D^* \frac{d^2c_3}{dz^2} - v_z \frac{dc_3}{dz} - k_{s3}c_3 + k_{s2}c_2 = 0 \qquad (3.46c)$$

where $k_{s1}^* = \left\{k_{s1} + \left(\frac{1-n}{n}\right) R\, k_{a1}\right\}$, k_{a1} is the reaction coefficient in the adsorbed phase, v_z, the pore-velocity (q_z/n) and $D^* = D/n$. We have assumed that NO_2^- and NO_3^- do not participate in ion-exchange reactions and that both the adsorbed and free NH_4^+ are subject to nitrification.

The solution of the ammonium equation, a homogeneous linear differential equation, is

$$c_1(z) = C_1e^{r_1z} + C_2e^{r_2z}$$

where C_1 and C_2 are constants of integration and r_1 and r_2 are roots of the auxiliary polynomial equation,

$$r^2 - \frac{v_z}{D^*} r - \frac{k_{s1}^*}{D^*} = 0$$

or

$$r = v_z/2D^* \pm 1/2 \left\{(v_z/D^*)^2 + \left[4 \frac{k_{s1}^*}{D^*}\right]\right\}^{1/2}$$

Assuming the boundary condition at the ground surface, $z = 0$, is c_1^0 and the boundedness of c as $z \rightarrow \infty$, then the steady-state ammonium concentration is

$$c_1(z) = c_1^0 \exp\left[\left(\frac{z}{2D^*}\right)\left(v_z - \sqrt{v_z^2 + 4D^*\, k_{s1}^*}\right)\right],\ 0 \leq z \leq \infty \qquad (3.47)$$

As discussed by Misra, et al. (1974a, b), the make-up of the nitrifier species and the soil type regulate the rate at which NO_2^- is oxidized. Assuming that NO_2^- is oxidized relatively quickly, the concentration of the constituent at any point in the soil is negligible. As a result, NH_4^+ oxidation and NO_3^- reduction can be considered as irreversible reactions described by the source/sink terms,

$$S_1^* = -k_{s1}c_1$$

$$S_3^* = -k_{s3}c_3 + k_{s1}c_1$$

The steady-state nitrate concentration can then be obtained by substituting the ammonium solution in the nitrate transport equation and solving the resulting ordinary differential equation. The solution may be expressed as

$$c_2(z) = c_3^0\left\{\exp\frac{z}{2D^*}\left[v_z - \sqrt{v_z^2 + 4D^*k_{s3}}\right]\right\} + \frac{k_{s1}^*c_1^0}{k_{s1}^* - k_{s3}}\cdot$$
$$\left\{\exp\left[\frac{z}{2D^*}\left(v_z - \sqrt{v_z^2 + 4D^*k_{s3}}\right)\right] - \exp\left[\frac{z}{2D^*}\left(v_z - \sqrt{v_z^2 + 4D^*k_{s1}^*}\right)\right]\right\} \tag{3.48}$$

where c_3^0 is the input NO_3^- concentration. The transient simulation of the nitrogen cycle in soils has been presented by Mariño (1976) using a Crank-Nicolson finite difference model. Analytical solutions have also been developed by Cho (1971).

3.6 MASS TRANSPORT IN PARTIALLY SATURATED FLOW SYSTEMS

The prediction of mass transport in the unsaturated zone is conceptually similar to the mass transport problem discussed in Section 3.1. However, the problem is intrinsically more difficult because of the nonlinearity of the flow equation, hysteresis effects, and the lack of information defining the variation of the dispersion parameters with the volumetric water content.

The mass transport equation of the unsaturated system can be developed from the conservation of mass equations presented in Section 3.1. However, for partially saturated flow conditions, $\mathbf{q}$ and D are now functions of θ, the volumetric water content. The equations describing the mass transport of a conservative constituent in the unsaturated system are Duguid and Reeves (1976) and Bresler (1973).

$$\nabla \cdot \theta\, \mathbf{D}(\theta)\, \nabla c - \nabla \cdot \mathbf{q}c = \frac{\partial}{\partial t}(\theta\, c) \tag{3.49}$$

where

$$\mathbf{q} = -\,\mathbf{K}\,\nabla\,(\psi + z)$$

and ψ is the capillary potential. In comparison with Equation 3.16 the mass transport problem is similar to the mass transport equation for saturated flow. For the unsaturated problem, however, the velocity and dispersion fields are nonlinearly related to θ, the volumetric water content.

3.7 REFERENCES

Bear, J. *Dynamics of Fluids in Porous Media.* New York: Elsevier, 1972.

Bolt, O. H. *Journal of Agricultural Science,* 15:81-103, 1967.

Bredehoeft, J. D., & Pinder, G. F. "Mass Transport in Flowing Groundwater." *Water Resources Research,* 9 (1):194-210, 1973.

Bresler, E. "Simultaneous Transport of Solutes and Water under Transient Unsaturated Flow Conditions." *Water Resources Research,* 9(4):975-986, 1973.

Bruch, J. C., & Street, R. L. "Two-dimensional Dispersion." *ASCE, J. Sanitary Engr. Div.,* 93(346):17-39, 1967.

Cameron, D. R., & Klute, A. "Convective-Dispersive Solute Transport with a Combined Equilibrium and Kinetic Adsorption Model." *Water Resources Research,* 13(1):183-188, 1977.

Chao, T. T., Harward, M. E., & Fang, S. C. "Adsorption and Desorption Phenomena of Sulfate Adsorption in Soils." *Proceedings of the Soil Science Society of America,* 26:234-237, 1962.

Cho, C. M. "Convective Transport of Ammonium With Nitrification in Soil." *Can. J. Soil Science,* 51:339-350, 1971.

Duguid, J. O., & Reeves, M. "Material Transport Through Porous Media: A Finite Element Galerkin Model." *Oak Ridge National Laboratory, ORNL-4928,* Oak Ridge, Tennessee, 1976.

Enfield, C. G, & Bledsoe, B. E., "Fate of Wastewater Phosphorus in Soil." *ASCE, J. Irrigation and Drainage Div.,* 101 (IR3):145-155, 1975.

Helfferich, F., *Ion Exchange.* New York: McGraw-Hill, 1962.

John, M. D. "Cadmium Adsorption Maxima of Soils as Measured by The Langmuir Isotherm." *Can. J. Soil Science,* 52:343-350, 1972.

Kamke, E. *Differentialglelehungen, Losungsmethoden and Losungen.* New York: Chelsea Publishing, 1949.

Li, W-H. *Differential Equations of Hydraulic Transients, Dispersion and Groundwater Flow.* Englewood Cliffs, N.J.: Prentice-Hall, 1972.

Lindstrom, F. T., & Boersma, L. "Theory of Chemical Transport with Simultaneous Sorption in a Water Saturated Porous Media." *Soil Science,* 110:1-9, 1970.

Mariño, M. A. "Longitudinal Dispersion in Saturated Porous Media." ASCE, *J. Hyd. Div.,* 100(HY1):151-157, 1974a.

Mariño, M. A. "Distribution of Contaminants in Porous Media Flow." *Water Resources Research,* 10(5):1013-1018, 1974b.

Mariño, M. A. "Simulation Model of Nitrogen Transport in Soils Systems." *Proc. Int. Fed. for Inf. Processing Working Conf. on Modeling of Env. Systems,* T. L. Kunii and Y. Kava. Eds. Tokyo, Japan, 1976.

McLaren, A. D. "Temporal and Vectorial Reactions of Nitrogen in Soil: A Review." *Can. J. Soil Science,* 50:97-109, 1970.

Misra, C., Nielsen, D. R., & Biggar, J. W. "Nitrogen Transformation in Soils During Leaching: I, Theoretical Considerations." *Proceedings of the Soil Science Society of America*, 38:289-293, 1974a.

Misra, C., Nielsen, D. R., & Biggar, J. W., "Nitrogen Transformation in Soil During Leaching: II. Steady-State Nitrification and Nitrate Reduction." *Proceedings of the Soil Science Society of America*, 38:294-299, 1974b.

Oddson, J. K., Letey, L., & Weeks, L. U. "Predicted Distribution of Organic Chemicals in Solution and Adsorbed as a Function of Position and Time for Various Chemicals and Soil Properties." *Proceedings of The Soil Science Society of America*, 34:412-417, 1970.

Scheidegger, A. D. "General Theory of Dispersion in Porous Media." *J. Geophysical Research*, 66(10):3273-3278, 1961.

Tanji, K. K. "A Computer Analysis on the Leaching of Boron from Stratified Soil Column." *Soil Science,* 110(1):44-51, 1970.

4. NUMERICAL METHODS IN GROUNDWATER MANAGEMENT

4.1 INTRODUCTION—THE RESPONSE EQUATIONS

Groundwater supply and water quality optimization models are based on the response or transfer equations of the aquifer system. These equations relate the state variables of the hydraulic or water quality problem and the management or planning alternatives. These alternatives determine, for example, the location, magnitude, and duration of groundwater pumping or artificial recharge, or the maximum waste input concentration in a conjunctively managed groundwater supply and quality system.

The development of the aquifer system's response equations is based on the analytical or numerical solution of the governing equations of the system. For idealized groundwater systems with simple boundary conditions and uniform parameter fields, analytical techniques can be used for the direct solution of the hydraulic and/or mass transport equations. Superposition, complex variable theory, and Laplace and similarity transforms are the common mathematical tools for the solution of the groundwater system's governing equations. Table 4.1 summarizes the more common analytical response equations for linear and nonlinear groundwater hydraulic problems. The development of these analytical models has been presented in the example problems in Chapters 2 and 3. Greater coverage of these methods may be found in Polubarinova-Kochina (1962), Walton (1970), Bear (1972), and Li (1972).

A more general approach for determining the response or transfer functions of a distributed parameter, time-varying groundwater system is through the use of numerical techniques such as finite-differences or finite-elements. These techniques transform the partial differential equations of the groundwater system into systems

TABLE 4.1 Common Analytical Response Equations

	System Equation	*Response Equation*	*Remarks*
1.	$\frac{\partial^2 h}{\partial r^2} = \frac{S}{T}\frac{\partial h}{\partial t}$ $h(r,0) = h_0$ $\lim_{r\to\infty} h = h_0$ $\lim_{r\to 0} r\frac{\partial h}{\partial r} = \frac{Q}{2\pi T}$	$h = h_0 - \frac{Q}{4\pi T}\int_u^\infty \frac{e^{-x}}{x}dx$ $u = \frac{r^2 S}{4Tt}$	Thesis (1935)
2.	$\frac{\partial^2 h}{\partial^2} + \frac{1}{r}\frac{\partial h}{\partial r} + \frac{K_a}{Tm_a}(H_a - h) = \frac{S}{T}\frac{\partial h}{\partial t}$ $h(r,0) = h_0$ $\lim_{r\to\infty} h = h_0$ $\lim_{r\to 0} r\frac{\partial h}{\partial r} = \frac{Q}{2\pi T}$	$h = h_0 - \frac{Q}{4\pi T}\int_u^\infty \frac{e^{h(x)}}{x}dx$ $h(x) = -x - \frac{r^2}{4b^2 x}$ $u = \frac{r^2 S}{4Tt}$ $b^2 = \frac{Tm_a}{K_s}$	Hantush & Jacob (1935) K_a = aquitard permeability m_a = aquitard thickness H_a = head in overlying aquifer
3.	$\frac{1}{r}\frac{d}{dr}\left(r\frac{dh}{dr}\right) = 0$ $\lim_{r\to 0} r\frac{dh}{dr} = \frac{Q}{2\pi T}$ $h = h_e, r = r_e$	$h = \frac{Q}{2\pi T}\ln\left(\frac{r}{r_e}\right) + h_e$	Theim (1906)
4.	$\frac{1}{r}\frac{d}{dr}\left(r\frac{dh^2}{dr}\right) = 0$ $\lim_{r\to 0} r\frac{dh^2}{dr} = \frac{Q}{\pi K}$ $h = h_e, r = r_e$	$h^2 = \frac{Q}{\pi K}\ln\left(\frac{r}{r_e}\right) + h_e^2$	Steady, unconfined flow
5.	$T_{xy}\frac{\partial^2 s}{\partial x^2} + 2T_{xy}\frac{\partial^2 s}{\partial x \partial y} + T_{yy}\frac{\partial^2 S}{\partial y^2}$ $- Q\delta(x - x_w, y - y_w) = S\frac{\partial s}{\partial t}$ $s(x,y,0) = 0$ $s(\pm\infty, y, t) = 0$ $s(x, \pm\infty, t) = 0$	$s = \frac{Q}{4\pi\sqrt{T_{xx}T_{yy}}}\int_{-u_{xy}}^{\infty} \frac{e^{-u_{xy}}}{u_{xy}}du$ $u_{xy} = \frac{S}{4t}\left[\frac{T_{xx}y^2 + T_{yy}x^2}{T_{xx}T_{yy}}\right]$	Papadopulos (1965) $x_w, y_w = 0$

6.	$\frac{\partial}{\partial x}\left(h\frac{\partial h}{\partial x}\right) = \frac{S_y}{K}\frac{\partial h}{\partial t}$ $h(x,0) = h_0(x)$ $h(0,t) = h_0(0) = 0$ $\left.\frac{\partial h}{\partial x}\right\rvert_{x=L} = 0$	$h(x,t) = h_0(x)\left[1 + \frac{\beta K h_0(L)}{S_y L^2 t}\right]$ $\alpha = \frac{\sqrt{\pi}\,\Gamma(\frac{2}{3})}{3\,\Gamma(\frac{7}{8})} \approx .862$ $\beta = \frac{3}{2}\alpha^2 \approx 1.12$	Boussinesq (1904)
7.	$\frac{\partial^2 h}{\partial x^2} = \frac{S_y}{K\bar{h}}\frac{\partial h}{\partial t}$ $h(x,0) = h_0, x > 0$ $h(0,t) = D, t > 0$	$h = D + (h_0 - D)\,\mathrm{erf}(\alpha)$ $\alpha = \sqrt{\frac{S_y x^2}{4K\bar{h}t}}$	Linearized Boussinesq equation
8.	$\frac{\partial^2 h}{\partial x^2} = \frac{S_y}{K\bar{h}}\frac{\partial h}{\partial t}$ $h(x,0) = h_0, 0 \leqslant x \leqslant \infty$ $Q = bt, x = 0, t \geqslant 0$	$h = h_0 - \frac{2}{3}\left\{\frac{bS_y}{(K\bar{h})^2}\right\}\left\{\frac{K\bar{h}t}{S_y}\right\}^{3/2} - [(2\alpha^3 + 3\alpha)(1 - \mathrm{erf}(\alpha)) - 2(\alpha^2 + 1)\pi^{-1/2}e^{-\alpha^2}]$ $\alpha = \left(\frac{S_y x^2}{4Tt}\right)^{1/2}$	Edlemann (1947)
9.	$K\bar{h}\frac{\partial^2 h}{\partial x^2} + W(x) = S_y\frac{\partial h}{\partial t}$ $h(x,0) = h_0(x)$ $W(x) = W, \lvert x\rvert \leqslant L$ $W(x) = 0, \lvert x\rvert > L$ $h(\pm\infty, t) = h_0$	$h = h_0 + \frac{W}{2S_y}\int_0^t \left[\mathrm{erf}\left(\frac{L - x}{2\alpha\sqrt{t - \tau}}\right) + \mathrm{erf}\left(\frac{L + x}{2\alpha\sqrt{t - \tau}}\right)\right] d\tau$ $\alpha = \left(\frac{S_y x^2}{4Tt}\right)^{1/2}$	Polubarinova-Kochina (1963)
10.	$T\left(\frac{\partial^2 h}{\partial r^2} + \frac{1}{r}\frac{\partial h}{\partial r}\right) = S\frac{\partial h}{\partial t} + D_i S_y \int_0^t \frac{\partial h}{\partial t} e^{-D_i(t-\tau)} d\tau$ $h(r,0) = h_0$ $\lim_{t\to\infty} h(r,t) = h_0$ $\lim_{r\to 0} r\frac{\partial h}{\partial r} = \frac{Q}{2\pi T}$	$h = h_0 - \frac{Q}{4\pi T}\int_0^\infty \frac{2}{x}\left[1 - e^{-u_1}[\cosh u_2 + \frac{D_i N(1 - x^2)t}{2u_2}\sinh u_2]\right] J_0\left(\frac{rx}{\eta D_t}\right) dx$ $u_1 = \frac{D_i t N(1 - x^2)}{2}$ $u_2 = \frac{D_i t\sqrt{N^2(1 + x^2)^2 - 4Nx^2}}{2}$ $\eta = \sqrt{\frac{N - 1}{N}} = \sqrt{\frac{S_y}{S + S_y}}$ $D_t = \sqrt{\frac{T}{D_i S_y}}$	Boulton (1963)

TABLE 4.1 Continued

	System Equation	*Response Equation*	*Remarks*
11.	$T\left(\frac{\partial^2 s}{\partial x^2} + \frac{\partial^2 s}{\partial y^2}\right) - \sum_{w\epsilon\pi} Q_w(t)\delta(x - x_w, y - y_w) = S\frac{\partial h}{\partial t}$ $s(x, y, 0) = 0$ $\lim_{x,y\to\infty} s(x, y, t) = 0$	$s(x, y, t) = \frac{1}{4\pi T}\sum_{w\epsilon\pi}\int_0^t Q_w(\tau)e^{r(\tau)}d\tau$ $r(\tau) = -\frac{R^2 S}{4T(t - \tau)}$ $R^2 = \{(x - x_w)^2 + (y - y_W)^2\}$	Maddock (1972)
12.	$T\left(\frac{\partial^2 s}{\partial x^2} + \frac{\partial^2 s}{\partial y^2}\right) - Q\delta(x - \eta, y - \xi) = S\frac{\partial s}{\partial t}$ $s(x, y, 0) = s_0$ $\left(-K\frac{\partial s}{\partial x} + \frac{K_a}{m_a}s\right)\Big\vert_{x=0} = 0 \; 0 \leqslant y \leqslant b, t > 0$ $\left(-K\frac{\partial s}{\partial y} + \frac{K_a}{m_a}s\right)\Big\vert_{y=0} = 0, 0 \leqslant x \leqslant a, t > 0$ $s(a, y, t) = 0,\ 0 \leqslant y \leqslant b, t > 0$ $s(x, b, t) = 0,\ 0 \leqslant x \leqslant a, t > 0$	$s(x, y, t) = \sum_{m=1}^{\infty}\sum_{n=1}^{\infty} e^{-\tau(\alpha_n^2 + \beta_m^2)t}D_m \cdot$ $\sin \alpha_n(a - x) \sin \beta_m(b - y) \cdot$ $\left[\frac{s_0}{\alpha_n\beta_m}(1 - \cos \alpha_n a)(1 - \cos \beta_m b)\right.$ $-q \sin \alpha_n(a - \eta) \cdot$ $\left.\sin \beta_m(b - \xi)\frac{1}{\tau(a_n^2 + \beta_m^2)}(e^{\tau(a_n^2+\beta_m^2)t} - 1)\right]$ $D_m^2 = \frac{4(\alpha_n^2 + 1)(\beta_m^2 + 1)}{[a(\alpha_n^2 + 1) + 1][b(\beta_m^2 + 1) + 1]}$ $\tau = \frac{T}{S}$ $\alpha_n \cot \alpha_n a = -1$ $\beta_m \cot \beta_m b = -1$	Corapcioglu, et al. (1983)
13.	$\frac{\partial h}{\partial t} = \frac{K}{S_y}\frac{\partial}{\partial x}\left(h\frac{\partial h}{\partial x}\right)$ $h(0, t) = h_1$ $h(x, 0) = h_0$	$\int_0^\infty (H - 1)d\eta = -2\mu\frac{dH}{d\eta}\Big\vert_{\eta=0}$ $\eta = x\left[\frac{Kh_0}{S_y}t\right]^{-1/2}$ $H = \frac{h}{h_0}$ $\mu = \frac{h_1}{h_0},\ h_1 > h_0$	Tolikas, et al. (1984)

14.	$\frac{d}{dx}\left(h\frac{dh}{dx}\right) + l\frac{dh}{dx} + \frac{R}{K} = 0$ $h = h_0, x = 0$ $h\frac{dh}{dx} + lh = \frac{q_0}{K}, x = 0$	$H = -uQ_0 \frac{f(u)}{f(u_0)}, 0 \leqslant aX \leqslant Q_0, u < 0$ $H = uQ_0 \frac{f(u)}{f(u_0)}, Q_0 < aX, u \geqslant 0$ $f(u) = \lvert au^2 + u + 1\rvert^{-0.5} e^{0.5g(u)}$ $g(u) = 2(4a - 1)^{-0.5} \arctan\left[\frac{2au + 1}{(4a - 1)^{0.5}}\right], a > \frac{1}{4}$ $g(u) = (1 - 4a)^{-0.5} \ln\left\lvert\frac{2au + 1 - (1 - 4a)^{0.5}}{2au + 1 + (1 - 4a)^{0.5}}\right\rvert, a < \frac{1}{4}$ $g(u) = \frac{-2}{(2au + 1)}, a = \frac{1}{4}$ $Q_0 = \frac{q_0}{(h_0 Kl)}$ $H = \frac{h}{h_0}$ $X = \frac{lx}{h_0}$ $u_0 = -\frac{1}{Q_0}$	Yates, et al. (1985)

TABLE 4.1 Continued

	System Equation	Response Equation	Remarks
15.	$\frac{\partial^2 s}{\partial r^2} + \frac{1}{r}\frac{\partial s}{\partial r} + \frac{\alpha_{\theta\theta}}{r^2}\frac{\partial_s^2}{\partial\theta^2} + \alpha_{zz}\frac{\partial^2 s}{\partial z^2} + \left(\frac{\alpha_{r\theta} + \alpha_{\theta r}}{r}\right)\frac{\partial^2 s}{\partial r\partial\theta} + (\alpha_{rz} + \alpha_{zr})\frac{\partial^2 s}{\partial r\partial z} = \frac{1}{\beta}\frac{\partial s}{\partial t}$ $s(r, \theta, z, 0) = 0$ $s(r, \theta, z, t) = 0 \text{ as } r \to \infty$ $\left(\frac{\partial s}{\partial r} + \frac{\alpha_{r\theta}}{r}\frac{\partial s}{\partial\theta}\right) + \gamma s = -Q, r = r_w, \forall t$	$s(r, \eta, \xi, t) = \frac{Q}{8\pi K_{rr}} \sum_{n=-\infty}^{\infty} \cos n\delta_{r\theta}(\xi - \acute{\xi}) \int_0^\infty C(k, r)C(k, \acute{r}) \cdot \left[e^{-k(\eta^2-\eta^2)^{1/2}} \operatorname{erfc}\left[\frac{1}{2}\left\{\frac{\eta^2 - \acute{\eta}^2}{\alpha(t - \acute{t})}\right\}^{1/2} - K[\alpha(t - \acute{t})]^{1/2}\right] - e^{k(\eta^2-\eta^2)^{1/2}} \operatorname{erfc}\left[\frac{1}{2}\left\{\frac{\eta^2 - \acute{\eta}^2}{\alpha(t - \acute{t})}\right\}^{1/2} + K(\alpha(t - \acute{t})]\right] dk$ $C(k, r) = \frac{J_{n\delta_{r\theta}}[kY'_{n\delta_{r\theta}} - \beta Y_{n\delta_{r\theta}}(kr_w)] - Y_{n\delta_{r\theta}}(kr)[kJ'_{n\delta_{r\theta}} - \beta J_{n\delta_{r\theta}}(kr_w)}{[[kY'_{n\delta_{r\theta}}(kr_w) - \beta Y_{n\delta_{r\theta}}(kr_w)]^2 + [kJ'_{n\delta_{r\theta}}(kr_w) - \beta J_{n\delta_{r\theta}}(kr_w)]^2]^{1/2}}$ $k[Y'_{n\delta_{r\theta}}(kr_w) + J'_{n\delta_{r\theta}}(kr_w)] + \beta[J_{n\delta_{r\theta}}(kr_w) + Y_{n\delta_{r\theta}}(kr_w)] = 0$ $\beta = \frac{K_{rr}}{S_s}$ $\xi = \frac{1}{\delta_{r\theta}}[\theta - (\alpha_{r\theta} + \alpha_{\theta r})\ln r]$ $\eta = \frac{1}{\delta_{rz}}[z - \alpha_{rz} + \alpha_{zr})r]$ $\delta_{rj} = [\alpha_{jj} - (\alpha_{rj} + \alpha_{jr})^2]^{1/2}, j = \theta, z$ $\alpha_{ij} = \frac{K_{ij}}{K_{rr}}, i, j = r, \theta, z$ γ *is a given parameter*	Falade (1981) J_s Bessel function of order s Y_s modified Bessel function of order s

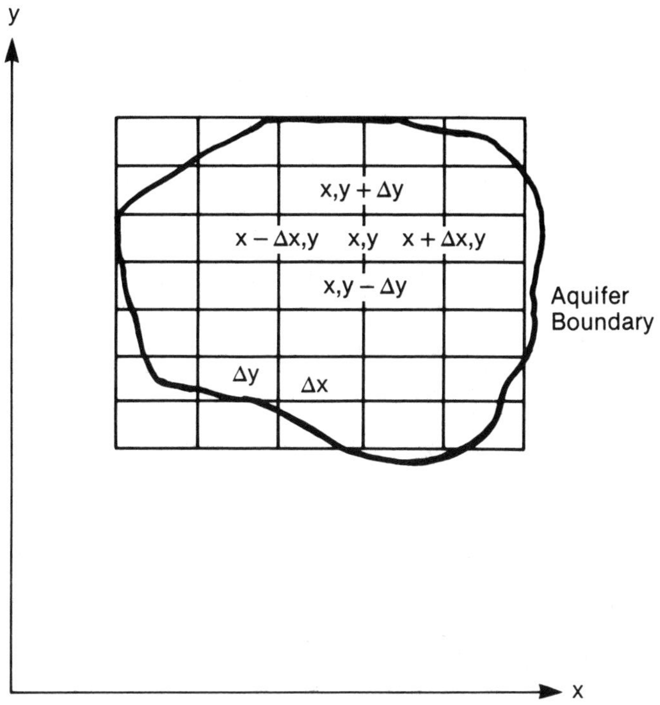

FIGURE 4.1 Finite-difference grid

of ordinary differential or algebraic equations. The solution of these equations determines the state variables at a predetermined set of discrete nodal points within the aquifer system. Again, the head or mass concentrations at these discrete points can be expressed as a function of possible pumping, injection, or other waste management policies. We begin by considering finite-difference methods.

4.2 FINITE-DIFFERENCE NUMERICAL MODELS

Finite-difference models of the groundwater flow and mass transport equations are based on a Taylor series representation of the time and spatial derivatives. In the discrete groundwater model, as shown in Figure 4.1, the head or mass concentration ϕ at the nodal points $x + \Delta x$ or $x - \Delta x$ can be expressed in terms of the state variables and all the derivatives evaluated at an adjacent nodal point, x, as

$$\phi(x + \Delta x) = \phi(x) + \left.\frac{\partial \phi}{\partial x}\right|_x \Delta x + \left.\frac{\partial^2 \phi}{\partial x^2}\right|_x \frac{\Delta x^2}{2!} + \left.\frac{\partial^3 \phi}{\partial x^3}\right|_x \frac{\Delta x^3}{3!} + \cdots \tag{4.1}$$

and

$$\phi(x - \Delta x) = \phi(x) - \left.\frac{\partial\phi}{\partial x}\right|_x \Delta x + \left.\frac{\partial^2\phi}{\partial x^2}\right|_x \frac{\Delta x^2}{2!} - \left.\frac{\partial^3\phi}{\partial x^3}\right|_x \frac{\Delta x^3}{3!} + \cdots \tag{4.2}$$

The only restrictions on the power series expansions are that the state variable functions are single-valued and continuous.

Finite-difference approximations for the first- and second-order derivatives of the groundwater flow and mass transport equations can be developed directly from the Taylor series approximations. For example, rearranging Equation 4.1, we have

$$\left.\frac{\partial\phi}{\partial x}\right|_x = \frac{\phi(x + \Delta x) - \phi(x)}{\Delta x} - \left.\frac{\Delta x}{2!}\frac{\partial^2\phi}{\partial x^2}\right|_x - \left.\frac{\Delta x^2}{3!}\frac{\partial^3\phi}{\partial x^3}\right|_x - \cdots$$

and

$$\left.\frac{\partial\phi}{\partial x}\right|_x = \frac{\phi(x) - \phi(x - \Delta x)}{\Delta x} + \left.\frac{\Delta x}{2!}\frac{\partial^2\phi}{\partial x^2}\right|_x - \left.\frac{\Delta x^2 \partial^3\phi}{3!\, \partial x^3}\right|_x + \cdots$$

The forward and backward first derivative approximations are then

$$\left.\frac{\partial\phi}{\partial x}\right|_x \cong \frac{\phi(x + \Delta x) - \phi(x)}{\Delta x} \tag{4.3a}$$

and

$$\left.\frac{\partial\phi}{\partial x}\right|_x \cong \frac{\phi(x) - \phi(x - \Delta x)}{\Delta x} \tag{4.3b}$$

The truncation error, or the error introduced by neglecting the higher-order terms of the Taylor series, is given as

$$\overline{E} = \pm \left.\frac{\Delta x}{2}\frac{\partial^2\phi}{\partial x^2}\right|_\xi = 0(\Delta x), \qquad \begin{cases} x \leq \xi \leq x + \Delta x \\ x - \Delta x \leq \xi \leq x \end{cases}$$

The error, $\overline{E}$, is on the order of Δx, which means there exists a positive constant δ that is independent of Δx, and that, $|\overline{E}| \leq \delta|\Delta x|$ for all sufficiently small Δx.

A central difference approximation to the first derivative can be obtained by differencing Equations 4.1 and and 4.2 and isolating the first-order derivative, or

$$\left.\frac{\partial\phi}{\partial x}\right|_x \cong \frac{\phi(x + \Delta x) - \phi(x - \Delta x)}{2\Delta x} \tag{4.3c}$$

The truncation error of the approximation is

$$\overline{E} = -\left.\frac{\Delta x^2}{3!}\frac{\partial^3\phi}{\partial x^3}\right|_{\xi} = 0(\Delta x^2), \quad x - \Delta x \leqslant \xi \leqslant x + \Delta x$$

An approximation to the second derivative can also be easily obtained by adding Equations 4.1 and 4.2, or

$$\left.\frac{\partial^2\phi}{\partial x^2}\right|_x \cong \frac{\phi(x + \Delta x) - 2\phi(x) + \phi(x - \Delta x)}{\Delta x^2} \tag{4.4}$$

The approximation is second-order correct; the truncation error is

$$\overline{E} = \left.\frac{\Delta x^2}{12}\frac{\partial^4\phi}{\partial x^4}\right|_{\xi} = 0(\Delta x^2), \; x - \Delta x \leqslant \xi \leqslant x + \Delta x$$

Higher-order derivative approximations can be developed using linear difference operators. The most common operators are defined in Table 4.2. To illustrate the use of these operators, consider again the Taylor series representation of $\phi(x + \Delta x)$, or

$$\phi(x + \Delta x) = \phi(x) + \Delta x \left.\frac{\partial\phi}{\partial x}\right|_x + \left.\frac{\Delta x^2}{2!}\frac{\partial^2\phi}{\partial x^2}\right|_x + \cdots$$

We can rewrite the Taylor series in terms of the differentiation operator, D, as

$$\phi(x + \Delta x) = \left[1 + \Delta x D + \frac{\Delta x^2}{2!}D^2 + \cdots\right]\phi(x) = e^{\Delta x D}\phi(x)$$

TABLE 4.2 Common Finite-Difference Operators

Operator	*Notation*	*Representation*
Forward	Δ	$\Delta\phi = \phi(x + \Delta x) - \phi(x)$
Backward	∇	$\nabla\phi = \phi(x) - \phi(x - \Delta x)$
Central	δ	$\delta\phi = \phi(x + \frac{\Delta x}{2}) - \phi(x - \frac{\Delta x}{2})$
Shift	E	$E\phi(x) = \phi(x + \Delta x)$
Average	μ	$\mu\phi(x) = \dfrac{\phi(x + \frac{\Delta x}{2}) + \phi(x - \frac{\Delta x}{2})}{2}$
Differentiation	D	$D\phi(x) = \dfrac{\partial\phi}{\partial x}$

Source: Lapidus & Pinder, 1982.

Since, however, $E\phi(x) = \phi(x + \Delta x)$ and $E = e^{\Delta x D}$, the differentiation operator can be expressed as

$$\Delta x D = \ln E = \left[\begin{array}{l} \ln (1 + \Delta) = \Delta - \frac{1}{2}\Delta^2 + \frac{1}{3}\Delta^3 \\ \\ -\ln (1 - \nabla) = \nabla + \frac{1}{2}\nabla^2 + \frac{1}{3}\nabla^3 \end{array} \right.$$

where

$$\Delta^p \phi (x) = \sum_{j=0}^{P} (-1)^j \binom{p}{j} \phi (x + (p - j)\Delta x),$$

$$\nabla^n \phi(x) = \nabla^{n-1} \phi(x) - \nabla^{n-1} \phi(x - \Delta x), \quad \binom{p}{j} = \frac{p!}{j!(p - j)!}$$

and we have used the infinite series expansion

$$\ln (1 \pm x) = \pm x - x^2/2 \pm x^3/3 - x^4/4, \qquad -1 \leqslant x \leqslant 1$$

However, we also have the relations,

$$\delta = 2 \sinh \left(\frac{\Delta x D}{2} \right)$$

and

$$\Delta x D = 2 \sinh^{-1} \left(\frac{\delta}{2} \right) = \delta - \frac{\delta^3}{2^2 3!} + \frac{3^2 \delta^5}{2^4 5!} - \ldots..$$

Using these relations, we can develop general expressions for the first- and second-order derivatives. As shown by Salvadori and Baron (1961), the first-order approximations may be written as

$$\Delta x \frac{\partial \phi}{\partial x} = \begin{cases} \left[\Delta - \frac{1}{2}\Delta^2 + \frac{1}{3}\Delta^3 - \right] \phi(x), \text{ or} \\ \left[\nabla + \frac{1}{2}\nabla^2 + \frac{1}{3}\nabla^3 - \right] \phi(x), \text{ or} \\ \left[\mu\delta - \frac{1}{3!}\mu\delta^3 + \frac{1}{30}\mu\delta^5 - \right] \phi(x) \end{cases} \tag{4.5}$$

The second-order derivative approximations are

$$\Delta x^2 \frac{\partial^2 \phi}{\partial x^2} = \begin{cases} \left[\Delta^2 - \Delta^3 + \frac{11}{12}\Delta^4 - \right]\phi(x), \text{ or} \\ \left[\nabla^2 + \nabla^3 + \frac{11}{12}\nabla^4 - \right]\phi(x), \text{ or} \\ \left[\delta^2 - \frac{1}{12}\delta^4 + \frac{1}{90}\delta^6 + \right]\phi(x) \end{cases} \tag{4.6}$$

By truncating these series expansions, a variety of approximations can be developed for the first- and second-order derivatives. The more important approximations and their associated truncation errors are summarized in Table 4.3.

The following example problems describe how these finite-difference approximations may be used in the development of the response equations of the groundwater system.

Example Problem 4.1 Two-Dimensional Semiconfined Flow

The aquifer system shown in Figure 4.2 is used as a source of domestic water supply. Well log data suggest that the aquifer is semiconfined, inhomogeneous, and anisotropic. A large river parallels the northern boundary of the system and the nodal points on the boundary are represented by Dirichlet boundary conditions, i.e., $h = \bar{h}$. The remaining boundaries are impermeable. Develop a finite difference model of the aquifer system.

We begin the analysis by superimposing on the groundwater basin the finite-difference grid network. As shown in Figure 4.2, the boundary of the aquifer system intersects the grid network only at the mesh points. This simplifies the difference approximations of the boundary conditions. In the event, however, this is impractical, the partial differential equation may be approximated in two ways. In one approach, the governing equations are approximated at every point internal to the boundary of the system. The approximations at these grid points are then modified to account for the boundary conditions and the nonuniform nodal spacing between the grid points and the boundaries. In the second approach, approximations are written for all points that are independent of the boundary nodes. The remaining nodes that are adjacent to the boundaries are required to satisfy the finite-difference approximation of the boundary condition. A further discussion of these approximations may be found in Remson, et al. (1971) and Lapidus and Pinder (1982).

The governing equation of the groundwater system is given by Equation 2.55,

$$\frac{\partial}{\partial x}\left(T_{xx}\frac{\partial h}{\partial x}\right) + \frac{\partial}{\partial y}\left(T_{yy}\frac{\partial h}{\partial y}\right) + \frac{K_a(H_a - h)}{m_a} \pm \sum_{w \in \Omega} Q_w \delta(x - x_w, y - y_w) = S\frac{\partial h}{\partial t}$$

TABLE 4.3 Common Finite-Difference Approximations

Derivative	*Approximation*	*Truncation Error*
$\frac{\partial\phi}{\partial x}$	$\frac{\phi(x + \Delta x) - \phi(x)}{\Delta x}$	$0(\Delta x)$
	$\frac{\phi(x) - \phi(x - \Delta x)}{\Delta x}$	$0(\Delta x)$
	$\frac{\phi(x + \Delta x) - \phi(x - \Delta x)}{2\Delta x}$	$0(\Delta x^2)$
	$\frac{-\phi(x + 2\Delta x) + 4\phi(x + \Delta x) - 3\phi(x)}{2\Delta x}$	$0(\Delta x^2)$
	$\frac{-\phi(x + 2\Delta x) + 8\phi(x + \Delta x) - 8\phi(x - \Delta x) + \phi(x - 2\Delta x)}{12\Delta x}$	$0(\Delta x^2)$
$\frac{\partial^2\phi}{\partial x^2}$	$\frac{\phi(x + \Delta x) - 2\phi(x) + \phi(x - \Delta x)}{\Delta x^2}$	$0(\Delta x^2)$
	$\frac{-\phi(x + 2\Delta x) + 16\phi(x + \Delta x) - 30\phi(x) + 16\phi(x - \Delta x) - \phi(x - 2\Delta x)}{12\Delta x^2}$	$0(\Delta x^4)$
$\frac{\partial^2\phi}{\partial x\partial y}$	$\frac{\phi(x + \Delta x, y + \Delta y) - \phi(x - \Delta x, y + \Delta y) - \phi(x + \Delta x, y - \Delta y) + \phi(x - \Delta x, y - \Delta y)}{4\Delta x\Delta y}$	$0(\Delta x^2 + \Delta y^2)$
$\frac{\partial\left(p(x)\frac{\partial\phi}{\partial x}\right)}{\partial x}$	$\frac{p(x + \frac{\Delta x}{2})(\phi(x + \Delta x) - \phi(x)) - p(x - \frac{\Delta x}{2})(\phi(x) - \phi(x - \Delta x))}{\Delta x^2}$	$0(\Delta x^2)$

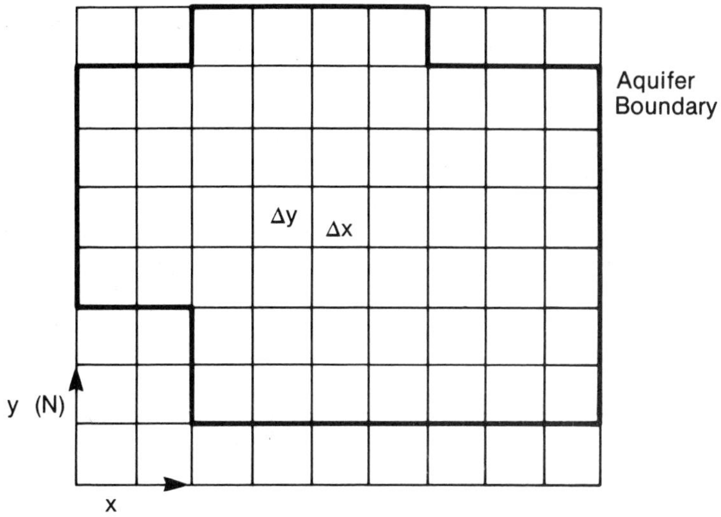

FIGURE 4.2 The discretized aquifer

and we have assumed that the x, y coordinate system is aligned with the principal axes of the transmissivity tensor. Assuming our objective is to minimize the truncation error of the approximation, then the finite difference representation of the semiconfined flow equation is for any point i, j (x, y)

$$\frac{1}{\Delta x^2}\{T_{xx}^{i+1/2,j}(h_{i+1,j} - h_{i,j}) - T_{xx}^{i-1/2,j}(h_{i,j} - h_{i-1,j})\} + \frac{1}{\Delta y^2}\{T_{yy}^{i,j+1/2}(h_{i,j+1} - h_{i,j})$$

$$- T_{yy}^{i,j-1/2}(h_{i,j} - h_{i,j-1})\} \pm \frac{Q_{w,i,j}}{\Delta x \Delta y} + \frac{K_{a,i,j}(H_{a,i,j} - h_{i,j})}{m_a} = S_{i,j}\dot{h}_{i,j} \qquad (4.7)$$

where $\dot{h}_{i,j}$ is the temporal derivative of h at point i,j. The truncation error of the approximations is $0(\Delta x^2 + \Delta y^2)$.

The transmissivity terms of the approximations are evaluated at the midpoints of the individual elements in the system. Three commonly used approximations for these terms are:

$$T_{\alpha,\beta}^{i+1/2,j} = \frac{T_{\alpha,\beta}^{i+1,j} + T_{\alpha,\beta}^{i,j}}{2} \quad \text{(arithmetic mean)} \qquad (4.8a)$$

$$T_{\alpha,\beta}^{i+1/2,j} = [T_{\alpha,\beta}^{i+1,j} \cdot T_{\alpha,\beta}^{i,j}]^{1/2} \quad \text{(geometric mean)} \qquad (4.8b)$$

$$T_{\alpha,\beta}^{i+1/2,j} = \frac{2\, T_{\alpha,\beta}^{i+1,j} \cdot T_{\alpha,\beta}^{i,j}}{T_{\alpha,\beta}^{i+1,j} + T_{\alpha,\beta}^{i,j}} \quad \text{(harmonic mean)} \qquad (4.8c)$$

where $\alpha, \beta = x, y$. All three approximations produce satisfactory results for most groundwater flow problems. However, the harmonic mean which heavily weights the smaller transmissivity values, is most appropriate for groundwater flow since the ability of the aquifer system to transmit water in a particular direction is dependent on the smallest transmissivity encountered in the direction of flow (Huntoon, 1974).

The boundary conditions may also be expressed in finite-difference form. If the index set Γ defines the boundary nodes of the aquifer system, then the Dirichlet and Neumann boundary conditions can be represented as ($\forall i, j \; \varepsilon \; \Gamma$)

$$h_{i,j} = \bar{h} \tag{4.9a}$$

$$\frac{h_{i,j} - h_{i-1,j}}{\Delta x} = 0, \tag{4.9b}$$

$$\frac{h_{i,j} - h_{i,j-1}}{\Delta y} = 0, \tag{4.9c}$$

Writing Equation 4.7 for all the internal nodes of the system, i.e., $i,j \notin \Gamma$, $i = 1,..m$, $j = 1,..n$, the resulting system of linear equations may be expressed in vector-matrix notation as

$$A\dot{\mathbf{h}} + B\mathbf{h} + \mathbf{g} = 0 \tag{4.10a}$$

$$\mathbf{h}(0) = \mathbf{h}_0 \tag{4.10b}$$

In this system of first-order, ordinary differential equations, the vector $\mathbf{h}$ is a $mnxl$ column vector of the unknown heads in the system, $\mathbf{h} = (\mathbf{h}_{11}..\mathbf{h}_{1m}..\mathbf{h}_{n1}..\mathbf{h}_{nm})^T$. The $mnxmn$ coefficient matrices A and B depend on the flow and storage parameters of the groundwater system. If the i and j indices are combined in a single index, $l = m(j - 1) + i$, then typical elements of the diagonal A matrix and the $mnxl$ column vector, $\mathbf{g}$, are

$$A_{l,l} = S_l, \quad l = 1, \ldots mn \tag{4.11}$$

$$g_l = \pm \frac{Q_i}{\Delta x \Delta y} + \frac{K_l H_l}{m_l} \tag{4.12}$$

The B matrix may be recursively generated for each point in the aquifer system using the boundary conditions and Equation 4.7. For an internal node in the system, the matrix elements are given as

$$B_{l,l} = -\frac{2T_{xx}^{l+1}T_{xx}^{l}}{(T_{xx}^{l+1} + T_{xx}^{l})\Delta x^2} + \frac{2T_{xx}^{l}T_{xx}^{l-1}}{(T_{xx}^{l} + T_{xx}^{l-1})\Delta x^2}$$

$$+ \frac{2T_{yy}^{l+1}T_{yy}^{l}}{(T_{yy}^{l+1} + T_{yy}^{l})\Delta y^2} + \frac{2T_{yy}^{l}T_{yy}^{l-1}}{(T_{yy}^{l} + T_{yy}^{l-1})\Delta y^2} + \frac{K_l}{m_l} \quad (4.13a)$$

$$B_{l,l+1} = \frac{2T_{xx}^{l}T_{xx}^{l-1}}{(T_{xx}^{l+1} + T_{xx}^{l})\Delta x^2} \quad (4.13b)$$

$$B_{l,l-1} = \frac{2T_{xx}^{l}T_{xx}^{l-1}}{(T_{xx}^{l} + T_{xx}^{l-1})\Delta x^2} \quad (4.13c)$$

$$B_{l+1,l} = \frac{2T_{yy}^{l+1}T_{yy}^{l}}{(T_{yy}^{l+1} + T_{yy}^{l})\Delta y^2} \quad (4.13d)$$

$$B_{l-1,l} = \frac{2T_{yy}^{l}T_{yy}^{l-1}}{(T_{yy}^{l} + T_{yy}^{l-1})\Delta y^2} \quad (4.13e)$$

In these dynamic response equations, the possible decision variables of the management model, the pumping or recharge rates and locations, are contained in the **g** vector. The solution of these equations, either analytically or using finite-difference approximations for the time derivatives, again expresses the hydraulic head as a linear function of the initial state of the system, and the pumping and/or injection decisions.

Example Problem 4.2 One-Dimensional Confined Flow

An inhomogeneous, anisotropic confined aquifer is located in an alluvial river basin. The aquifer's boundaries are two large rivers (see Figure 4.3) which are assumed to be unaffected by pumping. The initial piezometric head in the system is h_0. Develop the dynamic response equations of the system using finite-differences.

The governing equation of the aquifer system is the confined flow equation (Eq. 2.55). Assuming the coordinate system is again aligned with the principal directions of the conductivity tensor, and recognizing that the head is only a function of x and time (t), the flow equation is

$$\frac{\partial}{\partial x}\left(T_{xx}\frac{\partial h}{\partial x}\right) - \sum_{w=1}^{4} Q_w\delta(x - x_w) = S\frac{\partial h}{\partial t} \quad (2.55)$$

where Q_w is the discharge per unit thickness.

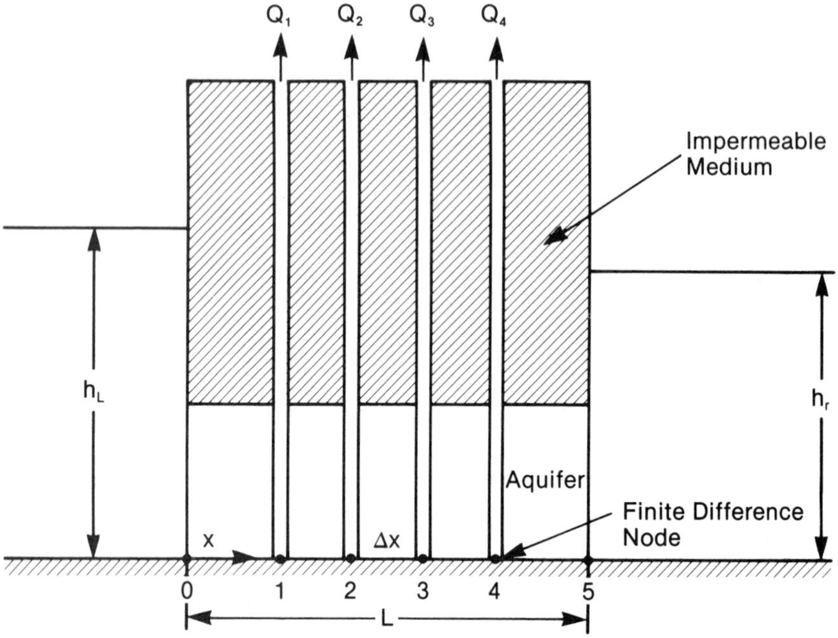

FIGURE 4.3

Assuming the water levels in the rivers are unaffected by pumping, the boundary and initial conditions for the problem are

$$h(0, t) = h_L \tag{4.14a}$$

$$h(L, t) = h_r \tag{4.14b}$$

$$h(x, 0) = h_0 \tag{4.14c}$$

We next discretize the aquifer system (an example is shown in Figure 4.3). There are a total of four internal nodes in the basin. Nodes 0 and 5 are known from the boundary conditions of the problem. The finite-difference approximation of the flow equation for any point i in the system is from Table 4.3,

$$\frac{1}{\Delta x^2}\{T_{xx}^{i+1/2}(h_{i+1} - h_i) - T_{xx}^{i-1/2}(h_i - h_{i-1})\} - \frac{Q_{w,i}}{\Delta x} = S_i\,\dot{h}_i \tag{4.15}$$

where $Q_{w,i}$ (L^2/T) is the pumping rate per unit width at node i. The average transmissivities will be evaluated using the harmonic mean, or

$$T_{xx}^{i+1/2} = 2T_{xx}^{i+1}\,T_{xx}^{i}/(T_{xx}^{i+1} + T_{xx}^{i})$$

The truncation error of the approximations is $0(\Delta x^2)$.

If the difference approximation (Equation 4.15) is written for all internal nodes in the aquifer, we again generate the dynamic response equations of the aquifer system, or

$$\mathbf{A}\dot{\mathbf{h}} + \mathbf{B}\mathbf{h} + \mathbf{g} = 0 \tag{4.16a}$$

where $\mathbf{h} = (h_1, \ldots, h_4)^T$. The A matrix is now a 4×4 diagonal matrix with the elements

$$A_{ij} = \begin{bmatrix} -S_i, & i = j \\ 0, & i \neq j \end{bmatrix} \tag{4.16b}$$

The B matrix is a tridiagonal matrix, or

$$\begin{bmatrix} -\frac{1}{\Delta x^2}\{T_{xx}^{1\frac{1}{2}} + T_{xx}^{\frac{1}{2}}\} & \frac{1}{\Delta x^2}\{T_{xx}^{1\frac{1}{2}}\} & 0 & 0 \\ \frac{1}{\Delta x^2}\{T_{xx}^{1\frac{1}{2}}\} & -\frac{1}{\Delta x^2}\{T_{xx}^{2\frac{1}{2}} + T_{xx}^{1\frac{1}{2}}\} & \frac{1}{\Delta x^2}\{T_{xx}^{2\frac{1}{2}}\} & 0 \\ 0 & \frac{1}{\Delta x^2}\{T_{xx}^{2\frac{1}{2}}\} & -\frac{1}{\Delta x^2}\{T_{xx}^{3\frac{1}{2}} + T_{xx}^{2\frac{1}{2}}\} & \frac{1}{\Delta x^2}\{T_{xx}^{3\frac{1}{2}}\} \\ 0 & 0 & \frac{1}{\Delta x^2}\{T_{xx}^{3\frac{1}{2}}\} & -\frac{1}{\Delta x^2}\{T_{xx}^{4\frac{1}{2}} + T_{xx}^{3\frac{1}{2}}\} \end{bmatrix}$$

The $\mathbf{g}$ vector contains the boundary conditions of the problem and the pumping or extraction rates,

$$\mathbf{g} = \begin{bmatrix} \frac{1}{\Delta x^2}\{T_{xx}^{1/2}\}\, h_L - Q_1/\Delta x \\ - Q_2/\Delta x \\ - Q_3/\Delta x \\ \frac{1}{\Delta x^2}\{T_{xx}^{4\frac{1}{2}}\}\, h_r - Q_4/\Delta x \end{bmatrix}$$

The initial conditions for the problem are $\mathbf{h}(0) = \mathbf{h}_0$

4.3 FINITE-ELEMENT METHODS

4.3.1 The Method of Weighted Residuals

The finite-element method is another numerical technique for the solution of groundwater flow and mass transport problems. For many groundwater problems, the finite-element method is superior to classical finite-difference models. Medium heterogeneities and irregular boundary conditions are handled naturally by the finite-element method. This is in contrast to difference approximations that require complicated interpolation schemes to approximate complex boundary conditions. Moreover, in the finite-element method, the size of the elements can be easily varied to reflect rapidly changing state variables or parameter values. The piecewise continuous representation of the dependent variables and, possibly, the parameters of the groundwater system can also increase the accuracy of the numerical approximations.

In the method of weighted residuals, the trial solution to the system's equations, which we express in terms of the differential operator L, as

$$L\{\phi\} = 0 \tag{4.17}$$

can be written as the finite series,

$$\phi \cong \hat{\phi}(\mathbf{x}, t) = N_1(\mathbf{x}) + \sum_{i=2}^{M} N_i(\mathbf{x})\tilde{\Phi}_i(t) \tag{4.18}$$

where $N_1(\mathbf{x})$ is selected to satisfy the essential boundary conditions and the remaining elements of the series, $N_i(\mathbf{x})$, also satisfy the given, homogeneous boundary conditions. The $N_i(\mathbf{x})$ are the basis or shape functions of the problem. As we will see in Section 4.3.2, these functions are defined by assumptions regarding spatial variation of the state variables in the solution domain. We can incorporate N_1 in the series through a specification on the $\tilde{\Phi}_i(t)$ and write the trial solution as,

$$\phi \cong \hat{\phi}(\mathbf{x}, t) = \sum_{i=1}^{M} N_i(\mathbf{x})\,\tilde{\Phi}_i(t) \tag{4.19}$$

Because the finite series is an approximation to the actual solution of the system's equations, there is a non-zero residual, R, associated with the approximation, or

$$L\{\hat{\phi}\} = R \neq 0 \tag{4.20}$$

The objective of the method of weighted residuals is to determine the nodal values, $\tilde{\Phi}_i$, in such a way so as to minimize the residual of the approximation. The residual is minimized by requiring the orthogonality of the weighting functions and the differential operator, L, over the domain D, or (Zienkiewicz, 1971)

$$\int_D W_k L\{\hat{\phi}\}\, dD = 0, \qquad k = 1, 2, \ldots n \tag{4.21}$$

where the W_k are weighting functions. These integral equations will yield n equations in the n unknown nodal values of the state variables.

The selection of the n weighting functions determines the type of weighted residual approximation. For example, if the weighting functions are the basis or shape functions ($W_k = N_k$), then the procedure is known as the *Galerkin method.* The integral equations are then given by

$$\int_D N_k L\{\hat{\phi}\}\, dD = 0, \quad \forall k \tag{4.22}$$

In the *subdomain method,* the weighting functions are chosen to be unity in the subregion where they are defined (V^*) and zero otherwise. In other words,

$$\int_D W_k L\{\hat{\phi}\}\, dD = 0, \quad \forall k \tag{4.23a}$$

where

$$W_k = \begin{bmatrix} 1, & \mathbf{x} \in V^* \\ 0, & \text{otherwise} \end{bmatrix} \tag{4.23b}$$

Note that the method is equivalent to requiring that the integral is zero over a sufficient number of subdomains to give the necessary number of simultaneous equations.

Alternatively, if the weights are the Dirac delta function, $W_k = \delta(\mathbf{x} - \mathbf{x}_k)$, then the orthogonality conditions in the *collocation method* are

$$\int_D \delta_k(\mathbf{x} - \mathbf{x}_k) L\{\hat{\phi}\} dD = 0, \quad \forall k \tag{4.24}$$

These conditions are equivalent to evaluating the residual at each point in the system where the weighting functions are defined. Although computationally straightforward, the collocation method is very sensitive to where the collocation points are located in the solution domain. However, the method has been particularly effective for certain classes of nonlinear flow problems (see, for example, Finlayson, 1972).

Although all of these weighted residual methods have been applied in water resources engineering, the Galerkin method will be used to represent the hydraulic and mass transport response equations of the aquifer system. This decision is based, in part, on the extensive applications of the Galerkin method to problems in reservoir engineering, groundwater systems, and coastal hydrodynamic problems (see, for example, Young, 1977; Pinder & Gray, 1977).

The following example problems illustrate how the Galerkin procedure may be used to develop the dynamic response equations for confined and unconfined aquifer systems.

Example Problem 4.3 Galerkin Finite-Element Model

Develop a Galerkin finite-element model for two-dimensional, semiconfined groundwater flow. Assume the aquifer is homogeneous and anisotropic, and that the principal axes of the transmissivity tensor are aligned with the x, y coordinate system.

We begin by writing the residual or orthogonality condition as

$$\sum_e \int_{D_e} N_k L\{h\}\, dD_e = 0, \quad \forall k \tag{4.25}$$

where $L\{h\}$ is given by Equation 2.55 and the summation is now, however, over all elements of the system (e). N_k are the basis functions and D_e is the elemental domain. Introducing the approximating equation, $h \cong \hat{h} = \mathbf{N}\,\tilde{\mathbf{h}}$, Equation 4.25 can be expressed for any element as

$$\int_{D_e} N_k L\{\mathbf{N}\,\tilde{\mathbf{h}}\}, \quad \forall k \tag{4.26}$$

where the $\tilde{\mathbf{h}}$ is the vector of unknown nodal head values. Substituting the semiconfined flow equation (Eq. 2.55) we have,

$$\int_{D_e} \left\{ \nabla \cdot \mathbf{T}\, \nabla(\mathbf{N}\,\tilde{\mathbf{h}}) \pm S^* + \frac{K_a}{m_a} H_a - \mathbf{N}\tilde{\mathbf{h}} - S\frac{\partial}{\partial t}(\mathbf{N}\tilde{\mathbf{h}}) \right\} N_k dD_e \tag{4.27}$$

Each term in these integral equations will generate an elemental matrix that is dependent on the basis functions and the system parameters. For example, if the storage coefficient is assumed constant within any element, the fourth term in Equation 4.27 becomes

$$-S \int_{D_e} \frac{\partial}{\partial t}(\mathbf{N}\,\tilde{\mathbf{h}}) N_k dD_e$$

If this equation is written for all the basis functions within the element, the elemental matrix A_e is

$$A_e = \begin{bmatrix} \int N_i{}^2 dD_e & \int N_i N_j dD_e & \int N_i N_k dD_e \ldots \\ \int N_j N_i dD_e & \int N_j{}^2 dD_e & \ldots \\ \cdot & \cdot & \\ \cdot & \cdot & \\ \cdot & \cdot & \\ \int N_k N_i dD_e & N_k N_j dD_e & \ldots \end{bmatrix}$$

The first part of the element equation can then be expressed as

$$-SA_e \dot{\tilde{\mathbf{h}}}$$

where $\dot{\tilde{\mathbf{h}}} = [\dot{\tilde{h}}_i, \dot{\tilde{h}}_j \ldots]^T$

For a point source or sink located at x_w, y_w the constant vector, $\mathbf{g}_e$, can be expressed as

$$\mathbf{g}_e = \pm Q \begin{bmatrix} N_i \\ N_j \\ N_k \end{bmatrix}_{\substack{x = x_w \\ y = y_w}}$$

where the basis functions are evaluated at x_w, y_w (the location of the well) and Q is the magnitude of the source or sink.

In a semiconfined aquifer, the $\mathbf{g}_e$ vector may also contain the leakage from the aquitard, or

$$\mathbf{g}_e = \begin{bmatrix} \int aN_i \, dD_e \\ \cdot \\ \cdot \\ \cdot \\ \int aN_k dD_e \end{bmatrix}$$

where $a = K_a H_a / m_a$. The remainder of the leakage term generates the elemental system of equations,

$$\frac{K_a}{m_a} \begin{bmatrix} \int N_i^2 dD_e & \int N_i N_j dD_e & \cdots\cdots \\ \int N_i N_j dD_e & \cdots\cdots & \\ \int N_i N_k dD_e & \cdots\cdots & \int N_k^2 dD_e \end{bmatrix} \begin{bmatrix} \tilde{h}_i \\ \tilde{h}_j \\ \tilde{h}_k \end{bmatrix}$$

Assuming the transmissivity is constant within an element, the transmissivity terms of Equation 4.27 may be written as

$$\int_{D_e} N_k \left\{ T_{xx} \frac{\partial^2 h}{\partial x^2} + T_{yy} \frac{\partial^2 h}{\partial y^2} \right\} dD_e \tag{4.28}$$

Green's theorem may be used to reduce the order of the equations to eliminate unnecessary continuity restrictions on the basis functions. Green's theorem replaces these second-order spatial derivatives with first-order derivatives and a surface integral evaluated on the boundary of the flow domain (Pipes, 1958), or

$$\int_D u \cdot \nabla^2 w dD = - \int_D u \cdot \nabla w dD + \int_S u \nabla w \, dS \tag{4.29}$$

where D is the domain and S is the external boundary of the domain; u and w are scalars.

To transform the equation, we introduce a new variable, $\bar{x} = x\left(\frac{T_{yy}}{T_{xx}}\right)^{1/2}$, and rewrite Equation 4.28 as

$$T_{yy}\int_{D_e}\left(N_k\frac{\partial^2 h}{\partial \bar{x}^2} + N_k\frac{\partial^2 h}{\partial y^2}\right)dD_e = T_{yy}\int_{D_e} N_k\,\nabla^2 h dD_e$$

where $\nabla^2 h$ is the Laplacian of the hydraulic head. Applying Green's theorem with $u = N_k$ and $w = h$,

$$-T_{yy}\int_{D_e}\left\{\frac{\partial N_k}{\partial \bar{x}}\frac{\partial h}{\partial \bar{x}} + \frac{\partial N_k}{\partial y}\frac{\partial h}{\partial y}\right\}dD_e + \int_S N_k\left\{\frac{\partial h}{\partial \bar{x}}, \frac{\partial h}{\partial y}\right\}\cdot d\mathbf{S}_e \qquad (4.30)$$

After recovering the original variables and substituting the approximating functions, the first term of the expression becomes

$$-\int_{D_e}\left\{T_{xx}\frac{\partial N_k}{\partial x}\mathbf{N}_x\,\tilde{\mathbf{h}} + T_{yy}\frac{\partial N_k}{\partial y}\mathbf{N}_y\,\tilde{\mathbf{h}}\right\}dD_e \qquad (4.31)$$

where $\mathbf{N}_x$ and $\mathbf{N}_y$ denote the vectors of the x, y partial derivatives of the basis functions.

The final term in Equation 4.30 represents the Neumann boundary conditions of the problem. Figure 4.4 shows a portion of the groundwater boundary S. If $d\mathbf{S}$,

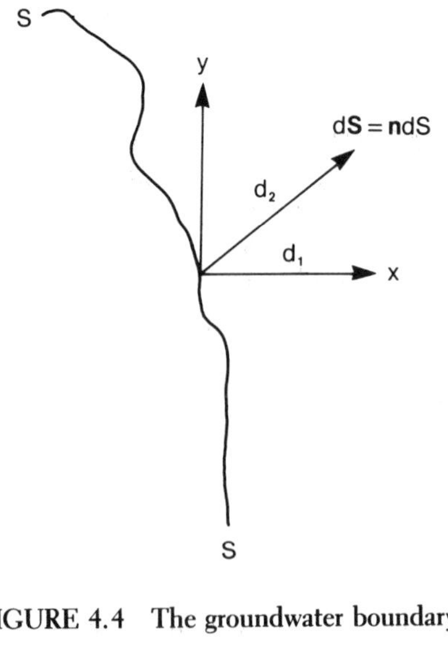

FIGURE 4.4 The groundwater boundary

the elemental boundary is expressed as

$$d\mathbf{S} = \mathbf{n}\, dS$$

Then, after a change of variables, the surface integral becomes

$$\int_S N_k \left\{ T_{xx} \frac{\partial \hat{h}}{\partial x} \mathbf{i} + T_{yy} \frac{\partial \hat{h}}{\partial y} \mathbf{j} \right\} \cdot \mathbf{n} dS = \int_S N_k \left\{ T_{xx} \frac{\partial \hat{h}}{\partial x} d_1 + T_{yy} \frac{\partial \hat{h}}{\partial y} d_2 \right\} dS \quad (4.32)$$

where d_1 and d_2 are the direction cosines associated with the boundary $d\mathbf{S}$ and $\mathbf{i}$ and $\mathbf{j}$ are unit vectors in the x, y directions. The argument of this integral is the normal flux $\mathbf{q}_n$ through the elemental boundary $d\mathbf{S}$.

$$\int_S N_k\, T_n \frac{\partial \hat{h}}{\partial \mathbf{n}}\, dS = \int_S N_k\, \mathbf{q}_n \cdot d\mathbf{S} \quad (4.33)$$

where T_n is the component of the transmissivity in the direction normal to the boundary. The boundary condition is only generated when non-zero fluxes are specified on the boundary. In the present example, this term is zero.

Combining the various matrix terms for the storage and transmissivity coefficients and the aquitard leakage, and dropping the ~ notation, the elemental equations may be expressed as

$$\mathrm{A}_e\, \dot{\mathbf{h}}_e + B_e\, \mathbf{h}_e + \mathbf{g}_e \quad (4.34)$$

The set of equations for the entire domain may be obtained by superimposing the individual elemental equations in the global matrix according to the node numbers of the elements or (see Example Problem 4.4)

$$\sum_e \int_{D_e} N_k\, L\{\hat{h}\} dD_e = 0, \qquad \forall_k$$

and, finally,

$$\mathrm{A}\dot{\mathbf{h}} + B\mathbf{h} + \mathbf{g} = 0 \quad (4.35a)$$

with

$$\mathrm{A} = \sum_e \mathrm{A}_e, \quad B = \sum_e B_e, \quad \mathbf{g} = \sum_e g_e \quad (4.35b)$$

The initial conditions are $\mathbf{h}\,(0) = g(\mathbf{x})$. The global equations may contain nodes where Dirichlet boundary conditions for groundwater flow or mass transport occur. In this event, the global matrices are partitioned to account for these passive nodes.

The dynamic response equations again relate the head, the initial state of the system, and the management of planning policies. As in the finite-difference examples, these equations are linear functions of the planning or management alternatives.

Example Problem 4.4 Global Matrix Generation

The example aquifer shown in Figure 4.5 has been discretized into five triangular elements. There are a total of six unknown nodal head values, and the elemental equations have been generated according to Equation 4.26. Assuming the boundary of the system is impermeable, generate the global coefficient matrices of the problem.

We begin by defining the global order of the unknown heads in the global coefficient matrices. In this example, each global matrix will be a 6 × 6 square matrix while the unknown heads are described by the 6 × 1 column vector $\mathbf{h} = (h_1, h_2, \ldots h_6)^T$. The assembly of each global matrix (A, B) is obtained by placing each elemental matrix in the global equation according to the node numbers of the elemental matrix (Figure 4.6). The node numbers are simply the position of the unknown head in the global vector $\mathbf{h}$. For example, in element 1, the unknown heads are h_1, h_2, h_3; the corresponding node numbers are then 1, 2, 3. The elemental matrix will then occupy the first three rows and columns of the global matrix. In element 4, the node numbers are 3, 5, and 6; the elemental matrix will then occupy the third, fifth, and sixth rows and columns of the global matrix. The total global matrix for the problem represents the superposition of all the elemental equations.

The global matrices typically have a large number of zero coefficients; that is, they are sparse matrices. All the non-zero coefficients are confined, however,

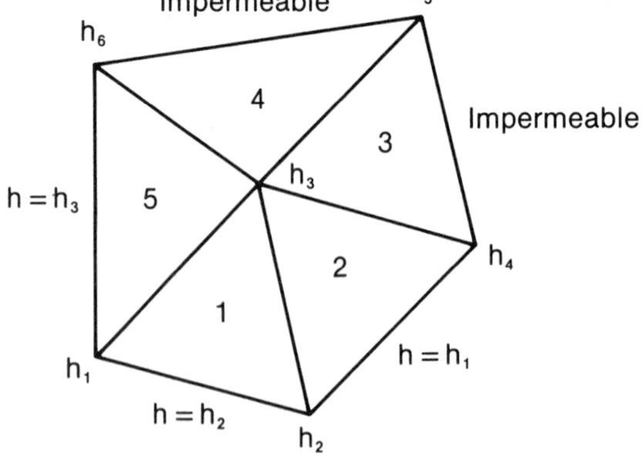

FIGURE 4.5 Example aquifer

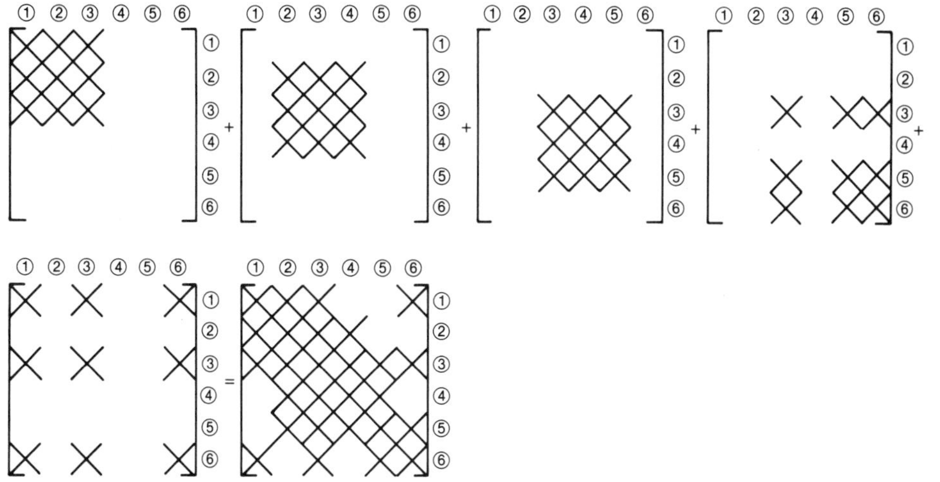

FIGURE 4.6 Global assembly of the finite-element equations

within a narrow band in the global equations. For triangular linear elements, for example, the bandwidth (BW) is (Segerlind, 1976),

$$\text{BW} = (R + 1) + NDOF$$

where R is the largest difference between the node numbers for all the elements of the assembly and $NDOF$ is the number of unknowns, the degrees of freedom, at each node. Similar formulas can be developed for other types of finite elements (see Pinder & Gray, 1977). The node numbering scheme is important because it affects the bandwidth of the global system of equations and, consequently, the overall size of the coefficient matrices.

The coefficient matrices for groundwater flow problems are also symmetric matrices. As a result, only the upper triangular portion of the matrices are actually stored during the computer solution of the finite-element equations. This, of course, minimizes computer storage and increases the computational efficiency of the procedure. In contrast, however, mass transport problems require substantially more computer resources because of the nonsymmetry of the convective, mass transport terms. Because all non-zero terms must be retained in core during the execution of the finite-element method, the solution of the mass transport problem is considerably more difficult than that of the hydraulic problem.

Example Problem 4.5 Soil Water Quality Model

A groundwater planner is analyzing the feasibility of the basin-spreading of municipal wastewaters. The disposal site overlies a shallow, unconfined aquifer and preliminary soil surveys indicate that the soil is homogeneous and isotropic. Assume that

1. The majority of the spreading cycle is spent under saturated soil water conditions.
2. Adsorption can be represented by a linear, equilibrium isotherm, A = Rc.
3. Biochemical reactions are first-order decay reactions.
4. One-dimensional (vertical) mass transport.

Develop a finite-element simulation model of contaminant transport in the soil system. The boundary and initial conditions for the problem are

$$-D\frac{\partial c}{\partial z} + q_z c = q_z c^*, \quad z = 0 \text{ (the ground surface)} \tag{4.36a}$$

$$\left.\frac{\partial c}{\partial z}\right|_{z=L} = 0 \quad \text{(soil-aquifer interface)} \tag{4.36b}$$

$$c(z, 0) = 0 \tag{4.36c}$$

where c^* is the input concentration in the spreading basin and L is the approximate distance to the water table.

We begin by considering the flow problem. Because the management model does not control the hydraulics of the spreading operation, the velocity distribution in the soil system can be found from the direct solution of the steady-state continuity equation (Eq. 2.12). The integration of the equation implies that the interstitial or pore velocity v_z is constant, or

$$v_z = q_z/n = W/n$$

where W is the given spreading rate and n is the porosity.

Introducing the adsorption relation in the mass transport equation (Eq. 3.40), the soil water quality may be expressed as,

$$\frac{\partial}{\partial t}(c + eRc) = D^*\frac{\partial^2 c}{\partial z^2} - v_z\frac{\partial c}{\partial z} - k_s c - k_a eRc \tag{4.37}$$

where $D^* = D/n$ and v_z is the pore velocity, R is the distribution coefficient relating the concentration of the constituent in the sorbed and solution phases, and $e = [(1 - n)/n]$ is the reciprocal of the void ratio. k_s and k_a are the reaction coefficients in the solution and sorbed phases. The transport equation may then be simplified to

$$\frac{\partial c}{\partial t} = \overline{D}\frac{\partial^2 c}{\partial z^2} - \overline{v}_z\frac{\partial c}{\partial z} - k^* c \tag{4.38}$$

where

$$\overline{D} = [D^*/(1 + eR)],\ \overline{v}_z = [v_z/(1 + eR)],\ k^* = [(k_s + k_a eR)/(1 + eR)]$$

The term $(1 + eR)$ is known as the *retardation factor* of the system. It effectively reduces the hydrodynamic dispersion and the velocity in the soil system.

The finite-element formulation of the mass transport model begins by assuming a trial solution to the mass transport equation (Eq. 4.38),

$$c \cong \hat{c} = \mathbf{N}\,\tilde{\mathbf{c}} \tag{4.39}$$

The orothogonality condition minimizes the residual associated with the finite-series approximations or

$$\int_0^L N_k L\{\hat{c}\}\, dz = 0, \quad \forall k \tag{4.40}$$

For an individual element, the orthogonality condition can be expressed as

$$\int_{D_e} N_k \left\{ \frac{\partial}{\partial t}(\mathbf{N}\tilde{\mathbf{c}}) - \overline{D}\frac{\partial^2}{\partial z^2}(\mathbf{N}\tilde{\mathbf{c}}) + \overline{v}_z \frac{\partial}{\partial z}(\mathbf{N}\tilde{\mathbf{c}}) + k^*(\mathbf{N}\tilde{\mathbf{c}}) \right\} dD_e, \quad \forall k \tag{4.41}$$

where $\tilde{\mathbf{c}}$ is the vector of the unknown nodal values of the concentration.

Again using Green's theorem, or for one-dimensional problems, integration by parts to eliminate the second derivatives, the transport equation becomes

$$\int_{D_e} N_k \{\mathbf{N}\,\dot{\tilde{\mathbf{c}}} + \overline{v}_z \mathbf{N}_z\,\tilde{\mathbf{c}} + k^*\,\mathbf{N}\,\tilde{\mathbf{c}}\} dD_e - \overline{D} N_k\,\mathbf{N}_z\,\tilde{\mathbf{c}}|_{S_e} + \overline{D}\int_{D_e} \frac{\partial N_k}{\partial z}\mathbf{N}_z\,\tilde{\mathbf{c}}\, dD_e$$

where $\mathbf{N}_z$ are the partial derivatives of the basis functions and S_e is the boundary of the element. This boundary term will be zero everywhere within the domain except at the ground surface, $z = 0$. Rewriting the flux boundary condition, the surface integral may be expressed as

$$-\overline{D}\mathbf{N}_z\,\tilde{\mathbf{c}} = \overline{v}_z\, c^* - \overline{v}_z \mathbf{N}\,\tilde{\mathbf{c}} = \overline{v}_z c^* - \overline{v}_z \tilde{c}_1$$

where $\tilde{c}_1$ is the concentration at the ground surface, and we have used the 0, 1 property of the basis functions ($N_i = 1$ at node i and 0 at any other node in the domain).

The elemental response equations can then be written as (dropping the ~ notation)

$$P_e\,\dot{\mathbf{c}} + R_e\mathbf{c} + \mathbf{f}_e \tag{4.42a}$$

where

$$P_e = \int_{D_e} N_k N_n dD_e \tag{4.42b}$$

$$R_e = \int_{D_e} \left\{ \overline{v}_z N_k \frac{\partial N_n}{\partial z} + \overline{D} \frac{\partial N_k \partial N_n}{\partial z \, \partial z} + N_k N_n k^* \right\} dD_e \tag{4.42c}$$

and $\mathbf{f}_e$ will be a zero vector, except for the first element.

Summing the elemental equations according to Equation 4.35, the global equations are

$$P\dot{\mathbf{c}} + R\mathbf{c} + \mathbf{f} = 0 \tag{4.43}$$

where $\mathbf{f}$ is a column vector with the elements,

$$f_k = \begin{cases} -\overline{v}_z (c_1 - c^*), & k = 1 \\ 0 & k \neq 1 \end{cases}$$

The initial conditions are given as $c(0) = g(z)$. The dynamic response equations are again a linear function of the waste input concentration, c^*.

4.3.2 Basis Functions

The accuracy and stability of the weighted residual model is dependent on the basis or approximating functions for the problem. In finite-element analysis, the dependent variables are approximated by polynomial functions that are defined in terms of the unknown nodal values of the state or dependent variables. These approximating or interpolating functions vary over each element of the system, although continuity of the function is maintained along the elemental boundaries. For one-dimensional hydraulic or mass transport problems, the basis function can be formulated as Lagrange or Hermitian polynomials. The Lagrange polynomials p_i^n are defined as

$$p_i^n = (\xi - \xi_0)\ldots(\xi - \xi_{i-1})(\xi - \xi_{i+1})\ldots(\xi - \xi_n)/\{(\xi_i - \xi_0)\ldots(\xi_i - \xi_{i-1})\cdot (\xi_i - \xi_{i+1})\ldots(\xi_i - \xi_n)\}, \quad -1 \leq \xi \leq 1 \tag{4.44}$$

where ξ is the dimensionless coordinate system. Hermitian cubic basis functions are given in local coordinates as

$$\begin{aligned} N_{0,1} &= \frac{1}{4}(\xi - 1)^2 \cdot (\xi + 2), \quad & N_{0,2} &= -\frac{1}{4}(\xi + 1)^2 \cdot (\xi - 2) \\ N_{1,1} &= \frac{1}{8}L^e(\xi + 1)\cdot(\xi - 1)^2, \quad & N_{1,2} &= \frac{1}{8}L^e(\xi + 1)^2 \cdot (\xi - 1) \end{aligned} \tag{4.45}$$

where L^e is the length of an element, and $\xi = 2[(x - x_1^e)/L^e] - 1$. These polynomials are cubic splines that have first-order derivative continuity between elements and second-order continuity over an element.

For higher-dimensional problems, the basis or approximating functions can be developed by forming the product of two one-dimensional basis functions, where each basis function is written for a particular coordinate direction. The approach can be applied whenever linear, quadratic, cubic, or Hermitian cubic bases are used to approximate the state variables of the aquifer system. Alternatively, serendipity elements can also be used to approximate the unknown functions. These elements are developed from the same degree polynomial Lagrangian elements, but do not require interior nodal points. Studies indicate that they are nearly as accurate as Lagrangian elements but require far fewer nodes (Pinder & Gray, 1977).

Isoparametric finite elements have also been widely used in groundwater simulation analysis. These elements, which may be deformed to match irregular boundaries, are obtained by first defining a simple element in one coordinate system and then distorting the element to a more convenient shape in another curvilinear coordinate system. The element's geometry can then be described by a set of functions that are independent of the parameter approximation; that is, the number of nodes used to define the element shape is equal to the number of nodes used in the definition of the interpolating equations. The development of the isoparametric basis function is illustrated in Example Problem 4.6. Mixed isoparametric elements, which allow the solution domain to be discretized into a number of different types of elements, are discussed in Pinder and Gray (1977).

Common one- and two-dimensional finite elements used in groundwater planning, and their associated basis functions, are summarized in Table 4.4.

Example Problem 4.6 Isoparametric Elements

Develop the basis functions for the linear isoparametric element in Figure 4.7.

We begin by considering Figure 4.7 which shows a local (ξ, η) and global Cartesian coordinate system (x, y). The element is irregularly shaped in x,y but is a square in the local system. The corners of the element are located at $\xi = \pm 1$ $\eta = \pm 1$. The relation between the local and global coordinates can be expressed as

$$x = \sum_{i=1}^{n} w_i x_i \tag{4.46a}$$

$$y = \sum_{i=1}^{n} w_i y_i \tag{4.46b}$$

where the w's are undetermined functions of η and x_i, and x_i, y_i are the nodal coordinates in x and y. To determine values of $\mathbf{w}$ that will provide the necessary variation in the head or mass concentration along the elemental sides, we require for the linear element

$$x = a + b\,\xi + c\eta + d\,\xi\,\eta \tag{4.47}$$

TABLE 4.4 Finite Elements in Subsurface Hydrology[1,2]

	Type of Element	*Physical Description*	*Basis Function*
1.	Linear simplex	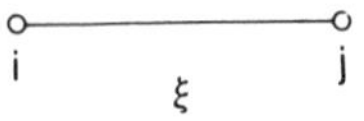	$N_i = \frac{1}{2}(1 - \xi)$ $N_j = \frac{1}{2}(1 + \xi)$
2.	Quadratic simplex	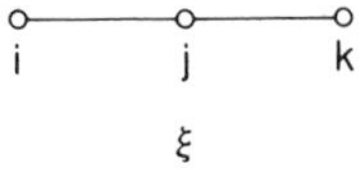	$N_i = -\frac{1}{2}\xi(1 - \xi)$ $N_j = 1 - \xi^2$ $N_k = \frac{1}{2}\xi(1 + \xi)$
3.	Cubic element	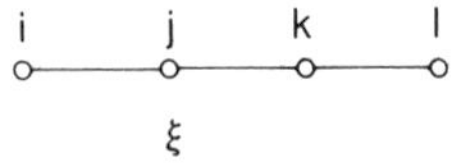	$N_i = -\frac{1}{16}(1 + 3\xi)(1 - 3\xi)(1 - \xi)$ $N_j = \frac{9}{16}(1 + \xi)(1 - 3\xi)(1 - \xi)$ $N_k = \frac{9}{16}(1 + \xi)(1 + 3\xi)(1 - \xi)$ $N_l = -\frac{1}{16}(1 + \xi)(1 + 3\xi)(1 - 3\xi)$
4.	Hermitian cubic polynomials[3]	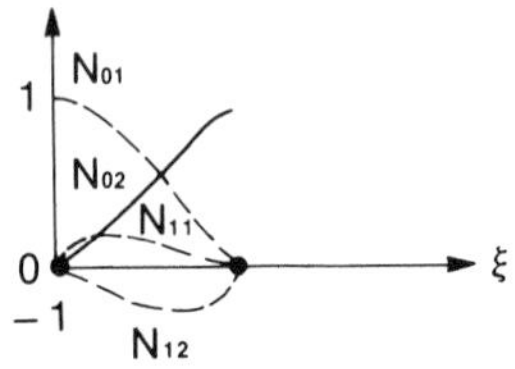	$N_{01} = \frac{1}{4}(\xi - 1)^2(\xi + 2)$ $N_{02} = -\frac{1}{4}(\xi + 1)^2(\xi + 2)$ $N_{11} = \frac{1}{8}L^e(\xi + 1)(\xi - 1)^2$ $N_{12} = \frac{1}{8}L^e(\xi + 1)^2(\xi - 1)$
5.	Triangular simplex	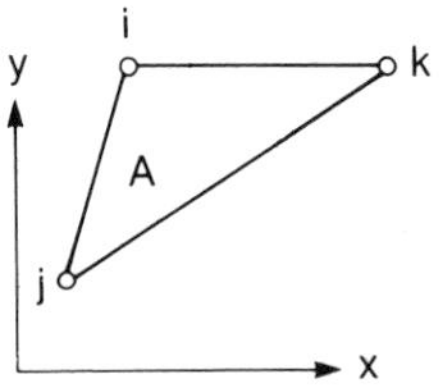	$N_i = [(x_j y_k - x_k y_j) + (y_j - y_k)x + (x_k - x_j)y]/2A$ $N_j = [(x_k y_i - x_i y_k) + (y_k - y_i)x + (x_i - x_k)y]/2A$ $N_k = [(x_i y_j - x_j y_i) + (y_i - y_j)x + (x_j - x_i)y]/2A$
6.	Linear Lagrangian element	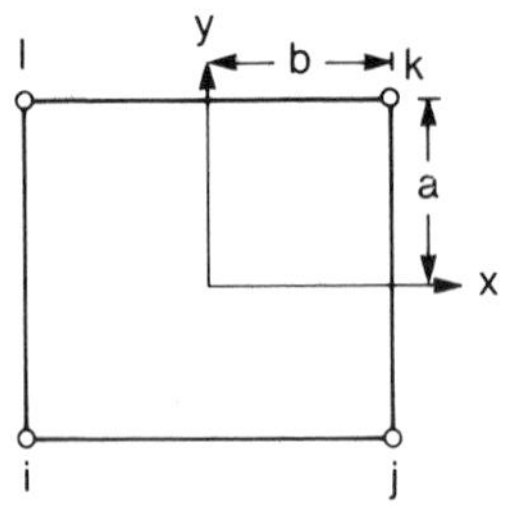	$N_i = \frac{1}{4ab}(b - x)(a - y)$ $N_j = \frac{1}{4ab}(b + x)(a - y)$ $N_k = \frac{1}{4ab}(b + x)(a + y)$ $N_l = \frac{1}{4ab}(b - x)(a + y)$ or $N_i = \frac{1}{4}(1 + \xi\xi_i)(1 + \eta\eta_i)\ \xi_i = \pm 1, \eta = \pm 1$

TABLE 4.4 Continued

	Type of Element	*Physical Description*	*Basis Function*
7.	Quadratic serendipity element	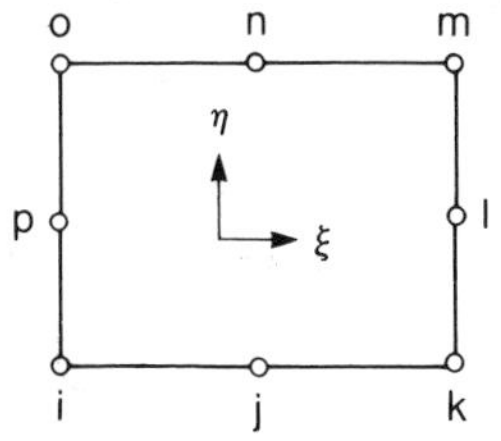 $1\epsilon,\eta 1 \le 1$	Corner nodes, $N_i = \frac{1}{4}(1+\xi\xi_i)(1+\eta\eta_i)(\xi\xi_i+\eta\eta_i-1)$ Midside nodes, $\xi_i = 0, N_i\ \frac{1}{2}(1-\xi)^2(1+\eta\eta_i)$ Midside nodes, $\eta_i = 0, N_i\ \frac{1}{2}(1+\xi\xi_i)(1-\eta^2)$
8.	Cubic serendipity element	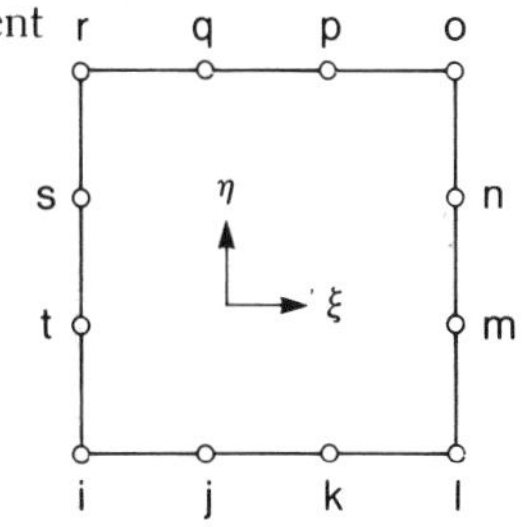	Corner nodes $N_i = \frac{1}{32}(1+\xi\xi_i)(1+\eta\eta_i)[9(\xi^2+\eta^2)-10]$ Midside nodes, $\xi_i = \pm 1, \eta_i = \pm\frac{1}{3}, N_i = \frac{9}{32}(1+\xi\xi_i)(1+\eta^2)(1+9\eta\eta_i)$ $\xi_i = \pm\frac{1}{3}, \eta_i = \pm 1, N_i = \frac{9}{32}(1+\eta\eta_i)(1-\xi^2)(1+9\xi\xi_i)$

[1] $-1 \le \xi \le 1$.
[2]. $-1 \le \eta \le 1$.
[3] $L^e = x_2 - x_1$; $\xi = 2[(x - x_i)/L^e] - 1$.

which is linear in ξ when $\eta = \pm 1$ and linear in η when $\xi = \pm 1$. We also have, from Figure 4.7, the following conditions,

$$x = x_1,\ \xi = \eta = -1 \qquad x = x_3,\ \xi = \eta = 1$$

$$x = x_2,\ \xi = 1, \eta = -1 \qquad x = x_4,\ \xi = -1, \eta = 1$$

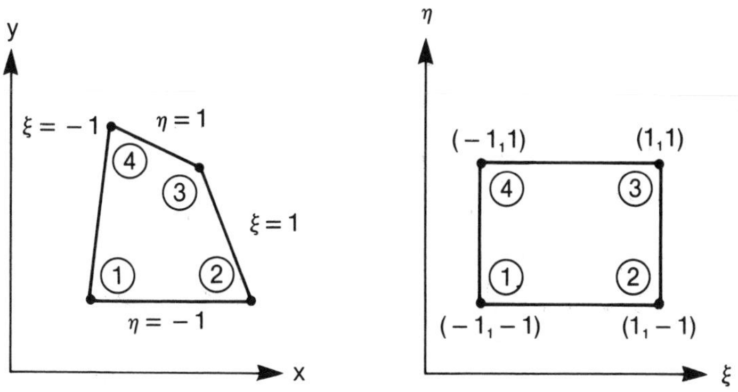

FIGURE 4.7 Linear isoparametric (quadrilateral) element

Substituting these restrictions into Equation 4.47, we have

$$\begin{bmatrix} x_1 \\ x_2 \\ x_3 \\ x_4 \end{bmatrix} = \begin{bmatrix} 1 & -1 & -1 & 1 \\ 1 & 1 & -1 & -1 \\ 1 & 1 & 1 & 1 \\ 1 & -1 & 1 & -1 \end{bmatrix} \begin{bmatrix} a \\ b \\ c \\ d \end{bmatrix}$$

or, in vector-matrix notation

$$\mathbf{x} = \mathbf{A}\mathbf{p}$$

where $\mathbf{p} = [a\ b\ c\ d]^T$. The parameters are then given as $\mathbf{p} = \mathbf{A}^{-1}\mathbf{x}$. However, from Equations 4.46 and 4.47, we have

$$x = [1\ \xi\ \eta\ \xi\eta]\ \mathbf{p} = \boldsymbol{\xi}\ \mathbf{p} = \xi\ \mathbf{A}^{-1}\mathbf{x} \quad \text{and} \quad \mathbf{w} = \boldsymbol{\xi}\ \mathbf{A}^{-1}$$

Expanding the equations, the transformation functions are defined by

$$\begin{aligned} w_1 &= \frac{1}{4}(1 - \xi)(1 - \eta) \qquad & w_3 &= \frac{1}{4}(1 + \xi)(1 + \eta) \\ w_2 &= \frac{1}{4}(1 + \xi)(1 - \eta) \qquad & w_4 &= \frac{1}{4}(1 - \xi)(1 - \eta) \end{aligned} \tag{4.48}$$

Examining Table 4.4 we see that the basis functions are identical to the shape functions defined for a rectangular element. What this implies is that the polynomials that are used to satisfy the coordinate transformations also fulfill the requirements of basis defined in local coordinates. Basis functions defined in this way satisfy the requirements for convergence of the finite element method, that is

1. The basis functions sum to one for any element.
2. The functions are continuous between the adjacent elements (Zienkiewicz, 1971)

Example Problem 4.7 Basis Functions for a Linear Triangular Element

The flow domain of a confined aquifer system has been discretized into a series of linear, triangular elements. Develop the basis functions for the problem.

The development of the basis functions assumes that within any element the head, h, is a linear function of the x, y coordinates, or

$$\hat{h} = a + bx + cy \tag{4.49a}$$

or

$$\hat{h} = [1\ x\ y] \begin{bmatrix} a \\ b \\ c \end{bmatrix} \tag{4.49b}$$

where a, b, c are the unknown parameters of the approximating function. These parameters may be determined from the nodal conditions of the system's finite elements. For the triangular two-dimensional element shown in Figure 4.8, we can write for each node (ijk)

$$h = \tilde{h}_i = a + bx_i + cy_i$$
$$h = \tilde{h}_j = a + bx_j + cy_j$$
$$h = \tilde{h}_k = a + bx_k + cy_k$$

where $\tilde{h}_i$ is the nodal value of the hydraulic head and (x_i, y_i) are the coordinates of the nodal points relative to the global coordinate system of the problem. The simultaneous equations may then be rewritten in vector-matrix notation as,

$$\tilde{\mathbf{h}} = \begin{bmatrix} \tilde{h}_i \\ \tilde{h}_j \\ \tilde{h}_k \end{bmatrix} = \begin{bmatrix} 1 & x_i & y_i \\ 1 & x_j & y_j \\ 1 & x_k & y_k \end{bmatrix} \begin{bmatrix} a \\ b \\ c \end{bmatrix} = \mathrm{Ap} \tag{4.50}$$

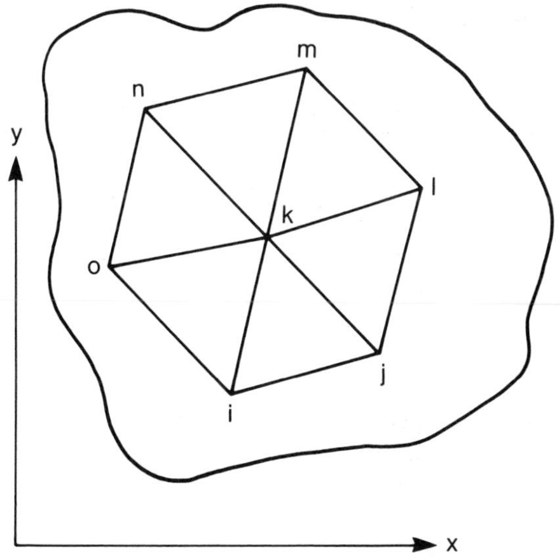

FIGURE 4.8 Triangular finite elements

where $\tilde{\mathbf{h}}$ is the vector of nodal values, A is the coefficient matrix, and $\mathbf{p}$ is the vector of the unknown parameters of the approximating function. The parameters are a function of the nodal coordinates and the nodal values of h, or

$$\mathbf{p} = A^{-1}\tilde{\mathbf{h}} \tag{4.51}$$

where A^{-1} is the matrix inverse of A. If the parameters are substituted into the interpolating function, Equation 4.49b, we have

$$\hat{h} = [1 \; x \; y]A^{-1}\tilde{\mathbf{h}} = \mathbf{N}\,\tilde{\mathbf{h}} \tag{4.52}$$

where the row vector $\mathbf{N}$ is composed of the basis or shape functions of the problem, or

$$N_i = \{(x_j y_k - x_k y_j) + (y_j - y_k)x + (x_k - x_j)y\}/2\Delta \tag{4.53a}$$

$$N_j = \{(x_k y_i - x_i y_k) + (y_k - y_i)x + (x_i - x_k)y\}/2\Delta \tag{4.53b}$$

$$N_k = \{x_i y_j - x_j y_i) + (y_i - y_j)x + (x_j - x_i)y\}/2\Delta \tag{4.53c}$$

and 2Δ is given as

$$2\Delta = \begin{vmatrix} 1 & x_i & y_i \\ 1 & x_j & y_j \\ 1 & x_k & y_k \end{vmatrix}$$

where Δ is the area of the triangular element and i, j, and k are arranged in the counterclockwise order.

Example Problem 4.8 Quadratic Basis Functions

Develop the basis functions for a quadratic finite element using Equation 4.44 to evaluate the basis functions.

The quadratic simplex element has three nodal points. The general representation of the head or concentration in any element is then given by

$$\hat{\phi} = a + bx + cx^2 \tag{4.54}$$

We use Equation 4.44 to evaluate the parameters of the approximating functions. For example, selecting those terms of the Lagrange polynomial containing $\xi_{i-1} = -1$, $\xi_i = 0$, $\xi_{i+1} = 1$, we have for the center node,

$$\begin{aligned} N_0 = P_i^2|_{\xi=0} &= (\xi - \xi_{i-1})(\xi - \xi_{i+1})/\{(\xi_i - \xi_{i-1})(\xi_i - \xi_{i+1})\} \\ &= (1 + \xi)(1 - \xi) = 1 - \xi^2 \end{aligned} \tag{4.55}$$

Similarly, the basis function, N_{-1}, involves values of ξ, corresponding to

$$\xi_i = -1, \quad \xi_{i+1} = 0, \quad \xi_{i+1} = 1$$

The basis function can then be written as

$$N_{-1} = P_i^2|_{\xi=-1} = \frac{(\xi - \xi_{i+1})(\xi - \xi_{i+2})}{(\xi_i - \xi_{i+1})(\xi_i - \xi_{i+2})} = \frac{\xi^2 - \xi}{2} \tag{4.56}$$

The basis function N_1 is given as

$$N_1 = P_i^2|_{\xi=1} = \frac{\xi + \xi^2}{2}$$

4.4 SOLUTION METHODS FOR THE DYNAMIC RESPONSE EQUATIONS

4.4.1 Linear Systems

The hydraulic or water-quality response equations of the groundwater system are obtained by numerically transforming the system's equations using finite-differences or finite-elements. These equations are systems of ordinary differential equations (in time) which can be expressed as

$$A\frac{d\boldsymbol{\phi}}{dt} + B\boldsymbol{\phi} + \mathbf{g} = 0 \tag{4.57a}$$

$$\boldsymbol{\phi}(0) = \boldsymbol{\phi}_0 \tag{4.57b}$$

The A and B coefficient matrices are a function of the system's hydraulic or quality parameters and the basis functions or difference approximations. The $\mathbf{g}$ vector contains the point sources and/or sinks, which are the control or policy variables for the problem, and the boundary conditions of the aquifer system. The $\boldsymbol{\phi}$ vector is a vector of the nodal values (grid points) of the state variables, the hydraulic head or mass concentrations. $\boldsymbol{\phi}_0$ are the initial conditions for the problem.

In many management or screening models (see Section 6.1, for example), the steady-state response equations are used to determine preliminary groundwater pumping or recharge alternatives. The steady-state response equations, relating the state variables and the management options, can be easily developed from Equation 4.57. For steady-state conditions, $d\boldsymbol{\phi}/dt = 0$ and, consequently, the response equations become

$$B\boldsymbol{\phi} + \mathbf{g} = 0 \tag{4.58a}$$

or

$$\boldsymbol{\phi} = B^{-1}\mathbf{g} \tag{4.58b}$$

The state variables are explicit (linear) functions of the decision variables contained in the **g** vector.

In time-varying simulation or optimization problems, the response equations may be solved using finite-difference approximations in time, or analytically using the matrix calculus. These approximations may be used to relate the policy or decision variables and the state of the system over the entire planning or design horizon.

If we approximate the time derivatives of the response equations using an explicit or forward-time approximation, the equations may be expressed as

$$\mathrm{A}\left\{\frac{\boldsymbol{\phi}^k - \boldsymbol{\phi}^{k-1}}{\Delta t}\right\} + B\boldsymbol{\phi}^{k-1} + \mathbf{g} = 0 \tag{4.59a}$$

where $\boldsymbol{\phi}^k$ is the vector of state variables at the kth time level within the planning horizon and Δt is the time step. The solution of the equations for any time step is

$$\boldsymbol{\phi}^k = \{I - \mathrm{A}^{-1} B\Delta t\}\, \boldsymbol{\phi}^{k-1} - \mathrm{A}^{-1}\Delta t\mathbf{g} \tag{4.59b}$$

where A^{-1} is the matrix inverse of A and I is the identity matrix. If finite-difference methods are used to generate the response equations, the A coefficient matrix is a diagonal matrix. The matrix inverse thus requires little computational effort. However, when finite elements are used to approximate the equations of the groundwater system, the A matrix will generally not be a diagonal form. The equations will then have to be solved using methods developed for large systems of equations (see, for example, Remson, et al., 1971).

However, it is possible to formulate Equation 4.59b so that A is diagonal. The most popular approach is to diagonalize A by lumping (Lapidus & Pinder, 1982). This means that the information contained in the off-diagonal elements is incorporated in the diagonal elements of A. In one approach, for example, the elements within any row are summed and allocated directly to the diagonal elements. Alternatively, different numerical integration schemes can be used to evaluate the elemental matrices. For example, if Simpson's rule rather than Gaussian quadrature is used with Lagrangian finite elements a diagonal form arises naturally (Gray & Van Genuchten, 1978; Young, 1977).

An alternative to the explicit scheme is the backward-in-time or implicit approximation. In this method, all the state variables are evaluated at the next or unknown time level, or

$$A\left\{\frac{\boldsymbol{\phi}^k - \boldsymbol{\phi}^{k-1}}{\Delta t}\right\} + B\boldsymbol{\phi}^k + \mathbf{g} = 0 \qquad (4.60a)$$

which can be rearranged to give

$$\boldsymbol{\phi}^k = \left\{\frac{A}{\Delta t} + B\right\}^{-1} \frac{A}{\Delta t}\boldsymbol{\phi}^{k-1} - \left\{\frac{A}{\Delta t} + B\right\}^{-1} \mathbf{g} \qquad (4.60b)$$

The matrix inversion is generally not performed directly. Rather, the set of simultaneous equations are solved using direct-solution algorithms, such as LU decomposition. Further details of these methods may be found in Lapidus and Pinder (1982), Remson, et al., (1971) and in Section 4.5. In contrast to the explicit scheme, the implicit scheme is unconditionally stable.

We can also generalize the difference approximation by writing the variable-weighted, implicit approximation as

$$A\left\{\frac{\boldsymbol{\phi}^k - \boldsymbol{\phi}^{k-1}}{\Delta t}\right\} + B\,\{\theta\boldsymbol{\phi}^k + (1 - \theta)\boldsymbol{\phi}^{k-1}\} + \mathbf{g} = 0 \qquad (4.61a)$$

where θ is the weighting parameter. The value of $\boldsymbol{\phi}^k$, the state variables, at the next time step, may be written as in terms of $\boldsymbol{\phi}^{k-1}$, and $\mathbf{g}$ as

$$\{A + \theta\Delta t B\}\boldsymbol{\phi}^k = \{A - (1 - \theta)\Delta t B\}\boldsymbol{\phi}^{k-1} - \mathbf{g}\Delta t \qquad (4.61b)$$

When θ is one, Equation 4.61b becomes the implicit or backward approximation. Similarly, when θ is zero, we have the forward or explicit approximation. The Crank-Nicolson approximation is obtained with $\theta = 0.5$,

$$A\left\{\frac{\boldsymbol{\phi}^k - \boldsymbol{\phi}^{k-1}}{\Delta t}\right\} + B\left\{\frac{\boldsymbol{\phi}^k + \boldsymbol{\phi}^{k-1}}{2}\right\} + \mathbf{g} = 0 \qquad (4.62a)$$

and

$$\boldsymbol{\phi}^k = \left\{\frac{A}{\Delta t} + \frac{B}{2}\right\}^{-1} \left\{\frac{A}{\Delta t} - \frac{B}{2}\right\}\boldsymbol{\phi}^{k-1} - \left\{\frac{A}{\Delta t} + \frac{B}{2}\right\}^{-1} \mathbf{g} \qquad (4.62b)$$

The Crank-Nicolson approximation minimizes the truncation error of the difference approximation and increases the convergence properties of the numerical approximations. This second-order correct scheme has a truncation error of $0\ (\Delta t^2)$.

An alternative to the finite-difference solution of the dynamic response equations is the direct analytical solution of these equations using the matrix calculus.

We rewrite Equation 4.57 as

$$\dot{\boldsymbol{\phi}} = P\boldsymbol{\phi} + \mathbf{z} \tag{4.63a}$$

$$\boldsymbol{\phi}(0) = \boldsymbol{\phi}_0 \tag{4.63b}$$

with $P = -A^{-1}B$, and $\mathbf{z} = -A^{-1}\mathbf{g}$. The solution of the equations is given as (Bellman, 1960)

$$\boldsymbol{\phi}(t) = e^{Pt}\boldsymbol{\phi}_0 + \int_0^t e^{P(t-\tau)}\,\mathbf{z}(\tau)d\tau \tag{4.64}$$

where e^{Pt} is the matrix exponential. An approach that can be used for the evaluation of the matrix exponential and the determination of error bounds is based on the diagonalization of the P coefficient matrix (Moler & Loan, 1978). Assuming the characteristic roots (eigenvalues) of P are distinct, P can be expressed as

$$P = RQR^{-1} \tag{4.65}$$

The R matrix contains the eigenvectors of P. The elements of the diagonal matrix Q are the eigenvalues of P, i.e., $Q_{jj} = \lambda_j$. The eigenvalues and eigenvectors are found from the solution of the equations,

$$P\boldsymbol{\phi} = \lambda\boldsymbol{\phi}$$

The exponential terms can then be written as

$$e^{Pt} = e^{RQR^{-1}t} = R\hat{Q}R^{-1}$$

where $\hat{Q}$ is now a diagonal matrix, i.e., $\hat{Q}_{jj} = e^{\lambda_j t}$. If we also can assume that the boundary conditions and/or the management decisions are constant over the period of integration, then the analytical solution of the equations is

$$\boldsymbol{\phi}(t) = R\hat{Q}R^{-1}\boldsymbol{\phi}_0 - P^{-1}(I - R\hat{Q}R^{-1})\mathbf{z} \tag{4.66}$$

or, after recovering the original matrices,

$$\boldsymbol{\phi}(t) = R\hat{Q}R^{-1}\boldsymbol{\phi}_0 - B^{-1}A\,(I - R\hat{Q}R^{-1})A^{-1}\mathbf{g} \tag{4.67a}$$

or

$$\boldsymbol{\phi}(t) = A_1\,(t)\boldsymbol{\phi}_0 + A_2(t)\mathbf{g} \tag{4.67b}$$

where now

$$A_1(t) = R\hat{Q}R^{-1}, \tag{4.68a}$$

$$A_2(t) = -B^{-1}A(I - R\hat{Q}R^{-1})A^{-1} \tag{4.68b}$$

Equations 4.67a and b are the general linear response equations for a finite, distributed parameter, groundwater flow or quality system.

Example Problem 4.9

In Example Problem 4.5, a Galerkin finite element model was developed to predict soil/water quality. Simulate the water quality variations using chapeau and cubic basis functions for a conservative, noninteracting constituent. Compare the results with a finite-difference solution of the mass transport equation.

From Table 4.4, the basis functions for the finite element model are given for the linear element as

$$N_i = \frac{(1 - \xi)}{2}$$

$$N_j = \frac{(1 + \xi)}{2}$$

where $-1 \leq \xi \leq 1$. And, for the cubic element,

$$N_i = -\frac{1}{16}(1 + 3\xi)(1 - 3\xi)(1 - \xi)$$

$$N_j = \frac{9}{16}(1 + \xi)(1 - 3\xi)(1 - \xi)$$

$$N_k = \frac{9}{16}(1 + \xi)(1 + 3\xi)(1 - \xi)$$

$$N_l = -\frac{1}{16}(1 + \xi)(1 + 3\xi)(1 - 3\xi)$$

Evaluating the elemental equations and assembling the global matrices according to Equation 4.35 produces the system of ordinary differential equations,

$$P\dot{c} + Rc + f = 0, \quad c(0) = 0$$

The solution of these equations for any time step, Δt, is approximated using the centered-in-time Crank-Nicolson scheme, or

$$P\left(\frac{\mathbf{c}^k - \mathbf{c}^{k-1}}{\Delta t}\right) + R\left(\frac{\mathbf{c}^k + \mathbf{c}^{k-1}}{2}\right) + \mathbf{f} = 0$$

and

$$\left\{\frac{P}{\Delta t} + \frac{R}{2}\right\}\mathbf{c}^k = \left\{\frac{P}{\Delta t} - \frac{R}{2}\right\}\mathbf{c}^{k-1} - \mathbf{f}$$

The truncation error of the approximation is $0(\Delta t^2)$.

Typical simulation results of the finite-element model are presented in Figures 4.9 and 4.10. For both linear and cubic basis functions, the results are in close agreement with Brenner's (1962) analytical solution, or

$$c/c_0 = \exp\,[P(2 - T)] \sum_{k=1}^{\infty} \frac{\lambda_k \sin 2\lambda_k}{(\lambda_k^2 + P^2 + P)} \exp\,[-\lambda_k^2 T/P] \qquad (4.69a)$$

where, referring to Equation 4.38, $T = \bar{v}_z t/L$, $P = 4q_z L/4\bar{D}$, and λ_n are the positive roots of

$$\tan 2\lambda = \frac{2\lambda P}{\lambda^2 - P^2} \qquad (4.69b)$$

For very low values of dispersion (large Peclet numbers), the chapeau functions exhibit oscillatory behavior. In contrast, the finite-difference solution produces severe oscillations for the two dispersion values illustrated in Figures 4.9 and 4.10.

The reasons for the poor performance of the finite-difference model are due in part to the types of approximations used in the numerical model. Consider the difference approximation of the mass transport equation (Eq. 4.38), assuming $k^* = 0$, and no adsorption,

$$\frac{\partial c}{\partial t} = \bar{D}\frac{\partial^2 c}{\partial z^2} - \bar{v}_z \frac{\partial c}{\partial z}$$

with $\bar{D} = D^* = D/n$ and $\bar{v}_z = q_z/n$. We can approximate the time derivative and convective terms using a variety of finite-difference approximations. For example, the time derivative could be differenced using a Crank-Nicolson ($\theta = 1/2$), explicit ($\theta = 0$), or an implicit ($\theta = 1$) scheme. The convective term could also be approximated using either a backward or central difference scheme. In

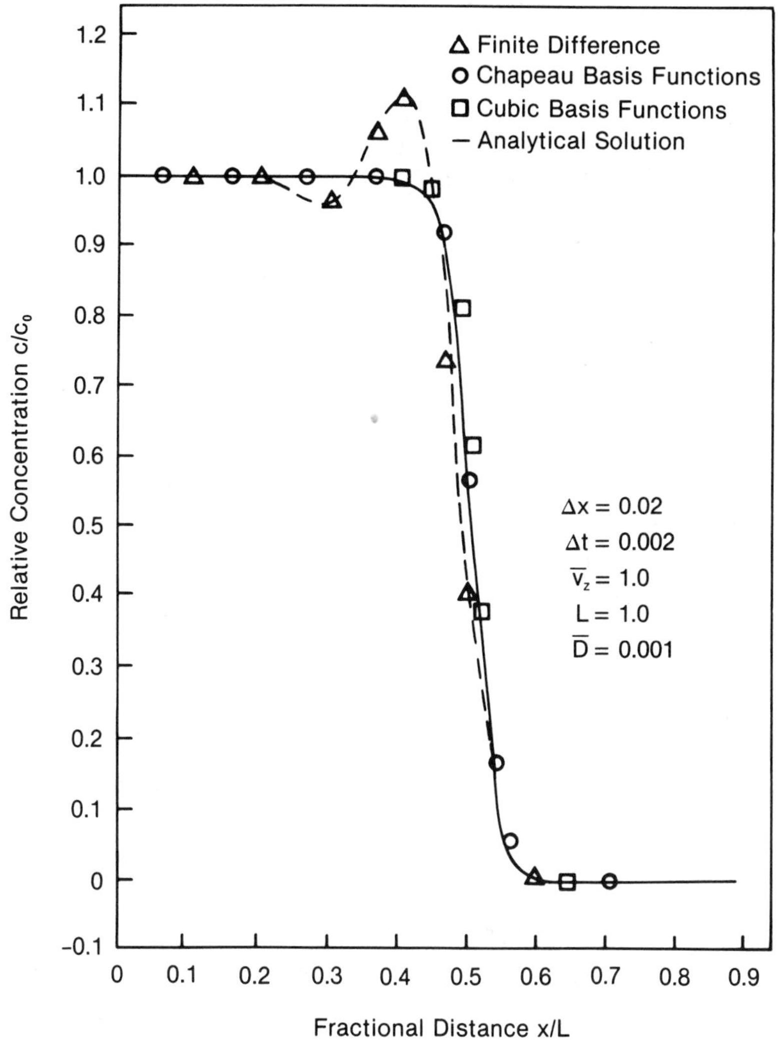

FIGURE 4.9 Comparison of finite-difference and finite-element solutions to the convective-diffusion equation for a dispersion coefficient of 0.0001 and a displaced pore volume of 0.5 (Adapted from D.B. Grove, "The Use of Galerkin Finite Element Methods to Solve Mass-Transport Equations." *Water Resources Investigation* 77-47. Denver: U.S.G.S., 1977, pp. 34–35)

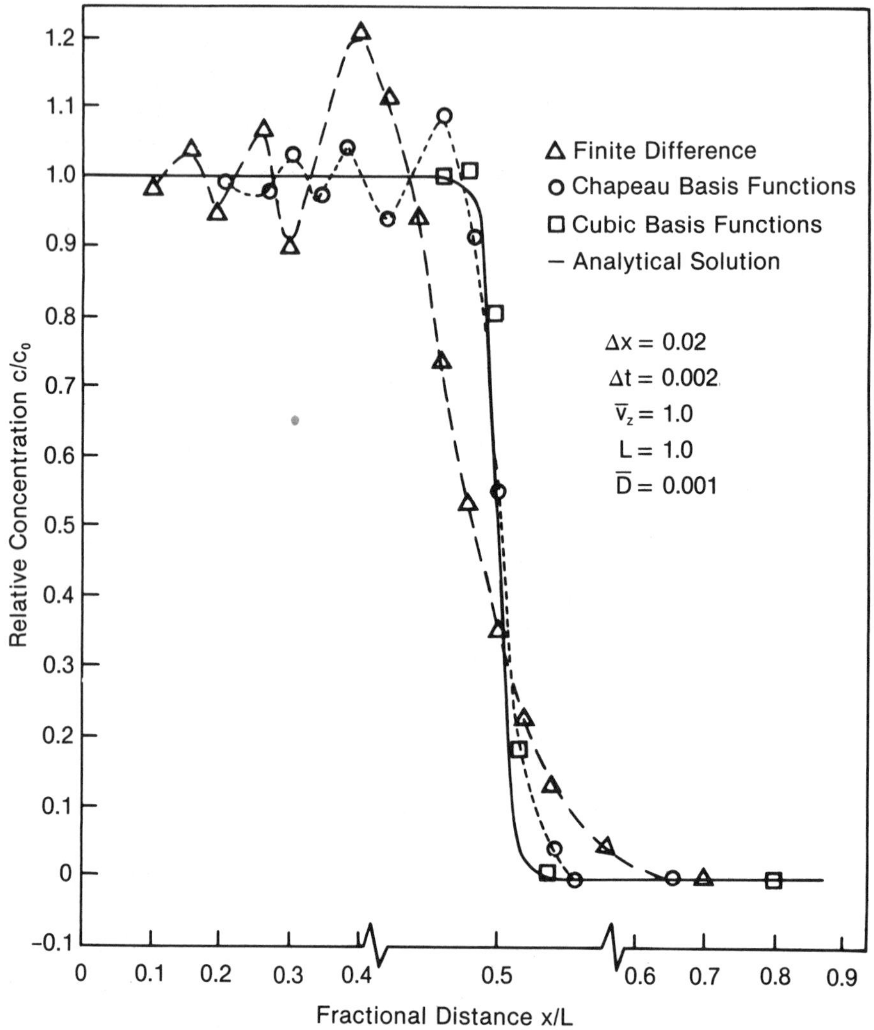

FIGURE 4.10 Comparison of finite-difference and finite-element solutions to the convective-diffusion equation for a dispersion coefficient of 0.0001 and a displaced pore volume of 0.5 (Adapted from D.B. Grove, "The Use of Galerkin Finite Element Methods to Solve Mass-Transport Equations." *Water Resources Investigation* 77-47. Denver: U.S.G.S., 1977, pp. 34–35)

general, for any point i and time k, the approximations can be written

$$c_i^{k+1} - c_i^k = \left(\frac{\Delta t}{\Delta z^2}\right) \overline{D}(c_{i+1}^{k+\theta} - 2c_i^{k+\theta} + c_{i-1}^{k+\theta})$$

$$-\frac{\Delta t}{2\Delta z}\,\overline{v}_z(c_{i+1}^{k+\theta} - c_{i-1}^{k+\theta}) \qquad (4.70)$$

where $$c_i^{k+\theta} = \theta c_i^{k+1} + (1 - \theta)c_i^k$$

Sufficient conditions for stability or nonoscillatory solutions of the difference model are given by Keller (1967) as

$$1 - (1 - \theta)\frac{2\overline{D}\Delta t}{\Delta z^2} \geq 0 \qquad (4.71a)$$

$$\frac{\overline{D}\Delta z}{2}\,|\overline{v}_z| \geq 0 \qquad (4.71b)$$

However, even though the solutions may appear stable, the approximations may still introduce artificial or numerical diffusion in the finite-difference model. This numerical diffusion for various types of approximations is summarized in Table 4.5 (Lantz, 1970). We note that, if the spatial variables are differenced centrally and time, explicitly, the dispersion coefficient must be larger than the numerical

TABLE 4.5 Numerical Diffusion Errors Caused by Various Space- and Time-Differencing of the Convective Term in the One-Dimensional Mass Transport Equation

Difference Form		
Spatial[1]	*Time*	*Truncation Error*[2]
BD	Explicit, $\theta = 0$	$(\overline{v}_z\Delta x - \overline{v}_z^2\Delta t)/2$
CD	Explicit	$-\overline{v}_z^2\Delta t/2$
BD	Implicit, $\theta = 1$	$(\overline{v}_z\Delta x + \overline{v}_z^2\Delta t)/2$
CD	Implicit	$\overline{v}_z^2\Delta t/2$
BD	C−N, $\theta = 1/2$	$\overline{v}_z\Delta x/2$
CD	C−N	0

[1]BD = backward difference CD = central difference C−N = Crank-Nicolson

[2]Second order, $\frac{\partial^2 c}{\partial x^2}$

dispersion or negative results may appear. Conversely, if a spatially centered difference and Crank-Nicolson schemes are used, there is no numerical diffusion in the difference approximation. These numerical problems will be fully explored in Chapter 7.

4.4.2 Nonlinear Systems

The hydraulic response of unconfined aquifer systems to pumping or natural or artificial recharge can be described by the Boussinesq equation, a nonlinear partial differential equation. The numerical transformation of the unconfined flow equation using finite-difference or finite-element methods produces a system of nonlinear ordinary differential equations. These equations may be expressed as

$$D \frac{d\mathbf{h}}{dt} + E(\mathbf{h})\,\mathbf{h} + \mathbf{r} = 0 \qquad (4.72)$$

where the coefficient matrices again contain the specific yield and the hydraulic conductivity. The operational or planning policies and the system's boundary conditions are contained in the $\mathbf{r}$ vector; $\mathbf{h}$ represents the vector of the hydraulic head at all nodal points in the system.

Although the response equations are nonlinear, a variety of algorithms have been developed for solution of the response equations. For example, Runge-Kutta methods may be used to numerically integrate the response equations of the system. To illustrate the application of these methods, we rewrite the response equations as

$$\dot{\mathbf{h}} = \mathbf{F}(\mathbf{h}, t) \qquad (4.73a)$$

with the initial conditions, $\mathbf{h}(0) = \mathbf{h}_0$. The vector-valued function, $\mathbf{F}$, defines the response equations, or

$$\mathbf{F} = -D^{-1}\,[E(\mathbf{h})\mathbf{h} + \mathbf{r}] \qquad (4.73b)$$

Runge-Kutta methods are designed to approximate a Taylor series approximation of the equations without requiring explicit definition or evaluation of the derivatives, other than the first. The method can be summarized by the recursive equations (Daniel & Moore, 1970).

$$\mathbf{h}_{i+1} = \mathbf{h}_i + \Delta t_i \sum_{j=1}^{J} b_j\,\mathbf{k}_j$$

with,

$$\mathbf{k}_j = \mathbf{F}\left(t_i + \alpha_j\,\Delta t_i,\ \mathbf{h}_j + \Delta t_i \sum_{l=1}^{J-1} d_{jl}\,\mathbf{k}_l\right),\quad j = 1,2..J \tag{4.74}$$

The constants of Equation 4.74, α_j, d_{jl}, b_j, are chosen such that the series expansion of the equations in powers of Δt_i matches the Taylor series expansions for $\mathbf{h}_{i+1}$ in Δt_i. The most widely used Runge-Kutta method is the fourth-order scheme,

where
$$\mathbf{h}_{i+1} = \mathbf{h}_i + \frac{1}{6}\Delta t_i\,(\mathbf{k}_1 + 2\mathbf{k}_2 + 2\mathbf{k}_3 + \mathbf{k}_4) \tag{4.75a}$$

and
$$\mathbf{k}_1 = \mathbf{F}(t_i, \mathbf{h}_i) \tag{4.75b}$$

$$\mathbf{k}_2 = \mathbf{F}\left(t_i + \frac{1}{2}\Delta t_i, \mathbf{h}_i + 1/2\Delta t_i\,\mathbf{k}_1\right) \tag{4.75c}$$

$$\mathbf{k}_3 = \mathbf{F}(t_i + 1/2\ \Delta t_i,\ \mathbf{h}_i + 1/2\ \Delta t_i\,\mathbf{k}_2) \tag{4.75d}$$

$$\mathbf{k}_4 = \mathbf{F}(t_i + \Delta t_i,\ \mathbf{h}_i + \Delta t_i\,\mathbf{k}_3) \tag{4.75e}$$

As Δt_i approaches zero, the integration scheme agrees with the Taylor series asymptotically through the fourth-order terms.

Several finite difference schemes can also be used to approximate the non-iterative solution of the response equations. In Lee's three-level approximation, the finite-difference equations may be expressed as (Lee, 1966)

$$D_k\left\{\frac{\mathbf{h}^{k+1} - \mathbf{h}^{k-1}}{2\Delta t}\right\} + \frac{1}{3}\{E_k(\mathbf{h}^{k+1} + \mathbf{h}^k + \mathbf{h}^{k-1})\} + \mathbf{r}_k = 0 \tag{4.76a}$$

or, after rearranging,

$$\left\{D_k + \frac{2\Delta t}{3}E_k\right\}\mathbf{h}^{k+1} = \left\{D_k - \frac{2\Delta t}{3}E_k\right\}\mathbf{h}^{k-1} - \frac{2\Delta t}{3}E_k\mathbf{h}^k - 2\Delta t\ \mathbf{r}_k \tag{4.76b}$$

The coefficient matrices are all evaluated at time level k and are known and can be solved by proceeding stepwise through time. It is assumed that the $\mathbf{r}_k$ vector can vary over the simulation period to reflect to changing boundary conditions and/or pumping schedules.

The implicit difference scheme introduced in Section 4.4.1 can also be used to solve the equations. Because the E matrix is time-dependent and the source of the nonlinearity, the matrix is evaluated at the previous time step, or

$$D_{k-1}\left\{\frac{\mathbf{h}^k - \mathbf{h}^{k-1}}{\Delta t}\right\} + E_{k-1}\mathbf{h}^k + \mathbf{r}_k = 0 \tag{4.77a}$$

and

$$\{D_{k-1} + \Delta t \mathbf{E}_{k-1}\}\mathbf{h}^k = D_{k-1}\mathbf{h}^{k-1} - \Delta t\ r_k \tag{4.77b}$$

The scheme has been shown to be superior to a centered-in-time approximation and Lee's three-level time approximation (Culham & Varga, 1970).

Douglas and Dupont (1970) introduced a series of predictor/corrector approximations for the solution of nonlinear finite difference equations. These techniques have been used to simulate unconfined groundwater flow, groundwater recharge, and the hydraulics of stream-aquifer systems (Mariño, 1975a, b). The approximations can be expressed as

$$D_{k-1}\left\{\frac{\mathbf{h}^{k,*} - \mathbf{h}^{k-1}}{\Delta t}\right\} + \frac{1}{2}E_{k-1}\{(1+\theta)\mathbf{h}^{k,*} + (1-\theta)\mathbf{h}^{k-1}\} = 0 \tag{4.78a}$$

and

$$D^*\left\{\frac{\mathbf{h}^k - \mathbf{h}^{k-1}}{\Delta t}\right\} + \frac{1}{2}E^*\{(1+\theta)\mathbf{h}^k + (1-\theta)\mathbf{h}^{k-1}\} = 0 \tag{4.78b}$$

$\mathbf{h}^{k,*}$ is a temporary solution (the predictor) and the D^* and E^* matrices are generated from the relation,

$$\frac{1}{2}\{(1+\theta)\mathbf{h}^{k,*} + (1-\theta)\mathbf{h}^{k-1}\}$$

Two sets of linear difference equations are produced for each time step. The method is second-order correct in time.

The computational efficiency of the difference scheme can be improved by solving the predictor equations using the matrix obtained from the correction equations from the previous time step. The equations can then be expressed as

$$D^1\left\{\frac{\mathbf{h}^{k,*} - \mathbf{h}^{k-1}}{\Delta t}\right\} + \frac{1}{2}E^1\{\mathbf{h}^{k,*} + \mathbf{h}^{k-1}\} = 0 \tag{4.79a}$$

and

$$D^2\left\{\frac{\mathbf{h}^k - \mathbf{h}^{k-1}}{\Delta t}\right\} + \frac{1}{2}E^2\{\mathbf{h}^k + \mathbf{h}^{k-1}\} = 0 \tag{4.79b}$$

The D^1 and E^1 coefficient matrices are calculated using

$$1/2\ \{\mathbf{h}^{k-1,*} + \mathbf{h}^{k-2,*}\}$$

Similarly, the D^2 and E^2 matrices are determined from

$$1/2\ \{\mathbf{h}^{k,*} + \mathbf{h}^{k-1}\}$$

Quasilinearization, an iterative approach, has been widely used in parameter estimation problems (Bellman & Kalaba, 1965). *Quasilinearization* is a technique that transforms boundary value problems into a sequence of more readily solvable, linear, initial-value problems. Moreover, the method also provides a sequence of functions, which in general converges monotonically and quadratically (if at all) to the solution of the original nonlinear equations (Lee, 1968).

Quasilinearization is a generalized Newton-Raphson method for functional equations. However, since the unknowns are now functions and not fixed values, the computational and theoretical aspects of the method are more complicated than the Newton-Raphson method. Here we examine how the method can be used for the solution of a general set of nonlinear response equations.

The quasilinearization method can be illustrated by considering the solution of the nonlinear response equations over the interval $0 \leqslant t \leqslant T$,

$$\dot{\mathbf{h}} = \mathbf{F}(\mathbf{h}, t) \tag{4.73}$$

where $\mathbf{h}$ is a vector of the hydraulic head and $\mathbf{F}$ defines the system of nonlinear equations. We represent the boundary conditions as the multipoint boundary value problem,

$$h_j(T) = h_j^T,\ j = 1, \ldots m \tag{4.80a}$$

$$h_k(0) = h_k^0,\ k = m + 1, \ldots n \tag{4.80b}$$

In quasilinearization, the nonlinear response equations are linearized about a trial solution at iteration $n, \mathbf{h}^n(t)$, for $0 \leqslant t \leqslant T$, or

$$\dot{\mathbf{h}}^{n+1} = \mathbf{F}(\mathbf{h}^n) + J(\mathbf{h}^n)\,(\mathbf{h}^{n+1} - \mathbf{h}^n) \tag{4.81}$$

where J is the Jacobian matrix. The transformed boundary conditions are

$$h_j^{n+1}(T) = h_j^T,\ j = 1, \ldots m \tag{4.82a}$$

$$h_k^{n+1}(0) = h_k^0,\ k = m + 1, \ldots n \tag{4.82b}$$

The recursive solution of these linear differential equations, for any iteration, can be obtained using the method of complementary functions. The general solution of these equations may be expressed using the principle of superposition,

$$\mathbf{h}^{n+1}(t) = \mathbf{h}_p^{n+1}(t) + \sum_{j=1}^{m} a_{j,n+1}\mathbf{h}_{h,j}^{n+1}(t) \qquad 0 \leqslant t \leqslant T \tag{4.83}$$

where the particular solution $\mathbf{x}_p$ is any solution of

$$\dot{\mathbf{h}}_p^{n+1} = \mathbf{F}(\mathbf{h}^n, t) + J(\mathbf{h}^n)(\mathbf{h}_p^{n+1} - \mathbf{h}_p^n) \tag{4.84a}$$

which satisfies the conditions,

$$h_{k,p}^{n+1} = h_k^0, \; k = m + 1, \ldots n \tag{4.84b}$$

The m vectors, $\mathbf{h}_{h,j}^{n+1}$ are homogeneous solutions that are any m sets of nontrivial and distinct solutions of the equations,

$$\dot{\mathbf{h}}_{h,j}^{n+1} = J(\mathbf{h}^n)\mathbf{h}_{h,j}^{n+1}, \quad j = 1, \ldots m \tag{4.85a}$$

which satisfy the conditions,

$$\sum_{j=1}^{m} a_{j,n+1} \, h_{k,h,j}^{n+1}(0) = 0, \quad k = m + 1, \ldots n \tag{4.85b}$$

where $a_{j,n+1}$ are m constants of integration. Note that the general solution uses only the given initial (boundary) conditions of the problem. The successive solution of these equations effectively identifies the unknown initial conditions ($j = 1, \ldots m$). In Chapter 8 we will explore how this approach may be used to identify boundary conditions and unknown system parameters from observational data.

A more restrictive formulation of the boundary value problem assumes that all the initial conditions for the problem are known. In other words,

$$\mathbf{h}(0) = \mathbf{h}_0$$

In this event, finite differences can be used for the direct solution of the linearized equations. For example, by rewriting the Boussinesq equations as

$$\begin{aligned} \dot{\mathbf{h}} &= -D^{-1}E(\mathbf{h})\mathbf{h} - D^{-1}\mathbf{r} \\ &= \mathrm{A}(\mathbf{h})\mathbf{h} + \mathbf{s} \end{aligned} \tag{4.86}$$

the linearized equations can be expressed for any iteration n as

$$\dot{\mathbf{h}}^{n+1} = \overline{\mathrm{A}}^n \, \mathbf{h}^{n+1} + \bar{\mathbf{s}}^n \tag{4.87a}$$

where

$$\overline{\mathrm{A}}^n = \frac{\partial}{\partial \mathbf{h}}(\mathrm{A}(\mathbf{h}^n)\mathbf{h}^n) \quad \text{and} \quad \bar{\mathbf{s}}^n = \mathrm{A}(\mathbf{h}^n)\mathbf{h}^n - \overline{\mathrm{A}}^n\mathbf{h}^n + \mathbf{s} \tag{4.87b}$$

Again, these equations may be solved recursively over the time interval, $0 \leq t \leq T$.

The quasilinearization method for the solution of nonlinear ordinary differential equations can be summarized by the following stages:

1. Assume a solution to the model, $\mathbf{h}^n$, for $0 \leq t \leq T$.
2. Generate the linearized response equations.
3. Solve the linearized response equations using the method of complementary function for multipoint boundary value problems or with finite differences if all the initial conditions are specified.
4. Compare the generated results with the results from the previous solution. If the difference at all times and spatial intervals,

$$|\mathbf{h}^{n+1}(t_k) - \mathbf{h}^n(t_k)| < \boldsymbol{\epsilon},\ k = 0, 1, 2$$

is less than $\boldsymbol{\epsilon}$ stop, the algorithm has converged. If not set $\mathbf{h}^{n+1} = \mathbf{h}^n$ and return to step 2.

The example problems describe how Runge-Kutta methods and quasilinearization can be used to approximate the solution of the dynamic response equations of the groundwater system.

Example Problem 4.10 Unconfined Groundwater Flow (after Yeh, 1970)

Determine the head distribution in the unconfined aquifer system shown in Figure 4.11 using the Runge-Kutta numerical integration method. Assume the aquifer is homogeneous and isotropic.

The mathematical model of the groundwater system is described by the Boussinesq equation,

$$K \frac{\partial}{\partial x}\left(h \frac{\partial h}{\partial x}\right) = S_y \frac{\partial h}{\partial t} \tag{4.88a}$$

and the initial and boundary conditions,

$$h(x, 0) = H,\ x > 0 \tag{4.88b}$$

$$h(0, t) = d,\ t > 0 \tag{4.88c}$$

$$h(\infty, t) = H,\ t > 0 \tag{4.88d}$$

Defining the dimensionless variables,

$$\phi = h/H,\ y = x/H,\ \tau = \frac{Kt}{S_y H}$$

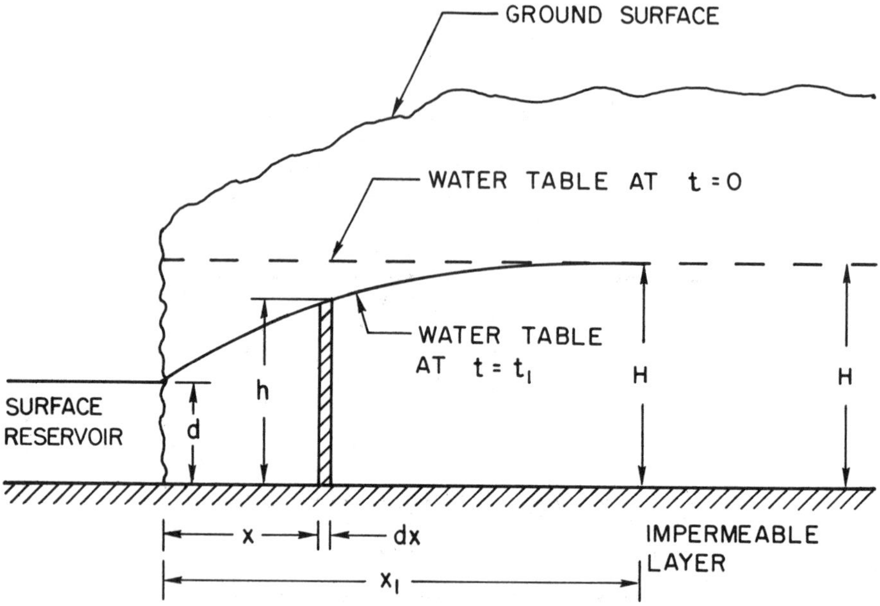

FIGURE 4.11 Groundwater flow to a surface reservoir in an unconfined aquifer (From W.W-G. Yeh, "Nonsteady Flow to Surface Reservoir." *ASCE, J. Hydr. Div.*, 96(HY3): 1970, p. 610. Reprinted with permission from the American Society of Civil Engineers)

the model can be expressed as

$$\frac{\partial \phi}{\partial \tau} = \frac{\partial}{\partial y}\left(\phi \frac{\partial \phi}{\partial y}\right) \tag{4.89a}$$

with the conditions

$$\tau = 0,\ \phi = 1,\ y > 0 \tag{4.89b}$$

$$y = 0,\ \phi = d/H,\ \tau > 0 \tag{4.89c}$$

$$y = \infty,\ \phi = 1,\ \tau > 0 \tag{4.89d}$$

Introducing the Boltzmann transformation, $\phi = y\tau^{-1/2}$, the model can be reduced to the ordinary differential equation,

$$-\frac{\phi}{2}\frac{d\theta}{d\phi} = \frac{d}{d\phi}\left(\theta \frac{d\theta}{d\phi}\right) \tag{4.90a}$$

with

$$\phi = 0, \quad \theta = d/H \tag{4.90b}$$

$$\phi = \infty, \quad \theta = 1 \tag{4.90c}$$

To apply the Runge-Kutta method, we rewrite the transformed flow equation as

$$\ddot{\theta} = -\frac{1}{\theta}(\dot{\theta})^2 - \frac{\phi}{2\theta}\dot{\theta} = F(\phi, \theta, \dot{\theta}) \tag{4.91a}$$

The boundary conditions are replaced by

$$\phi = 0, \quad \theta = d/H \tag{4.91b}$$

$$\phi > \phi_1, \quad \theta = 1 \tag{4.91c}$$

where ϕ_1 is defined as

$$\phi = \phi_1 = \frac{x_1}{\sqrt{\dfrac{HK}{S_y t_1}}} = \text{constant} \tag{4.91d}$$

and x_1 is the distance at t_1 beyond which the water table is effectively constant.

The solution of the two-point boundary value problem using Runge-Kutta methods requires an initial estimate of the slope, $\dot{\theta}|_{\phi=0}$. Figure 4.12 illustrates the

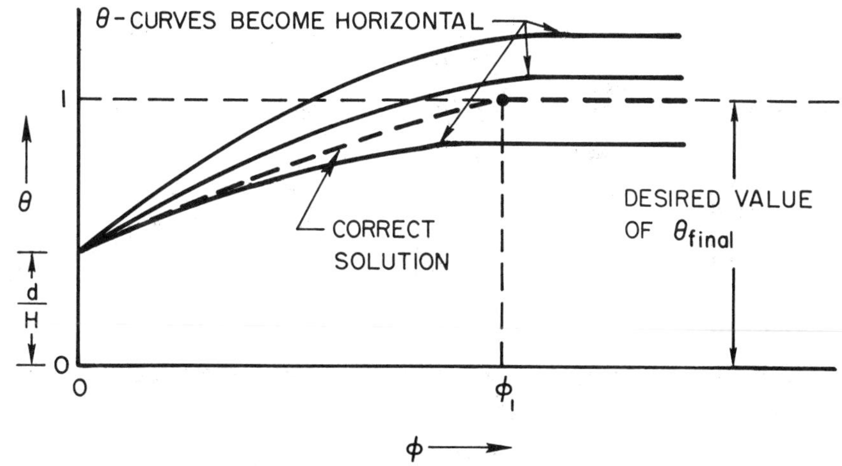

FIGURE 4.12 θ curves with different assumed initial slopes (From W.W-G. Yeh, "Non-steady Flow to Surface Reservoir." ASCE, *J. Hydr. Div.*, 96(HY3): 1970, p. 614. Reprinted with permission from the American Society of Civil Engineers)

effect of the initial gradient on the overall solution. This parametric information can be used to obtain a series of θ_{final} which are within a reasonable range of the desired θ value ($\theta = 1$). Lagrangian interpolation can then be used to identify the $\dot{\theta}|_{\phi=0}$ that gives the desired value of θ_{final}.

The simulated water table profiles are shown in Figure 4.13 for $d/H = 0.2$. The results are in close agreement with the Hele-Shaw experimental data of Todd (1954). In comparison the solution of the linearized Boussinesq model, h^*, (Eq. 2.83),

$$h^* = d + (H - d)\,\mathrm{erf}\left(\frac{x}{\alpha}\right) \tag{4.92a}$$

$$\alpha = \sqrt{\frac{4KHt}{S_y}} \tag{4.92b}$$

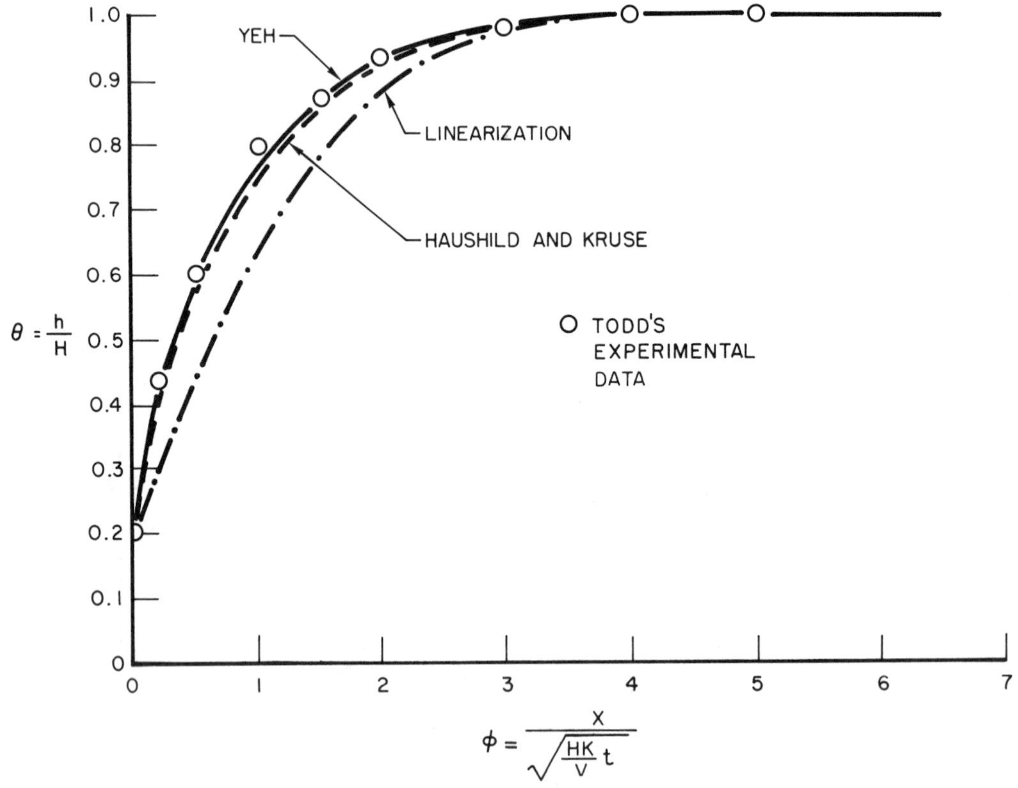

FIGURE 4.13 Computed dimensionless water table profile and experimental data (From W.W-G. Yeh, "Nonsteady Flow to Surface Reservoir." ASCE, *J. Hydr. Div.*, 96(HY3): 1970, p. 615. Reprinted with permission from the American Society of Civil Engineers)

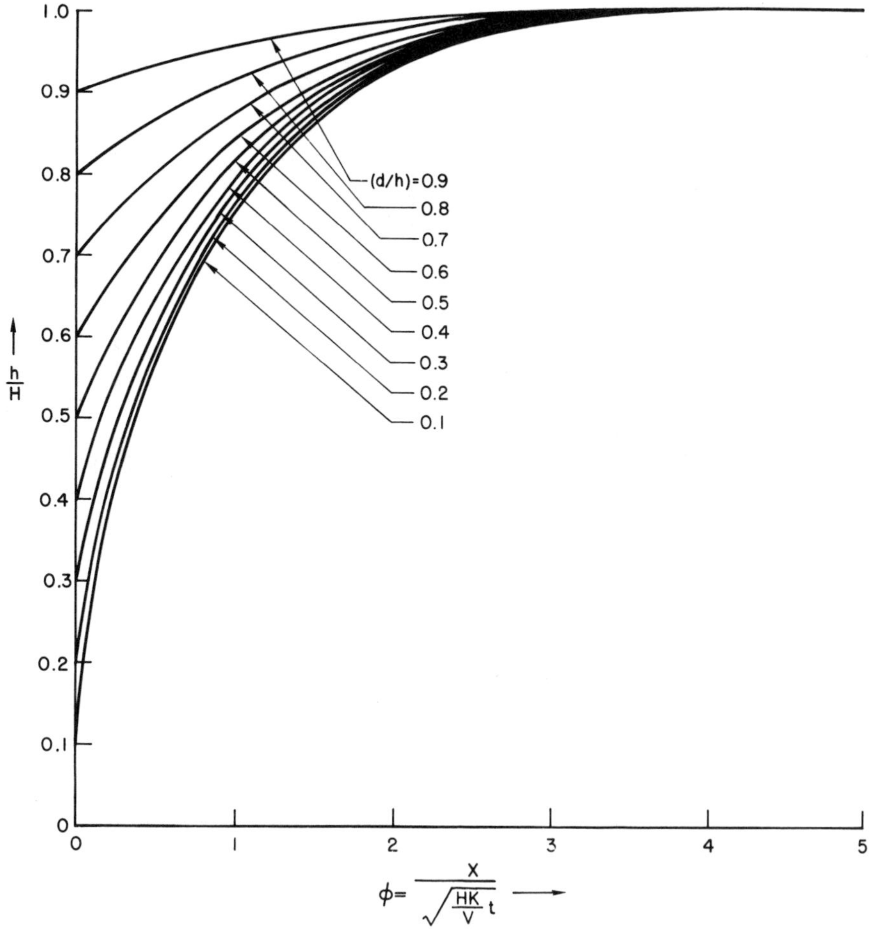

FIGURE 4.14 Dimensionless water table profiles for different boundary conditions (From W.W-G. Yeh, "Nonsteady Flow to Surface Reservoir." *ASCE, J. Hydr. Div.*, 96(HY3): 1970, p. 616. Reprinted with permission from the American Society of Civil Engineers)

deviates considerably from the experimental data. Figure 4.14 extends these results for a range of different boundary conditions.

Example Problem 4.11 Groundwater Recharge

A stream channel is used to recharge the aquifer system shown in Figure 4.15. The water table is initially at a constant height h_0 and for $t > 0$, water enters the aquifer at a constant rate F_0 per unit wetted area. Determine the head distribution

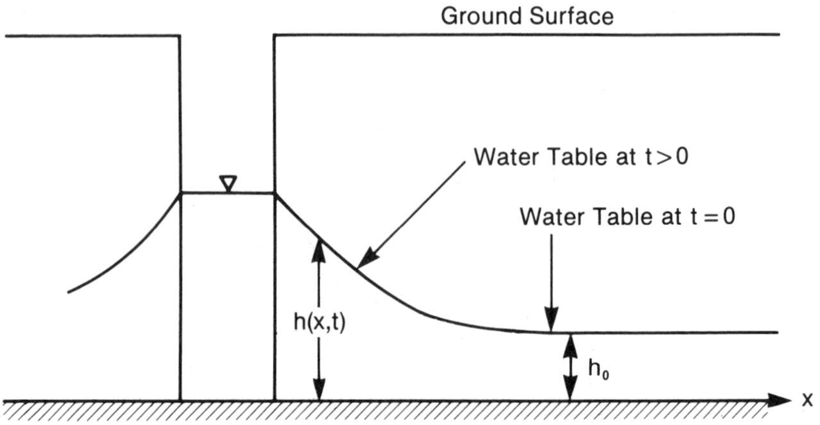

FIGURE 4.15 Cross-section of a ditch showing the water table position during the seepage (Adapted from A.B. Gureghian and E.G. Youngs, "Numerical Solutions of Boussinesq's Equation for Seepage Flow." In *Finite Elements in Water Resources*, W.G. Gray, G.F. Pinder, and C.A. Brebbia, Eds. London: Pentech Press, 1978, p. 3.188. Reprinted with permission)

in the aquifer using a predictor-corrector finite-difference model. Compare these results with a Galerkin finite-element solution of the Boussinesq equation.

The mathematical model of the stream-aquifer system is described by the equations

$$K\frac{\partial}{\partial x}\left(h\frac{\partial h}{\partial x}\right) = S_y\frac{\partial h}{\partial t} \tag{4.93a}$$

$$h = h_0, \quad 0 \leqslant x \leqslant \infty, \quad t = 0 \tag{4.93b}$$

$$-Kh\frac{\partial h}{\partial x} = F_0, \quad x = 0 \tag{4.93c}$$

$$h = h_0, \quad x \to \infty,\ t > 0 \tag{4.93d}$$

To apply the predictor-corrector method, the Boussinesq equation is expressed in quasilinear form as

$$\frac{\partial^2 h}{\partial x^2} = \frac{S_y}{Kh}\frac{\partial h}{\partial t} - \frac{1}{h}\left(\frac{\partial h}{\partial x}\right)^2 \tag{4.94}$$

For any node i at time k, the predictor equations are

$$\frac{h_{i+1,k} - 2h_{i,k} + h_{i-1,k}}{\Delta x^2} = \frac{S_y}{Kh_{i,k}}\left(\frac{h_{i,k+1/2} - h_{i,k}}{\Delta t/2}\right) - \frac{1}{h_{i,k}}\left(\frac{h_{i+1,k} - h_{i-1,k}}{2\Delta x}\right)^2 \tag{4.95a}$$

These equations are a tridiagonal system of equations in $h_{i,k+1/2}$. Similarly, the corrector equations are

$$1/2 \left\{ \frac{h_{i+1,k+1} - 2h_{i,k+1} + h_{i-1,k+1}}{\Delta x^2} + \frac{h_{i+1,k} - 2h_{i,k} + h_{i-1,k}}{\Delta x^2} \right\}$$

$$= \frac{1}{h_{i,k+1/2}} \left(\frac{h_{i,k+1} - h_{i,k}}{\Delta t} \right) - \frac{1}{h_{i,k+1/2}} (h_{i+1,k+1/2} - h_{i-1,k+1/2})^2 \quad (4.95b)$$

Again, however, the equations are tridiagonal. The unknowns are the head distribution at time $k + 1$. The truncation error of the approximations is $0(\Delta x^2 + \Delta t^{3/2})$.

The boundary conditions can be expressed as

$$-Kh_{0,k} \left(\frac{h_{1,k} - h_{0,k}}{\Delta x} \right) = F_0 \quad (4.96a)$$

$$h_{n,k} = h_0 \quad (4.96b)$$

where $h(0, t) = h_{0,k}$, $h(L,t) = h_{n,k}$, and L is the distance beyond which the head is equal to the initial head h_0.

The Galerkin finite-element equations are again given by the orthogonality conditions (Eq. 4.22), or

$$\int_0^L \mathrm{N}_\kappa \left\{ \frac{\partial}{\partial x} (K\mathbf{N}\tilde{\mathbf{h}} \frac{\partial}{\partial x} (\mathbf{N}\tilde{\mathbf{h}})| - S_y \frac{\partial}{\partial t} (\mathbf{N}\tilde{\mathbf{h}}) \right\} dx = 0, \forall \kappa \quad (4.97)$$

Discretizing the domain into a series of linear simplex elements, the elemental equations for the time-derivative terms can be expressed as

$$\frac{-S_y}{K} \begin{bmatrix} \int_0^{L^e} \mathrm{N}_i^2 dx & \int_0^{L^e} \mathrm{N}_i \mathrm{N}_j dx \\ \\ \int_0^{L^e} \mathrm{N}_j \mathrm{N}_i dx & \int_0^{L^e} \mathrm{N}_j^2 dx \end{bmatrix} \begin{bmatrix} \dot{h}_i \\ \\ \dot{h}_j \end{bmatrix} \quad (4.98)$$

where L_e is the length of a finite element.

Similarly, for the spatial derivatives, we have after integration by parts,

$$-\int_0^{L^e} \frac{\partial \mathrm{N}_i}{\partial x} \frac{\partial}{\partial x} (\mathbf{N}\tilde{\mathbf{h}})^2 dx + \mathrm{N}_i \frac{\partial}{\partial x} (\mathbf{N}\tilde{\mathbf{h}})^2 |_{\text{boundary}}$$

The elemental equations are then

$$2\tilde{h}_i\int^{Le}\frac{\partial N_i}{\partial x}\left(N_i\frac{\partial N_i}{\partial x}\right)dx + 2\tilde{h}_j\int^{Le}\frac{\partial N_i}{\partial x}\left(\frac{\partial N_i}{\partial x}N_j\right)dx \qquad 2\tilde{h}_i\int^{Le}\frac{\partial N_i}{\partial x}\left(N_i\frac{\partial N_j}{\partial x}\right) + 2\tilde{h}_j\int^{Le}\frac{\partial N_i}{\partial x}\left(N_j\frac{\partial N_j}{\partial x}\right)dx$$

$$2\tilde{h}_i\int^{Le}\frac{\partial N_j}{\partial x}\left(N_i\frac{\partial N_i}{\partial x}\right)dx + 2\tilde{h}_j\int^{Le}\frac{\partial N_j}{\partial x}\left(\frac{\partial N_i}{\partial x}N_j\right)dx \qquad 2\tilde{h}_r\int^{Le}\frac{\partial N_j}{\partial x}\left(N_i\frac{\partial N_j}{\partial x}\right) + 2\tilde{h}_j\int^{Le}\frac{\partial N_j}{\partial x}\left(N_j\frac{\partial N_j}{\partial x}\right)dx$$

The flux boundary condition at $x = 0$ can be expressed as

$$\left.\frac{\partial\,(\mathbf{N}\tilde{\mathbf{h}})^2}{\partial x}\right|_{x=0} = -\frac{2F_0}{K} \tag{4.100a}$$

or after evaluating the basis functions

$$\frac{1}{L_1}\tilde{h}_1^2 + \frac{\tilde{h}_1\tilde{h}_2}{L_1} = -\frac{F_0}{K} \tag{4.100b}$$

where L^1 is the length of element 1. The boundary condition is a nonlinear equation relating $\tilde{h}_1$ and $\tilde{h}_2$.

Assembling the elemental equations and incorporating the boundary conditions, the response equations may be expressed as the system of ordinary differential equations,

$$D\dot{\mathbf{h}} + E(\mathbf{h})\mathbf{h} + \mathbf{r} = 0 \tag{4.101}$$

Again we linearize these equations using the generalized Newton-Raphson method. Assuming the trial solution, $\dot{\mathbf{h}}^n(t)$ and $\mathbf{h}^{n+1}(t)$ (where n denotes the iteration), the equations may be expressed as

$$D\dot{\mathbf{h}}^n + E(\mathbf{h}^n)\mathbf{h}^n + \mathbf{r} + D\,(\dot{\mathbf{h}}^{n+1} - \dot{\mathbf{h}}^n) \tag{4.102}$$

$$+\frac{\partial}{\partial\mathbf{h}}\left\{E(\mathbf{h}^n)\mathbf{h}^n\right\}(\mathbf{h}^{n+1} - \mathbf{h}^n) = 0$$

The solution of these equations may be obtained again using either finite-difference methods or complementary functions. Here, however, finite-differences will be used to simulate the water table profiles. A good initial approximation of dh/dt over the initial time interval, Δt, can be obtained by using the analytical solution of the linearized Boussinesq equations with a constant surface flux (Carslaw &

Jaegar, 1959),

$$\frac{dh}{dt} \equiv 2H_0 \sqrt{\Delta\tau}\ ierfc\ \{\xi/\sqrt{2\Delta\tau}\}, \tag{4.103}$$

where $\xi = \dfrac{xF_0}{H_0^2K}$, $\eta = \dfrac{h}{h_0}$, $\Delta\tau = \dfrac{F_0^2t}{Kh_0^3S_y}$, and $ierfc\ (x) = \displaystyle\int_x^\infty \text{erfc}\ \xi d\xi$.

The results of the finite-difference and finite-element models are summarized in Figure 4.16. Also shown are the experimental results obtained with a Hele-Shaw model (Gureghian & Youngs, 1977). The numerical results agree quite closely with the experimental data. However, the finite-element solutions are superior to the finite-difference results, especially where the water table is the highest (e.g., $\tau = 5.90$ and $\tau = 12.52$). Computationally, however, the finite-difference method

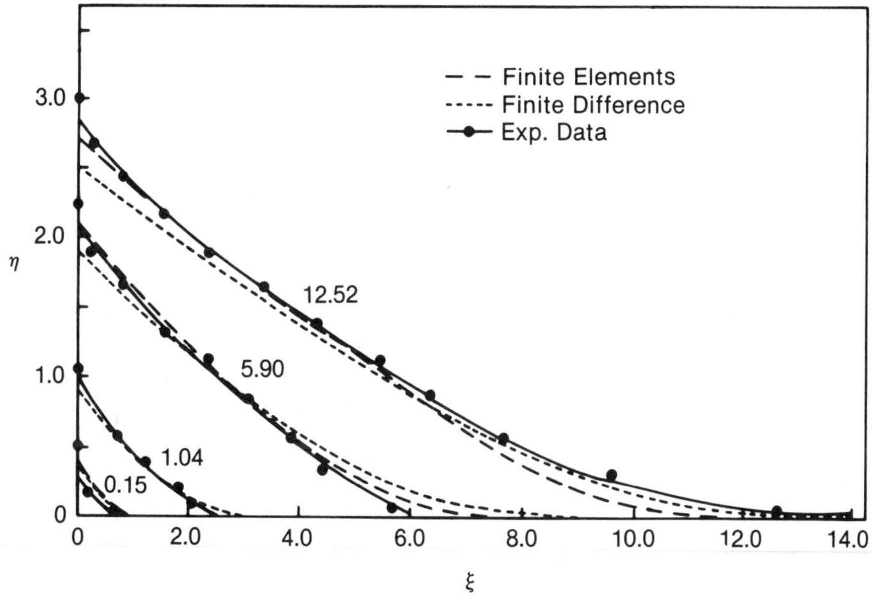

FIGURE 4.16 Comparison of solutions using finite-difference and finite-element methods with experimental results obtained in a Hele-Shaw analog for different values of τ, indicated by the curves (Adapted from A.B. Gureghian and E.G. Youngs, "Numerical Solutions of Boussinesq's Equation for Seepage Flow." In *Finite Elements in Water Resources*, W.G. Gray, G.F. Pinder, and C.A. Brebbia, Eds. London: Pentech Press, 1978, p. 3.188. Reprinted with permission)

may be preferable because the method requires substantially less CPU time. In this example, the finite-element method requires approximately 0.74 sec/time step in contrast to the finite-difference method, which takes only 0.073 sec/time step. The computations were performed on an IBM 370/165 (Gureghian & Youngs, 1977).

Example Problem 4.12 Quasilinearization Solution of the Boussinesq Equation (after Yeh and Tauxe, 1971)

The hydraulics of the stream-aquifer system shown in Figure 4.17 are described by the Boussinesq equation,

$$\frac{\partial}{\partial x}\left(Kh\,\frac{\partial h}{\partial x}\right) \pm S^* = S_y\,\frac{\partial h}{\partial t} \tag{4.104a}$$

and the boundary and initial conditions,

$$h(0, t) = h_0(t) \tag{4.104b}$$

$$\left.\frac{\partial h}{\partial x}\right|_{x=L} = 0 \tag{4.104c}$$

$$h(x, 0) = g(x) \tag{4.104d}$$

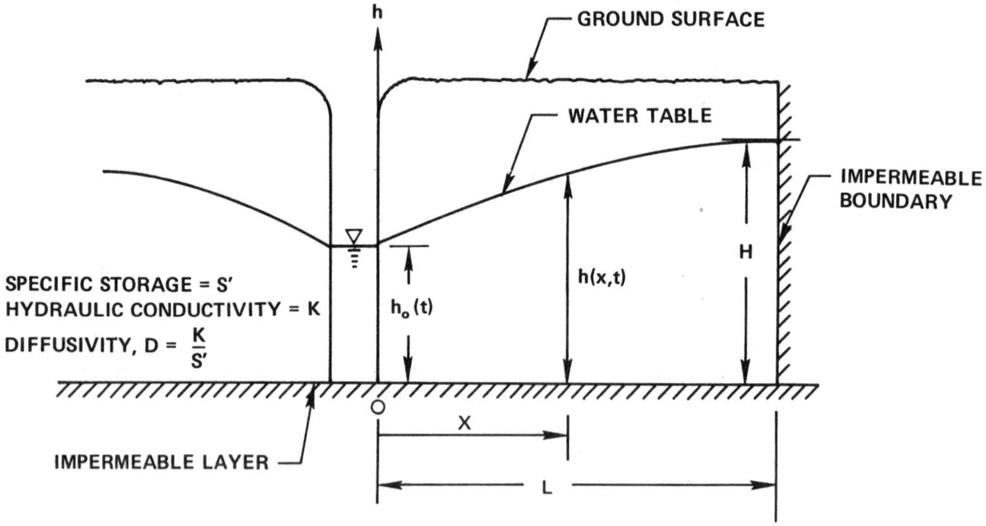

FIGURE 4.17 Unconfined aquifer and stream interaction configuration. ***S*** **is the specific storage;** ***K*** **is the hydraulic conductivity;** ***D*** **is diffusivity, equal to** ***K/S*** (From W.W-G. Yeh and G.W. Tauxe, "Optimal Identification of Aquifer Diffusivity Using Quasilinearization." *Water Resources Research*, 7(4): 1971, p. 956. Copyright © 1971 by the American Geophysical Union. Reprinted with permission)

where $g(x)$ is the initial head in the groundwater basin. Develop a finite-difference model of the stream-aquifer system and simulate the response of the aquifer to the change in stream stage, $h_0(t)$, using quasilinearization.

A second-order correct finite difference model of the Boussinesq equation can be obtained using the approximations in Table 4.3. A typical difference equation is

$$\frac{K}{\Delta x^2}\left\{\frac{h_{i+1}+h_i}{2}(h_{i+1}-h_i)-\frac{h_i+h_{i-1}}{2}(h_i-h_{i-1})\right\}\pm\frac{S_i^*}{\Delta x}=S_y\dot{h}_i \quad (4.105a)$$

or, after simplifying,

$$\frac{K}{2\Delta x^2}\{h_{i+1}^2-2h_i^2+h_{i-1}^2\}\pm S_i^*=S_y\dot{h}_i \quad (4.105b)$$

The arithmetic mean has been used to describe the heads at nodes $i+1/2$, $i-1/2$.

Assembling the nodal equations and incorporating the Dirichlet boundary conditions, the dynamic response equations may be expressed as

$$\dot{\mathbf{h}}=\mathrm{A}(\mathbf{h})\mathbf{h}+\mathbf{s} \quad (4.106a)$$

where

$$\mathrm{A}(h)=\frac{K}{2\Delta x^2 S_y}\begin{bmatrix}-2h_1 & h_2 & & \\ h_1 & -2h_2 & h_3 & \\ \cdot & \cdot & \cdot & \\ & \cdot & \cdot & \cdot \\ & & h_{n-1} & -h_n\end{bmatrix},$$

$$\mathbf{s}=\frac{1}{\Delta x S_y}\begin{bmatrix}-Q_1+\dfrac{Kh_0^2}{2\Delta x}\\ -Q_2\\ \cdot\\ \cdot\\ -Q_n\end{bmatrix} \quad (4.106b)$$

and the initial conditions are given as

$$\mathbf{h}(0)=\mathbf{h}_0 \quad (4.106c)$$

These nonlinear equations can be integrated using Runge-Kutta methods or quasilinearization. In the quasilinearization approach, we first assume an initial solution $\mathbf{h}^n(t)$, which satisfies the boundary conditions of the problem. Assuming there are five internal nodes in the system, the linearized response equations can be expressed as

$$\dot{\mathbf{h}}^{n+1} = \bar{\mathrm{A}}^n \mathbf{h}^{n+1} + \bar{\mathbf{s}}^n \tag{4.107a}$$

where

$$\bar{\mathrm{A}}^n = \frac{K}{S_y \Delta x^2} \begin{bmatrix} -2h_1 & h_2 & 0 & 0 & 0 \\ h_1 & -2h_2 & h_3 & 0 & 0 \\ 0 & h_2 & -2h_3 & h_4 & 0 \\ 0 & 0 & h_3 & -2h_4 & h_5 \\ 0 & 0 & 0 & h_4 & -h_5 \end{bmatrix}_n \tag{4.107b}$$

$$\bar{s}^0 = \frac{1}{\Delta x S_y} \begin{bmatrix} -Q_1 + \dfrac{Kh_0^2}{2\Delta x} \\ -Q_2 \\ -Q_3 \\ -Q_4 \\ -Q_5 \end{bmatrix} - \frac{K}{2\Delta x^2 S_y} \begin{bmatrix} -2 & 1 & 0 & 0 & 0 \\ 1 & -2 & 1 & 0 & 0 \\ 0 & 1 & -2 & 1 & 0 \\ 0 & 0 & 1 & -2 & 1 \\ 0 & 0 & 0 & 1 & -1 \end{bmatrix} \begin{bmatrix} h_1^2 \\ h_2^2 \\ h_3^2 \\ h_4^2 \\ h_5^2 \end{bmatrix}_n \tag{4.107c}$$

Since all the initial conditions are known for the problem, finite-difference methods can be used for the solution of the system of equations.

The results of the quasilinearization solution are summarized in Figure 4.18 and Table 4.6. Figure 4.18 is a comparison between the experimental data and the direct numerical integration of the finite-difference equations with the initial and boundary conditions,

$$h = H, \quad 0 \leq x \leq L, \ t = 0$$

$$h = .5H, \quad t > 0, \ x = 0$$

$$\left.\frac{\partial h}{\partial x}\right|_{x=L} = 0, \quad t > 0$$

where H is the maximum height of the water table above the impervious layer and L is the distance from the river to the water divide. The parameters of the finite-difference model are the diffusivity, $D = \dfrac{K}{S_y} = 1$ and $H/L^2 = 1$ (Ibrahim & Brutsaert, 1965). The results of the quasilinearization algorithm are shown in Table

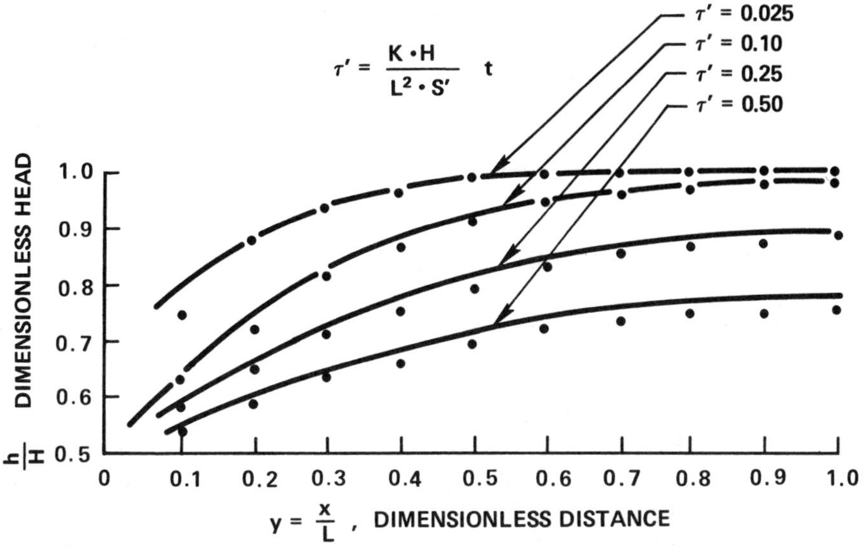

FIGURE 4.18 Comparisons between numerical solutions and experimental results obtained by Ibrahim and Brutsaert, 1965. The solid line represents the experimental data and the circles the numerical solutions (From W.W-G. Yeh and G.W. Tauxe, "Optimal Identification of Aquifer Diffusivity using Quasilinearization." *Water Resources Research*, 7(4): 1971, p. 958. Copyright © 1971 by the American Geophysical Union. Reprinted with permission)

TABLE 4.6 Results of Successive Approximations

	Zero	*First*	*Second*	*Third*	*Fourth*	*True Value*
D	0.10000	0.48200	0.92312	0.97667	1.00102	1.00000
$\theta_8(0)$	1.00000	1.00000	1.00000	1.00000	1.00000	1.00000
$\theta_8(0.1)$	0.99939	0.96261	0.91251	0.90733	0.90506	0.90518
$\theta_8(0.2)$	0.99428	0.90844	0.83805	0.83101	0.82789	0.82802
$\theta_8(0.3)$	0.98452	0.86675	0.78380	0.77567	0.77200	0.77215
$\theta_8(0.4)$	0.97256	0.83265	0.74057	0.73175	0.72787	0.72803
$\theta_8(0.5)$	0.96007	0.80334	0.70516	0.69609	0.69213	0.69229
$\theta_8(0.6)$	0.94779	0.77752	0.67578	0.66673	0.66280	0.66296
$\theta_8(0.7)$	0.93608	0.75449	0.65118	0.64234	0.63852	0.63867
$\theta_8(0.8)$	0.92500	0.73380	0.63044	0.62193	0.61827	0.61842
$\theta_8(0.9)$	0.91456	0.71511	0.61286	0.60475	0.60127	0.60142
$\theta_8(1.0)$	0.90473	0.69818	0.59788	0.59021	0.58694	0.58707
S_0	6.7566×10^{-1}	1.0712×10^{-1}	1.3257×10^{-2}	1.1503×10^{-4}	2.2010×10^{-7}	. . .

4.6. The values of the dimensionless head $\theta(h/H)$, and diffusivity, D, are shown for all iterations of the algorithm. Quasilinearization is also used in this example to identify the diffusivity using known observations on the head. This application of the approach will be more fully explored in Chapter 8. Convergence of the algorithm is quadratic.

4.5 MATRIX METHOD FOR THE SOLUTION OF LINEAR SYSTEMS OF EQUATIONS

The prediction of the hydraulic head in a confined or semiconfined groundwater system is based on the recursive finite-difference equations developed in Section 4.4. Solutions to these equations can be obtained using direct or iterative methods. Direct methods for solving linear equations are essentially variants of Gaussian elimination. Iterative methods, which are particularly suited to sparse, banded matrix systems, involve generating a sequence of approximations which, for certain conditions, converges to the solution of the system of equations.

In this section, we review the most common direct and iterative methods for the solution of the finite-difference or finite-element response equations.

4.5.1 Direct Methods

The response equations of the groundwater system can be expressed for any time step, using Equation 4.61b, as the system of linear equations,

$$\bar{A}\mathbf{x} = \mathbf{b} \tag{4.108}$$

where

$$\bar{A} = A + \theta\Delta tB,\ \mathbf{x} = \boldsymbol{\phi}_k,\ \text{and } \mathbf{b} = \{A - (1 - \theta)\Delta tB\}\ \boldsymbol{\phi}_{k-1} - \mathbf{g}\Delta t$$

The basis of all direct methods is the decomposition of the coefficient matrix, $\bar{A}$, into a lower triangular and an upper triangular matrix. That is, we require that

$$\bar{A} = LU \tag{4.109}$$

where L is defined as

$$L = \begin{bmatrix} 1 & & 0 & & \\ & 1 & & & \\ & & 1 & & \\ & L_{ij} & & 1 & \\ & & & & 1 \end{bmatrix} \tag{4.110}$$

and U is given as

$$U = \begin{bmatrix} U_{11} & & & \\ & U_{22} & & U_{ij} \\ & & \cdot & \\ & & & \cdot \\ 0 & & & \cdot\, U_{NN} \end{bmatrix} \tag{4.111}$$

Ignoring, for the moment, the problem of actually calculating L and U, the solution of the equations can be obtained by first introducing the unknown vector, $\mathbf{y}$, where

$$\mathbf{y} = U\mathbf{x} \tag{4.112a}$$

Equation 4.108 can then be expressed as

$$L\mathbf{y} = \mathbf{b} \tag{4.112b}$$

Equation 4.112b may be solved by forward substitution. That is, the first equation is solved for $\mathbf{y}_1$. This solution is then used in the second equation to calculate $\mathbf{y}_2$ and so on. With $\mathbf{y}$ known, the solution is completed by solving Equation 4.112a. Since U is upper triangular, these equations may be solved by backward substitution beginning with the last equation.

The algebraic basis of the matrix decomposition of $\overline{A}$ is based on the following theorem (Pinder & Gray, 1977):

> **Theorem:** Given a square matrix $\overline{A}$ of order N, let $\overline{A}_k$ denote the principal minor matrix made from the first k rows and columns. Assume that det $\overline{A}_k \neq 0$ for $k = 1, 2, \ldots N - 1$. Then there exists a unique lower triangular matrix L containing element L_{ij} with $L_{ii} = 1$, and a unique upper triangular matrix U with the element u_{ij} such that $LU = \overline{A}$ and det $\overline{A} = u_{11}\, u_{22} \ldots u_{NN}$.

Furthermore, if $\overline{A}$ is symmetric and positive definite, then it can be decomposed uniquely into the product GG^T, where G is a lower triangular matrix with positive diagonal elements (Forsythe & Moler, 1967).

Gaussian elimination is a form of LU decomposition since the matrix A becomes upper triangular through a sequence of elementary row operations. For example, the usual Gaussian elimination technique first eliminates the variable x_1 from all the equations except the first. Similarly, x_2 is eliminated from all but the second equation. The process continues until the original set of equations has been reduced to a triangular system of equations. Once the equations have been transformed, the last equation will identify the value of x_N. This back substitution procedure is equivalent to the decomposition of $\overline{A}$ into LU.

An important variation of LU decomposition is the *Crout method* (1941). Again, the coefficient matrix is decomposed into an upper and lower triangular matrices. However, in contrast to Gaussian elimination, the matrices can be calculated in a single operation rather than sequentially. To illustrate the approach, we formulate the augmented matrix of the system, i.e., the matrix A with the known right-hand side vector as the N + 1 column. Since A can be decomposed into an *LU* product, we write

$$\overline{A}\mathbf{b} = LU\mathbf{y} \tag{4.113}$$

where we now have a unit diagonal associated with the upper triangular matrix and **y** is a vector of intermediate values. Expanding Equation 4.113, the general equations for the algorithm can be summarized as (Salvadori & Baron, 1961; Pinder & Gray, 1977):

$$\begin{aligned} l_{ij} &= \overline{a}_{ij} - \sum_{r=1}^{j-1} l_{ir}\, u_{rj}, l_{i1} = \overline{a}_{i1}; \\ u_{ij} &= (1/l_{ii})\,(\overline{a}_{ij} - \sum_{r=1}^{i-1} l_{ir}u_{rj}), \quad u_{1j} = \overline{a}_{1j}/\overline{a}_{11} \end{aligned} \tag{4.114}$$

where $\overline{a}_{ij}$ and u_{ij} are the elements of the augmented matrices.

In comparison with Gaussian methods, the number of arithmetic computations are approximately the same for the Crout method. However, fewer operations are performed with the Crout method and the procedure also has smaller round-off error (Forsythe & Moler, 1967).

A related method for solving large systems of linear equations is Cholesky's *square root method*. The technique is applicable to symmetric, positive definite matrices as occur in the management of semiconfined aquifer systems. The coefficient matrix is expressed as the product of a lower triangular matrix with positive diagonal elements and its transpose or

$$\overline{A} = LL^T \tag{4.115}$$

Expanding these equations, we have for the main diagonal elements

$$\overline{a}_{jj} = l_{j1}^2 + l_{j2}^2 + \ldots + l_{jj}^2$$

and for below the diagonal,

$$\overline{a}_{ij} = l_{i1}l_{j1} + l_{i2}l_{j2} + \ldots\ l_{ij}l_{jj}\ (j < i)$$

Solving these equations, we have

$$l_{ij} = \left(\overline{a}_{ij} - \sum_{k=1}^{j-1} l_{ik}l_{jk}\right) / l_{jj} \tag{4.116a}$$

and for the diagonal elements,

$$l_{jj} = \left(\bar{a}_{jj} - \sum_{k=1}^{j-1} l_{jk}^2 \right)^{1/2} \qquad (4.116b)$$

(Pinder & Gray, 1977).

Cholesky's method has a small round-off error and, because of the symmetry of the matrices, only N(N + 1)/2 of the elements have to be stored during the solution of the equations.

The most commonly encountered matrix in groundwater analysis is the tridiagonal banded matrix:

$$\begin{bmatrix} e_1 & f_1 & & & 0 \\ d_2 & e_2 & f_2 & & \\ & d_3 & e_3\, f_3 & & \\ & & \cdot\ \cdot & \cdot & \\ & & & \cdot\ \cdot & \cdot \\ 0 & & & d_n & e_n \end{bmatrix}$$

The *LU* decomposition of the banded matrix follows from a theorem presented by Forsythe and Moler (1967).

Theorem: If a banded matrix, with bandwidth 2M + 1, has an *LU* decomposition, then L and *U* are triangular-banded matrices. In the tridiagonal case, $M = 1$, and the *LU* matrices may be expressed as

$$L = \begin{bmatrix} 1 & & & & 0 \\ l_2 & 1 & & & \\ & l_3 & 1 & & \\ & & & \cdot & \\ & & & & \cdot \\ 0 & & & & l_n 1 \end{bmatrix} \qquad U = \begin{bmatrix} u_1 & f_1 & \cdot & \cdot & \cdot & 0 \\ & u_2 & f_2 & & & \\ & & u_2 & f_3 & & \\ & & & \cdot & & \\ & & & & \cdot & \\ 0 & & & & & u_n \end{bmatrix}$$

Assuming $\bar{A}$ is factored in such a way that the diagonal of the upper triangular matrix is unity, the decomposition formulas are

$$\begin{aligned} l_1 &= e_1,\ u_i = f_i/l_i \quad (i = 1, 2, ..N - 1) \\ l_i &= e_i - d_i u_{i-1} \quad (i = 2, ..N) \end{aligned} \qquad (4.117)$$

The intermediate solution is given by

$$y_1 = b_1/l_1,\ y_i = (b_i - d_i y_{i-1})/l_i \quad (i = 2, 3, ..N) \qquad (4.118)$$

Back substitution is carried out according to the equations

$$x_n = y_n \tag{4.119a}$$

$$x_i = y_i - u_i x_{i+1}, \quad (i = N - 1, N - 2, \ldots 1) \tag{4.119b}$$

This decomposition algorithm is known as *Thomas' algorithm* (Thomas, 1949). Not only is the scheme extremely stable with respect to round-off errors (Douglas, 1959), but the total number of multiplication and division operations is approximately $5N - 4$ rather than $N^3/3$ for Gaussian elimination.

4.5.2 Iterative Methods

Iterative methods for solving large systems of linear, algebraic equations have been used extensively in groundwater simulation. Although the techniques were designed to solve the equations generated by finite-difference models, iterative methods continue to be developed for finite-element models with sparse, banded coefficient matrices. Here we consider point and block iterative solution schemes.

We begin by again considering the difference equations generated at any time step as

$$\overline{A}\mathbf{x} = \mathbf{b}$$

We assume that $\overline{A}$ is nonsingular and that the diagonal elements are non-zero. $\overline{A}$ can then be expressed as the difference of two new matrices, D and C as

$$\overline{A} = D - C \tag{4.120}$$

where D is a diagonal matrix and the C matrix contains the off-diagonal elements of $\overline{A}$ with opposite sign.

In the point-Jacobi method, the iterative equations can be expressed as

$$D\mathbf{x}^{n+1} = C\mathbf{x}^n + \mathbf{b} \qquad n = 0, 1, 2, \ldots \tag{4.121}$$

where $\mathbf{x}^0$ are the initial estimates of $\mathbf{x}$. Assuming D has non-zero diagonal elements, the nth iterative approximation to $\mathbf{x}$ is given as

$$\mathbf{x}^{n+1} = D^{-1}C\,\mathbf{x}^n + D^{-1}\mathbf{b} \tag{4.122}$$

where $D^{-1}C$ is the point-Jacobi matrix. As described by Pinder and Gray (1977), solution of the equations is obtained by expanding Equation 4.121 and isolating $\mathbf{x}_i^{n+1}$. These values can then be substituted in Equation 4.122 to determine $\mathbf{x}^{n+2}$.

A related iterative method is the point-Gauss-Seidel method. In the approach, the $\overline{A}$ matrix is expressed as

$$\overline{A} = D - L - U \tag{4.123}$$

where L and U are lower and upper triangular matrices that contain the elements of $\overline{A}$ but with opposite signs, that is

$$L = \begin{bmatrix} 0 & & & \\ -a_{2,1} & 0 & & \\ \cdot & \cdot & & \\ \cdot & & \cdot & \\ -a_{n,1} & \cdot & -a_{n,n-1} & 0 \end{bmatrix} \quad U = \begin{bmatrix} 0 & -a_{1,2} & \cdot & \cdot & -a_{i,n} \\ & 0 & & & \cdot \\ & & \cdot & & \cdot \\ & & & \cdot & \cdot \\ & & & & -a_{n-1,n} \\ & & & & 0 \end{bmatrix}$$

The iterative equations, which are the basis of the Gauss-Seidel method, are given as

$$D\mathbf{x}^{n+1} = L\mathbf{x}^{n+1} + U\mathbf{x}^{n} + \mathbf{b} \tag{4.125}$$

Since, however, $(D - L)$ is a nonsingular lower triangular matrix, Equation 4.125 can be expressed as

$$\mathbf{x}^{n+1} = (D - L)^{-1} U\mathbf{x}^{n} + (D - L)^{-1}\mathbf{b} \tag{4.126}$$

where $(D - L)^{-1}$ is now the point-Gauss-Seidel matrix. Expanding Equation 4.126 and dividing by the diagonal elements, we have

$$x_i^{n+1} = -\sum_{j=1}^{i-1}\left(\frac{\overline{a}_{ij}}{\overline{a}_{ii}}\right) x_j^{n+1} - \sum_{j=i+1}^{N}\left(\frac{\overline{a}_{ij}}{\overline{a}_{ii}}\right) x_j^{n} + \frac{b_i}{\overline{a}_{ii}} \tag{4.127}$$

A similar iterative approach is the *point-successive over relaxation iterative method*. In this method, the calculated values obtained at the $(n + 1)$st iteration are improved using a weighted mean of x_i^{n+1} and x_i^{n}. Introducing $\hat{\mathbf{x}}^{n+1}$ as a new vector defined by

$$D\hat{\mathbf{x}}^{n+1} = L\mathbf{x}^{n+1} + U\mathbf{x}^{n} + \mathbf{b} \tag{4.128}$$

$\mathbf{x}^{n+1}$ is obtained from the relationship,

$$\mathbf{x}^{n+1} = \mathbf{x}^{n} + \omega\,(\hat{\mathbf{x}}^{n+1} - \mathbf{x}^{n}), \quad \omega \geqslant 0 \tag{4.129}$$

or after rearranging,

$$\mathbf{x}^{n+1} = (1 - \omega)\,\mathbf{x}^{n} + \omega\hat{\mathbf{x}}^{n+1} \tag{4.130}$$

where ω, the parameter of the problem, is the relaxation factor. Substituting Equation 4.130 into Equation 4.129, we have

$$(D - \omega L)\mathbf{x}^{n+1} = [(1 - \omega)D + \omega U)]\mathbf{x}^n + \omega\mathbf{b} \tag{4.131}$$

Since $(D - L)$ is nonsingular for any ω, the equation can be solved uniquely for $\mathbf{x}^{n+1}$. The algorithm equations can then be expressed as (Smith, 1965),

$$x_i^{n+1} = (1 - \omega)x_i^n - (\omega/\bar{a}_{ii})\left\{\sum_{j=1}^{i-1} \bar{a}_{ij}x_j^{n+1} + \sum_{j=i+1}^{N} \bar{a}_{ij}x_j^n - b_i\right\} \tag{4.132}$$

Because these methods are iterative, an important consideration in implementation of these techniques is the convergence of the algorithms. Generally, the Gauss-Seidel and point-Jacobi methods are convergent provided the coefficient matrix, $\bar{\mathrm{A}}$, is strictly or irreducibly diagonally dominant (Varga, 1962). This, of course, holds regardless of the initial vector approximation.

Two important block-iterative schemes are *successive line over-relaxation* and the *alternating direction method.* These methods have been primarily designed for the grid networks encountered using finite differences.

In block-iterative methods, the coefficient matrix of the problem is partitioned into smaller block matrices. The original system of equations can then be expressed as

$$\begin{bmatrix} A_{11} & A_{12} & 0 & \cdot & & \cdot \\ A_{21} & A_{22} & A_{23} & & & \\ 0 & A_{32} & A_{33} & & & \\ \cdot & & & \cdot & & \\ \cdot & & & & \cdot & \\ & & & & A_{\bar{N}-1\bar{N}} & A_{\bar{N}\bar{N}} \end{bmatrix} \begin{bmatrix} x_1 \\ x_2 \\ \cdot \\ \cdot \\ x_{\bar{N}} \end{bmatrix} = \begin{bmatrix} b_1 \\ b_2 \\ \cdot \\ \cdot \\ b_{\bar{N}} \end{bmatrix} \tag{4.133}$$

where the matrix blocks are square, nonempty matrices and $\bar{N}$ is the number of blocks. As in the point-iterative schemes, we represent $\bar{\mathrm{A}}$ as

$$\bar{\mathrm{A}} = \mathrm{D} - \mathrm{L} - U \tag{4.134}$$

where again L and U are lower and upper triangular matrices. Since $\bar{\mathrm{A}}$ and D are symmetric, and $L^T = U$, the matrix $(D - \omega L)$ is nonsingular (we assume D is positive definite). The block successive overrelaxation is defined as (Pinder & Gray, 1977)

$$\mathbf{x}^{n+1} = (D - \omega L)^{-1}\{[\omega U + (1 - \omega)D]\}\mathbf{x}^n + \omega\mathbf{B} \tag{4.135}$$

where $\mathbf{B} = (\mathbf{b}_1, \mathbf{b}_2 \ldots \mathbf{b}_{\bar{N}})^T$ and $\mathbf{x}^{n+1} = (\mathbf{x}_1^{n+1}, \mathbf{x}_2^{n+1} \ldots \mathbf{x}_{\bar{N}}^{n+1})^T$. Rewriting Equation 4.135, we have

$$\mathbf{x}_i^{n+1} = (1 - \omega)\mathbf{x}_i^n - \omega\overline{A}_{ii}^{-1}\left\{\sum_{j=1}^{i-1} \overline{A}_{ij}\mathbf{x}_j^{n+1} + \sum_{j=i+1}^{\overline{N}} \overline{A}_{ij}\mathbf{x}_j^n - \mathbf{b}_i\right\} \qquad (4.136)$$

where $\overline{A}_{ii}^{-1}$ is the inverse of a diagonal block matrix.

The estimation of the relaxation factor in the algorithm's equations is given by

$$\omega = 2/[1 + (1 - \lambda^2)^{1/2}] \qquad (4.137)$$

where λ is the largest absolute value of the eigenvalues of the matrix $(D^{-1} A + I)$, the block-Jacobi iteration matrix. This eigenvalue is known as the spectral radius of the matrix, $\rho(A)$. Methods for determining the spectral radius may be found in Wachspress (1966).

In 1955, Peaceman and Rachford introduced the alternating direction procedure for the solution of matrix equations that occur in the discretization of partial differential equations with two or more spatial variables. In the approach, the coefficient matrix is again expressed as the sum of a nonnegative diagonal matrix, D, and two symmetric, positive definite matrices A_1 and A_2. These matrices have positive diagonal elements and nonpositive off-diagonal elements, or

$$\overline{A} = A_1 + A_2 + D \qquad (4.138)$$

We reformulate the matrix problem $\overline{A}\mathbf{x} = \mathbf{b}$ as

$$(A_1 + D + E_1)\,\mathbf{x} = \mathbf{b} - (A_2 - E_1)\,\mathbf{x} \qquad (4.139a)$$

$$(A_2 + D + E_2)\,\mathbf{x} = \mathbf{b} - (A_1 - E_2)\,\mathbf{x} \qquad (4.139b)$$

and assume that the matrix coefficients of $\mathbf{x}$ are nonsingular (Westlake, 1968). In the Peaceman-Rachford method, we define

$$E_1^n = \omega_n I$$

and

$$E_2^n = \hat{\omega}_n I$$

Then, for any iteration, we have two systems of equations,

$$(A_1 + D + \omega_n I)\,\mathbf{x}^{n+1/2} = \mathbf{b} - (A_2 - \omega_n I)\,\mathbf{x}^n \qquad (4.140a)$$

$$(A_2 + D + \hat{\omega}_n I)\,\mathbf{x}^{n+1} = \mathbf{b} - (A_1 - \hat{\omega}_n I)\,\mathbf{x}^{n+1/2} \qquad (4.140b)$$

Since these equations are tridiagonal, Thomas' algorithm can be used for the solution of the equations. Moreover, since $A_1 + D$ and $A_2 + D$ are derived from the finite-difference approximations to the spatial derivatives, the equations can be solved implicitly in one space dimension while using known values in the other coordinate direction. An application of the technique to groundwater recharge simulation has been presented by Mariño (1975c).

Again, convergence properties of the block-iterative schemes are well-defined. For example, if $\overline{A}$ is symmetric and the subblocks A_i are positive definite, then block-successive overrelaxation converges for $\omega = 1$, provided that $\overline{A}$ is positive definite. Furthermore, if A and D are Hermitian and positive definite then the method of block-successive overrelaxation converges for all $\mathbf{x}_0$ if $0 < \omega < 2$ (Westlake, 1968). The Peaceman-Rachford method is always convergent for $E_1 = E_2$ ($\omega = \hat{\omega}$) and when $(\omega I + \frac{1}{2}D)$ is positive definite and symmetric and $(A_1 + A_2 + D)$ is positive definite (Westlake, 1968). Examples of the convergence properties of the algorithms can be found in Forsythe and Wasow (1960).

4.6 FINITE-DIFFERENCE STABILITY ANALYSIS

Numerical models of groundwater hydraulics or mass transport are often based on the finite-difference equations developed in Section 4.2. These equations, because they are approximations of the aquifer system governing equations, introduce two types of errors in the numerical solutions. *Round-off error* occurs during the computer solution of the numerical model, primarily as the result of arithmetic operations. *Truncation error* is introduced in the approximation by neglecting the higher-order terms of the derivative approximations. The stability of the numerical model is determined by how these errors grow or decay during the solution of the finite-difference equations. In this section, we examine the stability properties of the approximating equations. We assume, without loss of generality, that the governing equation of the aquifer system can be represented as

$$L\{u(x, t)\} = 0 \tag{4.141}$$

where L is a differential operator and u is the dependent variable of the problem. Defining v_{ij} as the finite difference approximation to the true solution of the partial differential equation, u_{ij}, at node (i, j) in the $x - t$ plane (Δx and Δt are the discretized spatial and time intervals), then a finite difference scheme is convergent for a sequence

$$(\Delta x_n, \Delta t_n) \rightarrow 0$$

if for every $u_0 \in L^2$ and $t \geqslant 0$, we have

$$\| v(.,t_{u_n}) - u(.,t) \| \rightarrow 0 \quad \text{as} \quad n \rightarrow \infty \tag{4.142}$$

where, v_{0_n} is a sequence converging to u_0 in L^2 and u_n is a sequence of integers such that

$$t_{u_n} = u_n \, \Delta t_n \quad \text{as} \quad n \rightarrow \infty$$

where L^2 is a Banach space of all measurable functions $f(x), x \in R^0$, for which $\|f\| < \infty$, in which $\| \ \|$ denotes the L^2 norm which is defined as

$$\|f\| = (\textstyle\int_{R^0} |f|^2 \, dx)^{1/2}$$

Convergence of the numerical model and the stability of the solution are interrelated through Lax's equivalent theorem (Richtmeyer & Morton, 1967).

Theorem: Given a properly posed initial-value problem and a finite-difference approximation to it that satisfies the consistency condition, stability is the necessary and sufficient condition for convergence.

The ramifications of the theorem imply that a finite-difference scheme is convergent if:

1. The partial differential equation is well-posed.
2. The finite-difference scheme is consistent and stable.

The governing equations of the problem are well-posed in the sense that if the initial data are slightly perturbed, the partial differential equation provides solutions that deviate only slightly from the unperturbed solutions. Mathematically, a problem is well-posed in L^2 if

1. For every $u_0(x) \in D(R)$ (the set of C^∞ functions with compact support) there exists a unique classical solution in L^2.
2. There exists constants K and α, independent of $u_0\ (x)$, such that

$$\|u(.,t)\| \leqslant Ke^{\alpha t}\|u(.,0)\|, \tag{4.143}$$

for all $t \leqslant 0$ and for each classical solution $u(.,t)$.

Well-posedness of the problem can be shown using the Fourier method for linear partial differential equations with constant coefficients, and with the energy method, for equations with variable coefficients. (See Example Problem 4.14).

Assuming the problem is well-posed, a numerical scheme is consistent if it is accurate of order at least $(1, 1)$. A finite-difference scheme is accurate of order (Q_1, Q_2) for the particular solution $u(x, t)$ if there is a function $c(t)$ that depends on $u(x, t)$ and is bounded in each time interval, $0 \leqslant t \leqslant T$ such that for all Δx and

Δt sufficiently small, the local truncation error, $L_{\Delta x}u$ is given as

$$\|L_{\Delta x}u\| \leq c\Delta t\ (\Delta x^{Q_1} + \Delta t^{Q_2}) \tag{4.144}$$

where $\|\ \ \|$ is the norm of $L_{\Delta x}u$. The order of accuracy for a given finite difference scheme can be shown by a Taylor series expansion of $u(x, t)$.

If the difference scheme is consistent, stability of the numerical model is defined by the following theorem.

> **Theorem:** A finite-difference scheme is stable for a sequence $(\Delta x_n, \Delta t_n) \rightarrow 0$ as $n \rightarrow \infty$ if there exists constants α_s and K_s such that for all $t \geq 0$, with t being a multiple of Δt_n,

$$\|S_{\Delta x}(t,\ \Delta x_n, \Delta t_n)\| \leq K_s e^{\alpha_s t},\ \forall n \tag{4.145}$$

where $S_{\Delta x}$ is the solution operator of the finite-difference equations, or

$$v\ (.,t) = S_{\Delta x}(t,\ \Delta x_n,\ \Delta t_n)\ v\ (.,0),\ t \geq 0$$

The most widely used approach for determining the stability of a finite-difference model is *von Neumann stability analysis*. In the method, errors, which are represented by finite, Fourier series, are introduced in the difference equations. The growth or attenuation of these errors during the solution of the equations determines the stability of the numerical model. The method is strictly applicable, however, to initial value problems described by linear, constant coefficient, partial differential equations.

Fourier transformed finite-difference schemes can be written as

$$\hat{v}^{t+\Delta t} = G(\omega, \Delta t)\hat{v}^t$$

where $G(\omega, \Delta t)$ is the amplification matrix and $\hat{v}^{t+\Delta t}, \hat{v}^t$ are the Fourier transforms of $v^{t+\Delta t}$ and v^t. The Fourier transform of v is defined as

$$\hat{v}(\omega) = \frac{1}{\sqrt{2\pi}} \int_{-\infty}^{\infty} e^{-i\omega x} v(x) dx \tag{4.146}$$

and the von Neumann necessary condition for stability can be summarized as

$$\rho(G) \leq 1 + 0(\Delta t) \tag{4.147a}$$

or

$$|\lambda_j| \leq 1 + 0(\Delta t) \tag{4.147b}$$

where $\rho(G)$ is the spectral radius of G and λ_j, $j = 1, 2, \ldots, n$ are the eigenvalues of G. If G is a normal matrix, the von Neumann condition is necessary and sufficient for stability.

Stability of the numerical models can also be evaluated using matrix methods. In contrast to Fourier analysis, the matrix approach directly incorporates the boundary conditions in the stability analysis. To illustrate the method, we write the finite-difference equations for any time step as (see Equation 4.61b)

$$\boldsymbol{\phi}_{k+1} = \overline{B}\boldsymbol{\phi}_k \tag{4.148}$$

where $\overline{B} = (A + \theta\Delta tB)^{-1}(A - (1 - \theta)\Delta tB)$. In the stability analysis, we assume $\boldsymbol{\phi}_{k+1}$ is the sum of known value, and an error term, $\mathbf{E}_{k+1}$. The difference equations can now be written in terms of the error at time step k as

$$\mathbf{E}_{k+1} = \overline{B}\,\mathbf{E}_k$$

Taking the norm of both sides of the identity produces

$$\|\mathbf{E}_{k+1}\| \leqslant \|\overline{B}\|\;\|\mathbf{E}_k\|$$

Stability of the model requires that

$$\|\overline{B}\| < 1 \tag{4.149a}$$

This condition is equivalent to requiring that the spectral radius of $\overline{B}$, $\rho(\overline{B})$, cannot exceed one, or

$$\rho(\overline{B}) < 1 \tag{4.149b}$$

We also note that an upper bound on the spectral radius is given as $\rho(\overline{B}) \leqslant \|\overline{B}\|$ (Saul'yev, 1964).

The example problems demonstrate how these concepts can be used to evaluate the stability of common finite difference models.

Example Problem 4.13 Stability Analysis of the Explicit Difference Scheme

A finite-difference approximation is to be developed for the diffusion equation,

$$\frac{\partial^2 u}{\partial x^2} = \frac{\partial u}{\partial t} \tag{4.150a}$$

The model describes one-dimensional confined flow or dispersive mass transport.

The initial conditions for the problem are

$$u(x, 0) = u_0(x), \quad -\infty \leqslant x \leqslant \infty \tag{4.150b}$$

Investigate the stability of the difference approximations.

Defining $v_{i,j}$ as the approximation of the true solution of the partial differential equation, $u_{i,j}$ in the $x - t$ plane, then the explicit finite-difference approximation is from Table 4.3,

$$\frac{v_{i,j+1} - v_{i,j}}{\Delta t} = \frac{v_{i+1,j} - 2\, v_{i,j} + v_{i-1,j}}{\Delta x^2}$$

The order of accuracy of the approximation is (2, 1).

We approach the stability analysis of the difference approximation by again using Lax's equivalency theorem. Using Fourier methods, we write

$$\frac{d}{dt}\,\tilde{u}(\omega, t) = -\omega^2\tilde{u}(\omega, t), t \geqslant 0 \tag{4.151a}$$

$$\tilde{u}(\omega, 0) = \tilde{u}_0(\omega), \ \omega \epsilon R \tag{4.151b}$$

where $\tilde{u}$, the Fourier transform, is defined as

$$\tilde{u} = \frac{1}{\sqrt{2\pi}} \int_{-\infty}^{\infty} e^{-i\omega x} u(x, t) dx$$

Integrating Equation 4.150b, we have,

$$\tilde{u}(\omega, t) = e^{-\omega^2 t}\, \tilde{u}_0(\omega) \tag{4.152}$$

Taking the norm of Equation 4.152 yields,

$$\|\tilde{u}(\omega, t)\| = e^{-\omega^2 t}\|\tilde{u}_0(\omega)\| \tag{4.153a}$$

or

$$\|\tilde{u}(\omega, t)\| \leqslant \|\, \tilde{u}_0\|, \tag{4.153b}$$

since $e^{-\omega^2 t} \leqslant 1$.

By Parseval's relation,

$$\|u(x, t)\| \leqslant \|u_0(x)\| \tag{4.154}$$

which implies that $K = 1$ and $\alpha = 0$. Because these parameters exist, the problem is well-posed.

We can also determine the well-posedness of the model using the energy method. From the definition of energy, we have

$$\frac{d}{dt} \| u(.,t)\|^2 = \frac{d}{dt}(u, u) = (u_{xx}, u) + (u, u_{xx}) \tag{4.155}$$

Integrating by parts, we obtain

$$\frac{d}{dt}\|u(.,t)\|^2 = u_x u|_R - (u_x, u_x) + u\, u_x|_R - (u_x, u_x) = -2(u_x, u_x)$$

It has been assumed that the term $2u_x u|_R$ vanishes on the boundary.

Since $-2(u_x, u_x) \leq 0$

$$\frac{d}{dt}\|u(.,t)\|^2 \leq 0,$$

which implies

$$\|u(.,t)\|^2 \leq \|u(.,0)\|^2$$

or

$$\|u(.,t)\| \leq \|u_0(x)\| \tag{4.156}$$

Note that Equation 4.156 is identical to Equation 4.154; the results again imply the existence of the constants, $K = 1$ and $\alpha = 0$.

The next step in the analysis is to show that the approximating equations are consistent. We expand the approximation using a Taylor series or

$$L_{\Delta x}\, u = u_{i,j+1} - u_{i,j} - \frac{\Delta t}{\Delta x^2}(u_{i+1,j} - 2u_{i,j} + u_{i-1,j}) \tag{4.157a}$$

and

$$\begin{aligned} L_{\Delta x}\, u &= \Delta t \left\{ u_t + \frac{\Delta t}{2} u_{tt} + \frac{\Delta t^2}{6} u_{ttt} \right. \\ &\quad \left. - \left(u_{xx} + \frac{\Delta x}{12} u_{xxx} + \frac{\Delta x^4}{360} u_{xxx} + 0(\Delta x^6 + \Delta t^3) \right) \right\} \\ &= \Delta t \left\{ \left(\frac{\Delta t}{2} - \frac{\Delta x^2}{12} \right) u_{xxx} + \left(\frac{\Delta t^2}{6} - \frac{\Delta x^4}{360} \right) u_{xxx} + 0\,(\Delta x^6 + \Delta t^3) \right\} \end{aligned} \tag{4.157b}$$

If $\frac{\Delta t}{\Delta x^2} \neq \frac{1}{6}$ the scheme is on the order of (2,1). Similarly, if $\frac{\Delta x}{\Delta t^2} = \frac{1}{6}$, then the scheme is accurate of order (4,2). Therefore, the finite-difference scheme is consistent.

The final step in the stability analysis is the application of the von Neumann condition, the necessary and sufficiency condition for stability. We rewrite the difference equation as

$$v_{i,j+1} = \left[1 + \frac{\Delta t}{\Delta x^2}(E - 2 + E^{-1})\right] v_{i,j}$$

where

$$E(v_{i,j}) = v_{i+1,j} \quad \text{and} \quad E^{-1}(v_{i,j}) = v_{i-1,j}$$

The amplification matrix associated with the finite-difference approximation of Equation 4.149 is given as

$$G(\omega, \Delta t) = 1 - 4\frac{\Delta t}{\Delta x^2}\sin^2\frac{\xi}{2}$$

where $\xi = \omega\Delta x$. We observe that if $\frac{\Delta t}{\Delta x^2} \leqslant 1/2$ then $|G(\omega, \Delta t)| \leqslant 1$. The stability criterion for the explicit scheme can then be expressed as

$$\frac{\Delta t}{\Delta x^2} \leqslant 1/2$$

Of course, this in turn implies that the difference scheme is convergent. Further examples of stability analysis for parabolic equations may be found in Lapidus and Pinder (1982).

Example Problem 4.14 Matrix Stability Analysis

Investigate the stability of the one-dimensional finite-difference model developed in Example Problem 4.2. Using matrix methods, determine the stability of the explicit and Crank-Nicolson approximations. Assume the aquifer is homogeneous and isotropic with Dirichlet or Neumann no-flow boundary conditions.

The finite-difference equations arising from the solution of the confined flow model can be written for any time step as

$$\left\{\frac{\mathbf{h}^k - \mathbf{h}^{k-1}}{\Delta t}\right\} + B\,\{\theta\mathbf{h}^k + (1 - \theta)\mathbf{h}^{k-1}\} + \mathbf{g} = 0 \tag{4.158}$$

where, again, θ is a variable weighting factor and the **g** vector contains the boundary conditions. We drop this vector as it merely adds a constant vector to the manipulations.

For the explicit scheme we have ($\theta = 0$) and, for any time step k, the equations may be written recursively as

$$\mathbf{h}^k = \overline{B}\mathbf{h}^{k-1}, \quad \overline{B} = \{I - B\Delta t\}$$

Again, introducing the errors in the approximations, stability of the solution requires that

$$\|\overline{B}\| < 1$$

where $\overline{B}$ is given as

$$\overline{B} = I + \frac{\Delta t}{\Delta x^2} T \tag{4.159}$$

and T is a tridiagonal matrix with elements $(1, -2, 1)$. The stability condition, Equation 4.132, is again equivalent to requiring that the spectral radius of $\overline{B}$ be less than or equal to one, or

$$\rho(\overline{B}) \leqslant 1$$

Since $\overline{B}$ is tridiagonal, the eigenvalues of $\overline{B}$, λ_s, are given as (Lapidus & Pinder, 1982)

$$\lambda_s = 1 - 2\frac{\Delta t}{\Delta x^2}\left(1 - \cos\frac{s\pi}{n}\right)$$

For stability, we have

$$-1 \leqslant 1 - 4\frac{\Delta t}{\Delta x^2}\sin^2\left(\frac{s\pi}{2n}\right) \leqslant 1, s = 1, 2, \ldots n - 1$$

or, equivalently,

$$0 \leqslant \frac{\Delta t}{\Delta x^2} \leqslant 1/2$$

The explicit scheme is conditionally stable.

Similarly, for the Crank-Nicolson approximation, $\theta = 0.5$, the $\overline{B}$ matrix is given as

$$\bar{B} = \left(I + \frac{\Delta t}{2\Delta x^2} T\right)^{-1} \left(I - \frac{\Delta t}{2\Delta x^2} T\right) \tag{4.160}$$

where T is $(1, -2, 1)$. The eigenvalues of $\bar{B}$ are (Lapidus & Pinder, 1982)

$$\lambda_s = \frac{2 - 4\alpha \sin^2(s\pi/2n)}{2 + 4\alpha \sin^2 s(\pi/2n)}, \quad s = 1, 2, ..n - 1$$

where $\alpha = \Delta t/\Delta x^2$. Since the eigenvalues are less than one for $\alpha > 0$, the Crank-Nicolson is unconditionally stable.

4.7 FINITE-ELEMENT ANALYSIS

Convergence of the Galerkin finite-element method for linear problems is related to the completeness of the shape or basis functions of the numerical approximations. A sequence of these linearly independent basis functions N_i is complete with respect to the operator L if any admissible function ϕ can be approximated by a linear combination, $\sum_i N_i\tilde{\Phi}_i$, so that for every $\varepsilon > 0$

$$\int_D \left\{L\left(\phi - \sum_{i=1}^{N} N_i\tilde{\Phi}_i\right)\right\}\left(\phi - \sum_{i=1}^{N} N_i\tilde{\Phi}_i\right) dD < \epsilon \tag{4.161}$$

Functions satisfying Equation 4.161 are also complete in energy with respect to the operator L (Norrie & De Vries, 1973).

The basis functions are also complete in the sense of convergence to the mean, provided the operator is positive and bounded below (Norrie & de Vries, 1973). That is, for any operator, L, with the property

$$\int_D (L\phi)\phi dD \geqslant \gamma^2 \int_D (\phi)(\phi) dD, \quad \gamma > 0 \tag{4.162}$$

then,

$$\int_D \left\{\phi - \sum_{i=1}^{N} N_i\tilde{\Phi}_i\right\}^2 dD < \delta, \text{ for any } \delta > 0 \tag{4.163}$$

As a result, the approximating functions are capable of representing an arbitrary function to any degree of accuracy by increasing the number of terms in the series approximation.

Sufficient conditions for the convergence of the Galerkin method require more, however, than the completeness of the approximating functions. Convergence can only be demonstrated if the basis functions are also a complete set of orthonormal functions, or

$$\int_D N_i N_j \, dD = \begin{bmatrix} 0, & i \neq j \\ 1, & i = j \end{bmatrix} \tag{4.164}$$

A complete orthonormal set has the property that any function F, which is orthogonal to every member of the set, satisfies (Mikhlin, 1964),

$$\int_D F N_i dD = 0, \quad i = 1, 2, \ldots \infty \tag{4.165}$$

We observe that if the function is the residual of the approximation $L(\hat{\phi})$ then

$$\int_D L(\hat{\phi}) N_i dD = 0, \quad i = 1, 2, \ldots \infty \tag{4.166}$$

Comparing Equation 4.166 with the Galerkin equations (Equation 4.22), we see that the orthogonality condition constrains R to be zero as n approaches infinity. Again, since the complete orthonormal set is truncated, the residual is non-zero and the solution is an approximation.

Convergence and continuity of the finite-element method can also be analyzed by examining the behavior of the solution over an individual finite element. The solution of the finite-element model, for example, should converge toward the correct solution as the element size is decreased. Any arbitrary function tends to be represented exactly across an element as the element size approaches zero. However, these conditions are generally not sufficient in the finite-element method because there has to be nodal compatibility between the elements (Norrie & DeVries, 1973). *Nodal compatibility* implies that the nodal values of coincident nodes of adjacent elements are the same. For Lagrangian elements, the nodal values are the nodal values of the dependent variables. However, for Hermitian elements, the nodal parameters consist of the nodal values and the first derivatives of the functions. When nodal compatibility is required, the variable or derivative representation over an element must contain not only the first or constant terms of the complete set, but additional terms to satisfy the compatibility restriction. For example, if a complete polynomial sequence is being used, the next term group is the linear terms. The combination of these terms provides an accurate representation for a variable or derivative as the element size tends to zero. Nodal compatibility conditions are also satisfied.

Because the finite-element model consists of a set of piecewise continuous functions over the solution domain, the integration of these functions places certain restrictions on continuity between the elements. For example, the integral equations of the finite-element method can be expressed as

$$\int_D \frac{\partial^n \hat{\phi}}{\partial x^n}\, dD$$

For these integrals to be defined, ϕ must be continuous to the order $(N - 1)$ to ensure that only finite jump discontinuities exist in the nth derivative (Kaplan, 1952). If, for example, the governing equations of the problem contain second-order derivatives, the interpolation equations have to be continuous between the elements.

Additional theoretical convergence properties of the finite-element method may be found in Zienkiewicz (1971) and Pinder and Gray (1977).

Example Problem 4.15 Basis Functions and Convergence of the Finite-Element Method

Investigate the restrictions placed on the basis functions for convergence of the finite-element method. Assume the state variable can be represented as a polynomial approximation.

Convergence of the finite-element is based, in part, on the behavior of the solution as the element size decreases. We observe that the finite-element model will converge toward the correct solution as the element size decreases, if the basis or shape functions give a constant value throughout the element.

This restriction limits the types of basis functions that can be used in the approximations. For example, for any element, we have

$$\hat{\phi} = \sum_{i=1}^{n} N_i(x)\tilde{\Phi}_i$$

Since, however, in the limiting cases, $\tilde{\Phi}_i = \tilde{\Phi}_j = \tilde{\Phi}_k = \tilde{\Phi}_r \cdots \tilde{\Phi}_n$, we have

$$\left\{\sum_{i=1}^{n} N_i(x)\right\} \hat{\Phi} = \hat{\phi}$$

and the basis functions must tend to unity at every point within the element. Similar analyses of two- or three-dimensional Lagrangian or simplex elements will show that these basis functions also satisfy the convergence criteria (Segerlind, 1976).

4.8 REFERENCES

Bear, J. *Dynamics of Fluids in Porous Media*. New York: Elsevier, 1972.

Bellman, R.E. *Introduction to Matrix Analysis*. New York: McGraw-Hill, 1960.

Bellman, R.E., & Kalaba, R.E. *Quasilinearization and Nonlinear Boundary-Value Problems*. New York: Elsevier, 1965.

Boulton, N.S. "Analysis of Data from Nonequilibrium Pumping Tests Allowing for Delayed Yield from Storage." *Proc. Inst. Civil Engr.*, 26, No. 6693, 1963.

Boussinesq, J. "Recherches Théoriques sur L'Écoulement des Nappes D'Eau Infiltrées dans le Sol et Sur Débit de Sources." *C.R.H. Acad. Sci.*, *J. Math Pures Appl.*, 11:363–394, 1904.

Brenner, H. "The Diffusion Model of Longitudinal Mixing in Beds of Finite Length: Numerical Values." *Chem. Engr. Sci.* 17:229–243.

Carslaw, H.S., & Jaeger, J.C. *Conduction of Heat in Solids*. Fair Lawn NJ: Oxford University Press, 1959.

Corapcioglu, M.Y., Borekci, O., & Haridas, A. "Analytical Solutions for Rectangular Aquifers with Third-Kind (Caucky) Boundary Conditions." *Water Resources Research*, 19(2):523–528, 1983.

Culham, W.E., & Varga, R.S. "Numerical Methods for Time Dependent Nonlinear Boundary Value Problems." Soc. Pet. Engr. Paper 2806, 2nd Symposium on Numerical Simulation of Reservoir Performance, Dallas, Texas, 1970.

Daniel, J.W., & Moore, R.E. *Computation and Theory in the Ordinary Differential Equations*. San Francisco: W.H. Freeman and Co., 1970.

DeWiest, R.J.M. *Geohydrology*. New York: John Wiley, 1965.

Douglas, J., Jr. "Roundoff Error in the Numerical Solution of the Heat Equation." *J. Assoc. Computing Mach*, 6:48–58, 1959.

Douglas, J. Jr., & Dupont, T. "Galerkin Methods for Parabolic Equations." *SIAM J. Numer. Analy.*, 7:575, 1970.

Edelman, J.H. Over de Berekening Van Groundwaterstomingen. Thesis. Delft, The Netherlands, 1947.

Falade, G.K. "Mathematical Analysis of Fluid Flow in Porous Media with General Anisotropy." *Water Resources Research*, 17(4):1071–1074, 1981.

Finlayson, B. *The Method of Weighted Residuals and Variational Principles*. New York: Academic Press, 1972.

Forsythe, G.E., & Wasow, W.R. *Finite Difference Methods for Partial Differential Equations*. New York: John Wiley, 1960.

Forsythe, G.E., & Moler, C.B. *Computer Solution of Linear Algebraic Systems*. Englewood Cliffs NJ: Prentice-Hall, 1967.

Forsythe, G.E., Malcolm, M.A., & Moler, C.B. *Computer Methods for Mathematical Computations*. Englewood Cliffs NJ: Prentice-Hall, 1977.

Glover, R.E. *Transient Groundwater Hydraulics*. Fort Collins CO: Water Resources Publications, 1974.

Gray, W.G., & Van Genuchten, M. Th. "Economical Alternatives to Gaussian Quadrature or Isoparametric Quadrilaterals." *Int. J. Numer. Meth. Engr.*, (12), 1978.

Grove, D.B. "The Use of Galerkin Finite Element Methods to Solve Mass-Transport Equations." *U.S.G.S., Water Resources Investigation* 77-47, Denver, Colorado, 1977.

Gureghian, A.B., & Youngs, E.G. "Numerical Solutions of Boussinesq's Equation for

Seepage Flow." In *Finite Elements in Water Resources*, W.G. Gray, G.F. Pinder, & C.A. Brebbia, Eds. London: Pentech Press, 1978.

Hantush, M.S., & Jacob, C.E. "Non-Steady Radial Flow in an Infinite Leaky Aquifer." *Trans. Am. Geo. Union*, 36(1), 1955.

Huntoon, P.W. *Finite Difference Methods as Applied to the Solution of Groundwater Flow Problems*. Wyoming Water Resources Research Institute, Laramie, 1974.

Ibrahim, H.A., & Brutsaert, W. "Inflow Hydrographs from Large Unconfined Aquifers." *J. Irrigation and Drainage Division, ASCE*, 91(IR2):21-38, 1965.

Kaplan, W. *Advanced Calculus*. Reading MA: Addison-Wesley, 1952.

Keller, H.B. "The Numerical Solution of Parabolic Partial Differential Equations." In *Mathematical Methods for Digital Computers*, vol. 1. New York: Wiley, 1967.

Lantz, R.B. "Quantitative Evaluation of Numerical Diffusion (Truncation Error)." *J. Soc. Pet. Eng.*, September 1971, pp. 315–320.

Lapidus, L., & Pinder, G.F. *Numerical Solution of Partial Differential Equations in Science and Engineering*. New York: Academic Press, 1982.

Lee, E.S. *Quasilinearization and Invariant Imbedding*. New York: Academic Press, 1968.

Lees, M. "A Linear Three-Level Difference Scheme for Quasilinear Parabolic Equations." *Math. Compu.* 20:516, 1966.

Li, W.-H. *Differential Equations of Hydraulic Transients, Dispersion and Groundwater Flow*. Englewood Cliffs, NJ: Prentice-Hall, 1972.

Maddock, T., III. "Algebraic Technological Function from a Simulation Model." *Water Resources Research*, 8(1):129–134, 1972.

Mariño, M.A. "Digital Simulation Model of Aquifer Response to Stream Stage Fluctuation." *J. of Hydrology*, 25:51–78, 1975a.

Mariño, M.A. "Artificial Groundwater Recharge, I, Circular Recharge Area." *J. of Hydrology*, 25:201–208, 1975b.

Mariño, M.A. "Artifical Groundwater Recharge, II, Rectangular Recharge Area." *J. of Hydrology*, 26:29–37, 1975c.

Mikhlin, S.G. *Variational Methods of Mathematical Physics*. Oxford: Pergamon Press, 1964.

Moler, C., & Loan, C.V. "Nineteen Dubious Ways to Compute the Exponential of a Matrix," *SIAM Review*, 20(4):801–836, 1978.

Norrie, D.H., & deVries, G. *The Finite Element Method*. New York: Academic Press, 1973.

Papadopulos, I.S. "Nonsteady Flow to a Well in an Infinite Anisotropic Aquifer." Symposium Int. Ass. Sci. Hydrology, Dubrovnik, 1965.

Peaceman, D.W., & Rachford, H.H., Jr. "The Numerical Solution of Parabolic and Elliptic Differential Equations," *J. Soc. Ind. Appl. Math.*, 3:28–41, 1955.

Pinder, G.F., & Gray, W.G. *Finite Element Simulation in Surface and Subsurface Hydrology*. New York: Academic Press, 1977.

Pinder, G.F., Frind, E.O., & Celia, M.A. "Groundwater Flow Simulation Using Collocation Finite Elements." In *Finite Elements in Water Resources*, W.G. Gray, G.F. Pinder, and C.A. Brebbia, Eds. London: Pentech Press, 1978.

Pipes, L.A. *Applied Mathematics for Engineers and Physicists*. New York: McGraw-Hill, 1958.

Polubarinova-Kochina, P. Ya. *Theory of Groundwater Movement*. Princeton NJ: Princeton University Press, 1962.

Remson, I., Hornberger, G.M., & Molz, F.J. *Numerical Methods in Subsurface Hydrology*. New York: John Wiley, 1971.

Richtmeyer, R.D., & Morton, K.W. *Difference Methods for Initial-Value Problems*. New York: John Wiley, 1967.

Saul'yev, V.K. *Integration of Equations of Parabolic Type by the Method of Nets*. Oxford: Pergamon Press, 1964.

Salvadori, M.G., & Baron, M.L. *Numerical Methods in Engineering*. Englewood Cliffs NJ: Prentice-Hall, 1961.

Segerlind, L.J. *Applied Finite Element Analysis*. New York: John Wiley, 1976.

Smith, G.D. *Numerical Solution of Partial Differential Equations*. London: Oxford University Press, 1965.

Tewarson, R.P. *Sparse Matrices*. New York: Academic Press, 1973.

Theim, G. *Hydrologische Methoden*. Leipzig: Gebhardt, 1906.

Theis, C.V. "The Relocation Between the Lowering of the Piezometric Surface and the Rate and Duration of a Well Using Groundwater Storage." *Trans. Amer. Geo. Union*, 16:519–524, 1935.

Thomas, L.H. "Elliptic Problems in Linear Difference Equations over a Network." Report of Watson Science Computing Laboratory, Columbia University, New York, 1949.

Todd, D.K. "On Steady Flow in Porous Media by Means of a Hele-Shaw Viscous Fluid Model." *Trans. Amer. Geo. Union*, 35(6):905–916, 1954.

Tolikas, P.K., Sidiropoulos, E.G., & Tzimopoulos, C.D. "A Simple Analytical Solution for the Boussinesq One-Dimensional Groundwater Flow Equation." *Water Resources Research*, 20(1):24–28, 1984.

Varga, R.S. *Matrix Iterative Analysis*. Englewood Cliffs NJ: Prentice-Hall, 1962.

Wachspress, E.L. *Iterative Solution of Elliptic Systems*. Englewood Cliffs NJ: Prentice-Hall, 1966.

Walton, W.C. *Groundwater Resource Evaluation*. New York: McGraw-Hill, 1970.

Westlake, J.R. *A Handbook of Numerical Matrix Inversion and Solution of Linear Equations*. New York: Wiley, 1968.

Yates, S.R., Warrick, A.W., & Lonen, D.O. "Hillside Seepage: An Analytical Solution to a Nonlinear Dupuit-Forcheimer Problem." *Water Resources Research*, 21(3):331–336, 1985.

Yeh, W.W-G. "Nonsteady Flow to Surface Reservoir." *ASCE, J. Hydr. Div.*, 96(HY3):609–618, 1970.

Yeh, W.W-G., & Tauxe, G.W. "Optimal Identification of Aquifer Diffusivity Using Quasilinearization." *Water Resources Research*, 7(4):955–962, 1971.

Young, L.C. "A Preliminary Comparison of Finite Element Methods for Reservoir Simulation." In *Advances in Computer Methods for Partial Differential Equations*, vol. 2, R. Vichenevetsky, Ed. IMACS, 1977.

Zienkiewicz, O.C. *The Finite Element Method in Engineering Science*. New York: McGraw-Hill, 1971.

5. OPTIMIZATION METHODS FOR GROUNDWATER MANAGEMENT

5.1 INTRODUCTION

The management models that are presented in Chapters 6 through 8 are optimization models of the groundwater system. These mathematical models consist of a set of economic, hydraulic, water quality, or environmental objectives and the management or decision variables that control the groundwater pumping, recharge, or waste injection schedules. The management alternatives are constrained by the groundwater system's hydraulic or water quality response equations and possible well capacity, hydraulic gradient, or water demand requirements.

The objective of this chapter is to review the most common mathematical programming techniques for the solution of deterministic or stochastic, linear or nonlinear groundwater optimization models. We begin with a brief overview of mathematical programming.

5.2 MATHEMATICAL PROGRAMMING PRELIMINARIES

The general mathematical programming (MP) problem is to choose a control or decision vector, $\mathbf{x}$, from a given constraint or opportunity set, $\overline{\mathbf{X}}$, so as to maximize the system's objective function, $F(\mathbf{x})$, or

$$\max z = F(\mathbf{x}) \tag{5.1a}$$

subject to,

$$\mathbf{x} \in \overline{\mathbf{X}}$$

We assume that F is a real-valued function of $\mathbf{x}$ and is continuous and differentiable. In the general nonlinear programming problem, the constraint set can be expressed as

$$\overline{\mathbf{X}} = \{\mathbf{x} \mid \mathbf{g}(\mathbf{x}) \leqslant \mathbf{b},\ \mathbf{x} \geqslant 0\} \tag{5.1b}$$

where $\mathbf{g}$ is a $m \times 1$ vector of constraint equations and $\mathbf{b}$ is a $m \times 1$ column vector. The $\mathbf{g}$ vector is also assumed to be continuous and differentiable.

A global maximum of the MP problem, $\mathbf{x}^*$, is a feasible vector yielding a value of the objective function larger than or equal to that obtained by any other feasible vector, or

$$F(\mathbf{x}^*) \geqslant F(\mathbf{x}),\ \forall \mathbf{x} \in \overline{\mathbf{X}} \tag{5.2a}$$

$$\mathbf{x}^* \in \overline{\mathbf{X}}$$

$\mathbf{x}^*$ is a strict global maximum if

$$F(\mathbf{x}^*) > F(\mathbf{x}),\ \forall \mathbf{x} \in \overline{\mathbf{X}},\ \mathbf{x} \neq \mathbf{x}^* \tag{5.2b}$$

A solution may be a local maximum or minimum, even if it is not a global maximum or minimum. A solution, $\mathbf{x}^*$, is a local maximum if it is a feasible solution and if it yields a value of the objective function larger than or equal to that obtained by any feasible vector sufficiently close to it. Equivalently, there exists an $\epsilon > 0$, such that,

$$\mathbf{x}^* \in \overline{\mathbf{X}} \text{ and } F(\mathbf{x}^*) \geqslant F(x),\ \forall \mathbf{x} \in \overline{\mathbf{X}} \cap N_\epsilon(\mathbf{x}^*) \tag{5.3a}$$

where $N_\epsilon(\mathbf{x}^*) = \{\mathbf{x} \mid \|\mathbf{x} - \mathbf{x}^*\| \leqslant \epsilon\}$, where $\|\cdot\|$ is a suitable norm. Similarly, a strict local maximum is defined as

$$F(\mathbf{x}^*) > F(\mathbf{x}),\ \forall \mathbf{x} \in \overline{\mathbf{X}} \cap N_\epsilon(\mathbf{x}^*),\ \mathbf{x} \neq \mathbf{x}^* \tag{5.3b}$$

The Weierstrass theorem provides sufficient conditions for the existence of a global maximum. If the opportunity set is closed and bounded (compact) and nonempty, and the objective $F(\mathbf{x})$ is continuous on $\overline{\mathbf{X}}$, then $F(\mathbf{x})$ has a global maximum either in the interior or on the boundary of $\overline{\mathbf{X}}$.

The local-global theorem of mathematical programming gives sufficient conditions for a local maximum to be a global maximum. A local maximum is a global maximum if the feasible region, $\overline{\mathbf{X}}$, is a nonempty, compact, convex set and $F(\mathbf{x})$ is a continuous, concave function. If $F(\mathbf{x})$ is also strictly concave, then the solution is unique. It can also be shown that the set of points at which the maximum is obtained is also a convex set.

Although the Weierstrass and local-global theorems provide information regarding the existence of solutions and the type of optima, the theorems do not characterize or identify optimal solutions to mathematical optimization problems. These necessary conditions are given by the Kuhn-Tucker theorem of mathematical programming. The Kuhn-Tucker approach to the general nonlinear programming problem is based on the Lagrangian function of the optimization model. The Lagrangian, $L(\mathbf{x},\mathbf{y})$, is defined as

$$L(\mathbf{x},\mathbf{y}) = F(\mathbf{x}) + \mathbf{y}(\mathbf{b} - \mathbf{g}(\mathbf{x})) \tag{5.4}$$

where $\mathbf{y}$ is a m $\times$ 1 row vector of Lagrange multipliers. The Kuhn-Tucker conditions that characterize an optimal solution, $\mathbf{x}^*$, $\mathbf{y}^*$, are expressed in terms of the Lagrangian as (Intriligator, 1971)

$$\frac{\partial L}{\partial \mathbf{x}} \leqslant 0 \quad , \quad \frac{\partial L}{\partial \mathbf{y}} \geqslant 0 \tag{5.5a}$$

$$\frac{\partial L}{\partial \mathbf{x}} \mathbf{x} = 0 \quad , \quad \mathbf{y}\frac{\partial L}{\partial y} = 0 \tag{5.5b}$$

$$\mathbf{x} \geqslant 0 \quad , \quad \mathbf{y} \geqslant 0 \tag{5.5c}$$

These restrictions are necessary and sufficient for a (strict) global maximum if the objective is (strictly) concave, the constraint set is convex, and a constraint qualification condition is satisfied, or

$$\exists\ \mathbf{x}' \text{ such that } \mathbf{x}' \geqslant 0 \text{ and } \mathbf{g}(\mathbf{x}') < \mathbf{b} \tag{5.5d}$$

In other words, there exists a point in the opportunity set $\mathbf{x}'$ which satisfies all the inequality constraints as strict inequalities.

If $\mathbf{x}^*$, $\mathbf{y}^*$ can satisfy the Kuhn-Tucker conditions, then $\mathbf{x}^*$, $\mathbf{y}^*$ is a saddle point of the Lagrangian, or

$$L(\mathbf{x}, \mathbf{y}^*) \leqslant L(\mathbf{x}^*, \mathbf{y}^*) \leqslant L(\mathbf{x}^*, \mathbf{y}), \ \forall\ \mathbf{x} \geqslant 0,\ \mathbf{y} \geqslant 0 \tag{5.6}$$

According to the Kuhn-Tucker theorem, $\mathbf{x}^*$ solves the general nonlinear programming problem if $(\mathbf{x}^*, \mathbf{y}^*)$ solves the saddle point problem. If there exists a $\mathbf{y}^*$ such that $(\mathbf{x}^*, \mathbf{y}^*)$ solves the saddle point problem, $\mathbf{x}^*$ solves the nonlinear programming program. Furthermore, the Lagrange multipliers, $\mathbf{y}^*$, can be interpreted as the change in the optimal value of the objective as the constraint right-hand sides change, or

$$\mathbf{y}^* = \frac{\partial F(\mathbf{x}^*)}{\partial \mathbf{b}} \tag{5.7}$$

5.3 LINEAR PROGRAMMING

Linear programming models are optimization models that have a linear constraint set and a linear objective function. The general linear programming (LP) problem can be expressed as

$$\min_{\mathbf{x}} z = \mathbf{c}\,\mathbf{x} \tag{5.8}$$

subject to

$$A\,\mathbf{x} \leq \mathbf{b}, \quad \mathbf{x} \geq 0$$

where $\mathbf{c}$ is a $1 \times n$ cost coefficient, row vector, $\mathbf{x}$ is a $n \times 1$ column vector of decision variables, $\mathbf{b}$ is a $m \times 1$ right-hand side vector, and A is the $m \times n$ coefficient matrix defining the constraints of the optimization problem. Provided that the feasible region is nonempty and bounded, the Weierstrass and local-global theorems ensure that any local maximum to the LP problem is the global solution.

Associated with every LP problem is a dual programming problem. The dual problem of the model presented in Equation 5.8 can be expressed as

$$\max_{y} w = \mathbf{y}\,\mathbf{b} \tag{5.9}$$

subject to

$$\mathbf{y}\,A \geq \mathbf{c}, \quad \mathbf{y} \geq 0$$

where $\mathbf{y}$ are the dual decision variables (a $1 \times m$ row vector). Necessary and sufficiency conditions for the existence of solutions to the LP problem are based on the duality theorems of linear programming. If the feasible regions of both the primal and dual problems are nonempty, solutions exist for both problems. The strong duality theorem of linear programming prescribes the necessary and sufficient conditions for any feasible decision vector to represent the solution to a LP problem. The theorem requires that there exist feasible solutions for both primal and dual problems such that the values of the primal and dual objectives are equal, or

$$\min z = \max w \tag{5.10}$$

$$\mathbf{c}\,\mathbf{x}^* = \mathbf{y}^*\,\mathbf{b}$$

The dual variables, $\mathbf{y}^*$, are related to the primal shadow prices, $\partial z/\partial \mathbf{b}$, at the optimum, $\partial z/\partial b_i = y_i^*, i = 1, \ldots, m$.

The necessary and sufficiency conditions for feasible vectors, $\mathbf{x}^*$, $\mathbf{y}^*$, to solve the primal-dual problems are that the vectors satisfy the complementary slackness conditions, or

$$(\mathbf{c} - \mathbf{y}^*\mathrm{A})\,\mathbf{x}^* = 0 \tag{5.11a}$$

$$\mathbf{y}^*\,(\mathrm{A}\,\mathbf{x}^* - \mathbf{b}) = 0 \tag{5.11b}$$

The interpretation of these conditions is that if a constraint, at optimality, is a strict inequality, the corresponding dual variable is zero, i.e., if a resource is not limiting at the optimal solution, we will pay nothing for additional units of that resource. Conversely, if a decision variable is nonzero at the optimal solution, the corresponding inequality constraint in the dual problem is satisfied as a strict equality.

Linear programming problems are solved using the simplex algorithm, an algebraic iterative method. The algorithm begins from a vertex or corner point of the feasible region and moves to a neighboring vertex in a direction that decreases the objective function. The algorithm continues from vertex to vertex until a corner point is reached, where the objective function increases by a move to a neighboring vertex. The corner point solution is a global solution. If the movement to a neighboring vertex increases the objective, the solution is unique. If, however, a move to an adjacent vertex does not increase the objective, then the solution is obviously nonunique and all such vertices (and intermediate points) are solutions. Because there are a finite number of corner point solutions (an upper bound on the number of solutions is $(m + n)!/(m!n!)$, the simplex algorithm will identify the optimal solution or determine if the objective function is unbounded, or the problem infeasible, in a finite number of iterations. Commercially available computer codes, such as IBM's MPSX and CDC's APEX, are capable of solving LP problem with several thousand constraints and decision variables.

5.4 STOCHASTIC LINEAR PROGRAMMING

Groundwater management problems are often characterized by random variables. The water demand, groundwater recharge, or transmissive and storage parameters may be random variables or processes. LP methods have been extended to deal with the problem of optimizing, in some sense, an objective function, subject to constraints that include random variables. The stochastic LP problem can be expressed as

$$z = \min_{\mathbf{x}} (\mathbf{c}\,\mathbf{x} + E[\min_{\mathbf{y}} (\mathbf{g}\,\mathbf{y})]) \tag{5.12}$$

subject to

$$\mathrm{A}\mathbf{x} = \mathbf{b} \tag{5.13}$$

$$T\mathbf{x} + W\mathbf{y} = \mathbf{p} \tag{5.14}$$

$$\mathbf{x} \geqslant 0,\ \mathbf{y} \geqslant 0 \tag{5.15}$$

where $E\{\cdot\}$ is the expectation operator and

A is a $m \times n$ matrix
T is a $m \times n$ matrix
W is a $m \times n$ matrix
b, **c**, and **g** are $m \times 1$, $1 \times n$, and $1 \times n$ vectors
x and **y** are $n \times 1$ and $n \times 1$ vectors of decision variables
p is an $m \times 1$ random vector

A computational procedure for the solution of the model was presented by Dantzig (1955). In the two-stage LP model, the activity levels are determined in the first stage, followed by a corrective action in the second stage. In the first problem, a decision **x** (feasible) is chosen to satisfy

$$\begin{aligned} \mathrm{A}\mathbf{x} &= \mathbf{b} \\ \mathbf{x} &\in K \end{aligned} \tag{5.16}$$

where K is the set of **x** vectors such that there exists at least one **y** satisfying Equation 5.14 (whenever **p** is realized as a certainty). This will generate a **y** vector and cost **g y** whenever the random event **p** is observed with certainty. In the second stage, a recourse action **y** is found from solution of the model

$$\min_{\mathbf{y}} \mathbf{g}\,\mathbf{y} \tag{5.17}$$

subject to

$$W\mathbf{y} = \mathbf{p} - T\,\mathbf{x} \tag{5.18}$$

$$\mathbf{y} \geqslant 0$$

The optimium solution of Equations 5.17 and 5.18 is a function of **p** and **x**. Defining

$$\phi(\mathbf{x}, \mathbf{p}) = \min_{\mathbf{y}} \{\mathbf{gy}\,|W\mathbf{y} = \mathbf{P} - T\mathbf{x}, \mathbf{y} \geqslant 0\}$$

then $E(\phi(\mathbf{x}, \mathbf{p}))$ can be interpreted as the arithmetic mean of the optimum solutions. If $\mathbf{p}$ has a discrete distribution, we can obtain the first-stage and the second-stage optimum solutions simultaneously (Dantzig & Madansky, 1961).

Stochastic programming with recourse is conceptually straightforward but presents some difficulties in applications. Complete analysis by a stochastic programming with recourse model requires that the consequences of recourse actions be modeled and computed for all possible realizations of the random variables. Recourse actions are evaluated by an adequate estimation of losses resulting from random variation, which is difficult, if not impossible. And, it is computationally expensive.

Chance-constrained programming models are probabilistic models that incorporate randomness in the system constraints. In these models, the problem of estimating the loss function may be partially alleviated, even if the establishment of probability levels for constraints is a difficult task (Hogan, et al., 1981). The chance-constrained programming model may be expressed as

$$\min_{\mathbf{x}} z = \mathbf{c}\,\mathbf{x} \tag{5.19}$$

subject to

$$A\mathbf{x} = \mathbf{b} \tag{5.20}$$

$$\Pr\{T\mathbf{x} \geqslant \mathbf{p}\} \geqslant \alpha \tag{5.21}$$

$$\mathbf{x} \geqslant 0 \tag{5.22}$$

where $\Pr\{\cdot\}$ denotes the probability, α is an $m \times 1$ constant vector with components α_i $(0 \leqslant \alpha_i \leqslant 1)$ and $\mathbf{c}$, $\mathbf{x}$, A, $\mathbf{b}$, T, and $\mathbf{p}$ are defined as in Equations 5.12 through 5.15.

The deterministic equivalent of the probabilistic constraints can be developed by considering the constraint represented by Equation 5.21. Assuming that α is a scalar, $(0 \leqslant \alpha \leqslant 1)$, and that the probability distribution function of the random variable p is known, the probabilisitc constraint can be converted to a deterministic equivalence by using the cumulative probability distribution function of the random variable, p, F_p.

Since

$$\Pr\{p \leqslant T\mathbf{x}\} \geqslant \alpha \tag{5.23a}$$

and

$$F_p\,(T\mathbf{x}) \geqslant \alpha \tag{5.23b}$$

the resulting deterministic equation is

$$T\mathbf{x} \geq F_p^{-1}(\alpha) \tag{5.24}$$

where $F_p^{-1}(\alpha)$ is the inverse of the cumulative probability function with the given value of α. If α is chosen to be 0.9, then there will be a 0.1 or 10 percent probability that the probabilistic constraints will not be met.

A mathematical equivalence between the stochastic programming with recourse and chance-constrained programming can only be established for certain limited cases with simple linear recourse (Gartska, 1974). The mathematical equivalence means that for a given ($\mathbf{g}$ $\mathbf{y}$) and W of the stochastic programming with recourse as represented by Equations 5.12, 5.13, 5.14, and 5.15, it is possible to find an α for the chance-constrained programming problem. The solution of the chance-constrained programming model with this α is also, however, optimal for the stochastic programming with recourse. Conversely, for the chance-constrained programming problem and some recourse matrix W, it is possible to find ($\mathbf{g}$ $\mathbf{y}$) for the stochastic programming with recourse that has the same optimal solution as the corresponding chance-constrained programming. However, the equivalence seems purely mathematical and the penalty structure assumed by the chance-constrained programming is not equivalent to the stochastic programming with recourse. Chance-constrained formulations of stochastic programming neither penalize explicitly the constraint violations nor provide recourse action to correct the constraint violations. Hogan, et al., (1981) warn that the practical usefulness of the chance-constrained programming as a modeling technique is seriously limited and it should not be regarded as a substitution for the stochastic programming with recourse.

5.5 QUADRATIC PROGRAMMING

A quadratic programming (QP) problem is a nonlinear programming problem characterized by linear constraints and an objective function that is the sum of a linear and a quadratic form. The model can be expressed as:

$$\max_{\mathbf{x}} z = \mathbf{c}^T\mathbf{x} + 1/2\ \mathbf{x}^T D\mathbf{x} \tag{5.25}$$

subject to

$$A\mathbf{x} \leq \mathbf{b} \tag{5.26}$$

$$\mathbf{x} \geq 0$$

where

$$\begin{aligned}
\mathbf{x} &= n\text{-dimensional column vector} \\
A &= (m \times n) \text{ matrix} \\
\mathbf{b} &= m\text{-dimensional column vector} \\
\mathbf{c} &= n\text{-dimensional column vector} \\
D &= (n \times n) \text{ negative semidefinite symmetric matrix}
\end{aligned}$$

Several algorithms have been developed for the solution of the quadratic programming problem. The methods include Lemke's dual programming algorithm (Lemke, 1962), active-set null-space methods (Gill, et al., 1984), Beale's algorithm (1959), which is based on classical calculus, and Wolfe's (1959) linear programming approach. We will illustrate the development of Wolfe's algorithms because the approach is predicated on the Kuhn-Tucker conditions of the quadratic programming problem. These conditions can be expressed as

$$\mathbf{c} + \mathbf{x}^T D - \mathbf{y}A \leq 0, \qquad \mathbf{b} - A\mathbf{x} \geq 0 \tag{5.27a}$$

$$(\mathbf{c} + \mathbf{x}^T D - \mathbf{y}A)\mathbf{x} = 0, \qquad \mathbf{y}(\mathbf{b} - A\mathbf{x}) = 0 \tag{5.27b}$$

$$\mathbf{x} \geq 0, \qquad \mathbf{y} \geq 0 \tag{5.27c}$$

Introducing the slack variables, $\mathbf{v} > 0$, in Equations 5.27, the optimality conditions can be expressed as

$$\mathbf{c} + \mathbf{x}^T D - \mathbf{y}A + \mathbf{y} = 0, \qquad \mathbf{b} - A\mathbf{x} \geq 0 \tag{5.28a}$$

$$\mathbf{v}\,\mathbf{x} = 0, \qquad \mathbf{y}(\mathbf{b} - A\mathbf{x}) = 0 \tag{5.28b}$$

$$\mathbf{x}, \mathbf{v} \geq 0, \qquad \mathbf{y} \geq 0 \tag{5.28c}$$

The solution of the quadratic programming problem is reduced to identifying $\mathbf{x}^*$, $\mathbf{y}^*$, and $\mathbf{v}^*$ such that Equations 5.28 are satisfied. If it were not for the complementary slackness relations, the solution of the programming problem would be a simple matter. Feasible solutions to the Kuhn-Tucker conditions could be found, for example, using a Phase I linear programming algorithm. Fortunately, the nonlinear constraints are of a special type and the simplex algorithm with restricted basis entry (ensuring that the condition $\mathbf{xv} = 0$ is satisfied) can be used to identify feasible points that satisfy the Kuhn-Tucker conditions. Provided the coefficient matrix, D, is negative definite, the algorithm generates a sequence of feasible solutions that converge to the optimal solution of the quadratic program within a finite number of iterations.

Recent advances in mathematical programming have produced efficient algorithms to solve quadratic programs with a general Hessian matrix (positive and negative eigenvalues). Algorithms have been presented by Gill, et al. (1981) and

Fletcher (1981). An application of the algorithm to water resources management is given by Mariño and Loaiciga (1985).

5.6 DYNAMIC PROGRAMMING

Dynamic programming (DP) is used extensively in the optimization of groundwater resource systems (Buras, 1966). The popularity and success of this technique can be attributed in part to its efficiency in incorporating nonlinear constraints and objectives and stochastic or random variables in the DP formulation of the management or planning problem. DP effectively decomposes highly complex problems with a large number of decision variables into a series of subproblems that can be solved recursively. Dynamic programming, a method formulated largely by Richard Bellman (1957) is a procedure for optimizing multistage decision processes. Figure 5.1 illustrates a typical multistage optimization problem.

In the DP formulation of the planning or operational model, the optimization model is described or characterized by the state variables of the system, the system stages, and the control or decision variables. In groundwater systems, the state variables may represent the hydraulic head or mass concentrations, or the amount of groundwater in storage at any time. The decision variables control, in any time period, the pumping, recharge, or waste injection schedules. The time element of the planning problem defines the stages of the DP model.

An important element of any DP model is the state transition equations of the system. The response or transfer equations, which can be developed using finite-difference or finite-element methods, define how the system state variables change over successive stages or time periods. Defining $\mathbf{x}_n$ as the vector of state variables at stage n, the transition functions can be expressed as

$$\mathbf{x}_{n-1} = \mathbf{T}_n(\mathbf{x}_n, \mathbf{d}_n) \tag{5.29}$$

where $\mathbf{d}_n$ is the vector of decision variables at stage n, and $\mathbf{T}_n$ defines the system transition equation.

Dynamic programming models are based on the principle of optimality (Bellman, 1957). According to the principle, an optimal policy has the property that, whatever the initial state and decisions are, the remaining decisions must constitute an optimal policy with regard to the state resulting from the first decisions. To illustrate the application of the principle, define $f_n(\mathbf{x}_n)$ is the optimal return with n stages remaining given that the system is in state $\mathbf{x}_n$. Assuming the objective function is a separable and monotonic function of the decisions at any stage, the dynamic programming recursive equation can be expressed as

$$\begin{aligned} f_n(\mathbf{x}_n) &= \max_{\mathbf{d}_n}\{r_n(\mathbf{x}_n, \mathbf{d}_n) + f_{n-1}(\mathbf{x}_{n-1})\} \\ \mathbf{x}_{n-1} &= \mathbf{T}_n(\mathbf{x}_n \mathbf{d}_n) \end{aligned} \tag{5.30}$$

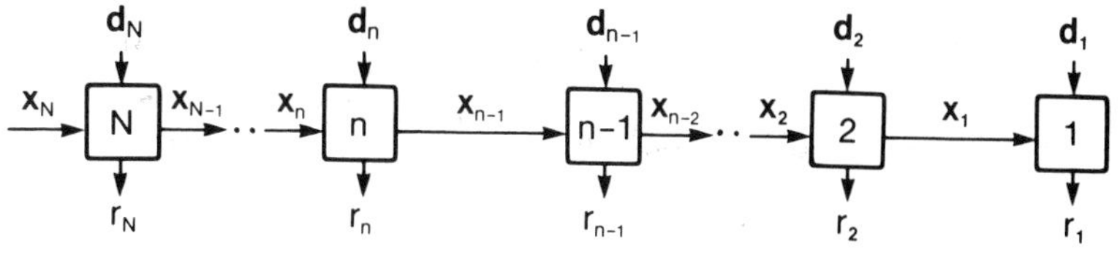

FIGURE 5.1 Serially connected system

where $r_n(\mathbf{x}_n, \mathbf{d}_n)$ is the return for stage n, the system transition function, $\mathbf{T}_n$ is again given by the response equations, and $f_n(\mathbf{x}_n)$ is known for all initial states.

In groundwater management problems, DP is particularly useful for nonconvex and stochastic, operational or planning problems because the algorithm can identify global solutions (within the discretization error). Also, in contrast to classical optimization methods, constraints can actually reduce the computational resources that are needed to solve the model. The major limitation of the approach in groundwater management problems remains the number of state variables that can be incorporated into the recursive equations. Since computer storage increases exponentially with the number of state variables, dynamic programming is most useful in planning studies where the groundwater system can be represented as a lumped parameter, homogeneous system. Design objectives and variables are also difficult to incorporate into dynamic programs because of the separability restrictions of the model.

As a simple example of a DP model, consider the groundwater system shown in Figure 5.2. The aquifer is homogeneous and isotropic and is a linear system. The stage-to-stage transformation is then described by the continuity equation, which for a lumped parameter, linear system can be expressed as

$$S_{n-1} = S_n + R_n - P_n - E_n \tag{5.31}$$

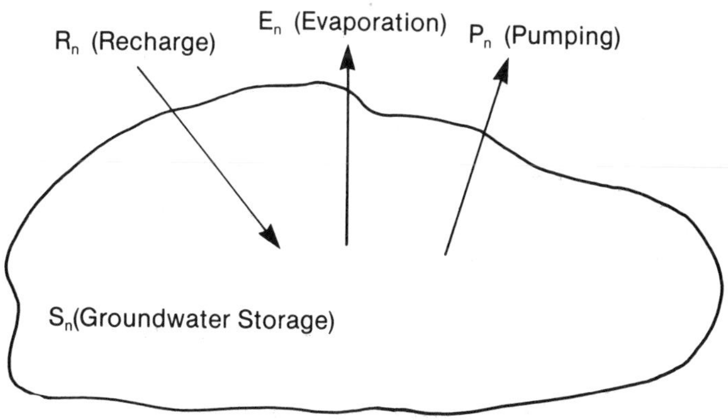

FIGURE 5.2 Example groundwater system

where S_n is the storage at the beginning of time period or stage n, P_n is the groundwater extraction during stage n, R_n is the groundwater recharge, and E_n is the subsurface outflow, leakage, or evaporation that occurs during stage n.

Assuming that the system objective, $r(S, P)$, is a function of the state variable, the groundwater storage, and the pumping occurring in any stage or time period, the DP recursive equation can be expressed as

$$f_n(S_n) = \max_{P_n} \{r(P_n, S_n) + f_{n-1}(S_{n-1})\}, \qquad n = 1, \ldots, n \tag{5.32}$$

where $f_n(S_n)$ denotes the optimal return function for stage n given the system is in state S_n and there are n stages remaining in the optimization. S_n is the given initial groundwater storage. Additional constraints of the model include maximum and minimum storage bounds to limit the groundwater extractions or to prevent possible flooding of the land surface, or

$$S_{\min} \leqslant S_n \leqslant S_{\max}, \ \forall n \tag{5.33}$$

The solution of the recursive equation requires that the state variable, the aquifer storage, be discretized into a number of feasible states as shown in Figure 5.3. For each stage and state variable level, the optimal policy can be determined from the recursive equation. The stage-to-stage transformation equation is given by the continuity equation. Since the solutions are imbedded in the discretized states, the infeasible transitions can be discarded in the solution process. The optimal solution(s) can ultimately be traced throughout the stages to identify the overall optimal extraction policy and groundwater storages. Note that n is numbered backwards with respect to real time.

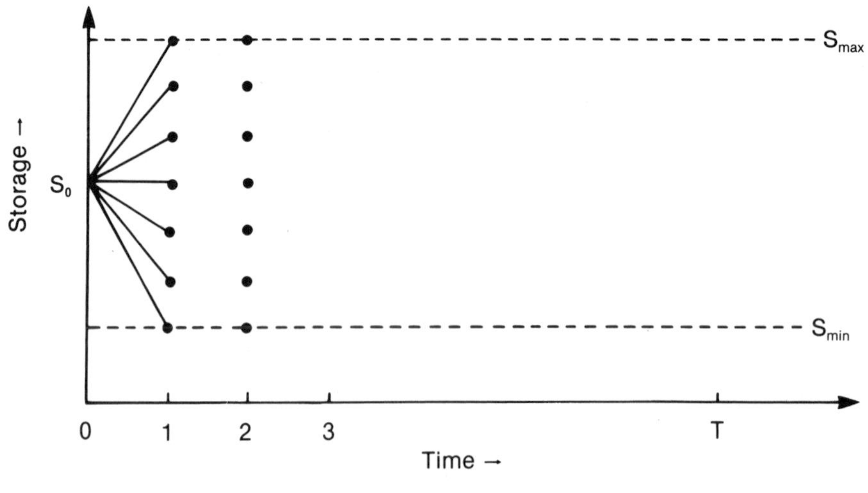

FIGURE 5.3 State space

5.7 STOCHASTIC DYNAMIC PROGRAMMING

An interesting application of DP involves the stochastic, optimal operation of an aquifer system. If, for example, the groundwater recharge is a Markov process (say, first-order), then the state of the system at any stage depends only on the state of the system at the previous stage and on known probabilities (Nemhauser, 1966). If there are a finite number of states at each stage and a finite number of stages, the probability of transition from a state at one stage to a state at the next stage exists. Howard (1960) introduced the concept of returns or rewards, corresponding to the probability transition matrix, into a Markov process. In a multistage Markovian model, the objective is generally to maximize the expected return. Continuing the example presented in Section 5.6, a stochastic DP model of the problem can be expressed as

$$f_n(S_n, R_{n+1}) = \max_{P_n}\left[\sum_{R_n=0}^{R_n=R_{n,\max}} p[R_n|R_{n+1}]\cdot[r(P_n) + f_{n-1}(S_{n-1}, R_n)]\right], \quad (5.34)$$

subject to

$$S_{n-1} = S_n + R_n - P_n - E_n, \quad (5.35)$$

$$f_1(S_1, R_2) = \max_{P_1}\left[\sum_{R_1=0}^{R_1=R_{1,\max}} p[R_1 \mid R_2]\cdot[r(P_1)]\right] \quad (5.36)$$

where $f_n(S_n, R_{n+1})$ = expected return from the optimal operation of the system with n stages remaining in the planning period

S_n = storage at the beginning of time period n

r = the return (or net benefit) function

$p[R_n|R_{n+1}]$ = the transition probabilities relating the recharge R_n in the nth time period or stage with the recharge R_{n+1} in time period $(n + 1)$

E_n = subsurface flow or loss during stage n

n = time period index, e.g., months numbered from the end of planning horizon

The recursive equation is based on the conditional probabilities, $p[R_n|R_{n+1}]$, relating the recharge R_n during the current time period and the recharge during the preceding time period R_{n+1}. The right-hand side of the recursive equation states that, for a given starting storage level S_n at stage n and a given recharge R_{n+1} for time period $n + 1$, and extraction rate "P" can be determined such that the expected value of the sum of the immediate return and future return is maximized. The expected value is taken over all possible recharge values ($R_n = 0$ to $R_n =$

$R_{n,\max}$) during the current stage, n, with conditional probability $p[R_n|R_{n+1}]$. Since the extraction during the current time period is a function of S_n and R_{n+1}, it is possible, with given values of S_n and R_{n+1}, to search over all possible values of "P" and select the pumping policy that maximizes the expected return.

5.8 NONLINEAR PROGRAMMING

Nonlinear programming (NLP) has not enjoyed the popularity that LP has had in groundwater systems analysis. The reasons are essentially computational: the optimization process is usually slow and requires large amounts of computer storage and time when compared with linear programming. The mathematics involved in the formulation and solutions of nonlinear models is much more complicated than in the linear case, and nonlinear programming, unlike DP, cannot easily accommodate stochastic or random variables. However, nonlinear programming can effectively handle nonseparable objective functions and nonlinear constraints which many programming techniques cannot. Furthermore, NLP, such as quadratic programming or separable programs, can be used iteratively as a master program or subprograms in large-scale system problems.

The general nonlinear programming problem can be stated as

$$\min_{\mathbf{x}} z = f(\mathbf{x}) \tag{5.37}$$

subject to

$$\mathbf{g}(\mathbf{x}) \leqslant \mathbf{b}, \quad \mathbf{x} \geqslant 0 \tag{5.38}$$

in which $\mathbf{x}$ is an n-dimensional vector of decision variables, and $f(\mathbf{x})$ and $g(\mathbf{x})$ are real-valued, continuous, and differentiable functions, respectively.

The Kuhn-Tucker conditions of nonlinear programming (Equation 5.5) again characterize the optimal solution of the optimization problem. Without the proper convexity assumptions, however, the Kuhn-Tucker equations are necessary conditions only for the nonlinear programming problem.

The existence of an optimal solution is related to the existence of a saddle point for the problem. A saddle point $(\mathbf{x}, \mathbf{y})$ of the Lagrangian exists if and only if

$$\mathbf{x} \text{ minimizes } L(\mathbf{x}, \mathbf{y}) \tag{5.39a}$$

$$\mathbf{g}(\mathbf{x}) \leqslant \mathbf{b} \tag{5.39b}$$

$$\mathbf{y}(\mathbf{b} - \mathbf{g}(\mathbf{x})) = 0 \tag{5.39c}$$

Furthermore, if $(\mathbf{x}, \mathbf{y})$ is a saddle point for L, then $\mathbf{x}$ solves the primal problem described by Equations 5.37 and 5.38.

Any vector of Lagrange multipliers, for which the minimum of the Lagrangian exists, solves a problem closely related to the primal, nonlinear programming problem. As shown by Everett (1963), if $\mathbf{x}(\mathbf{y})$ solves the Lagrangian problem,

$$\min_{\mathbf{x} \geq 0} L(\mathbf{x}, \mathbf{y}) \tag{5.40}$$

with $\mathbf{y} \geq 0$, then $\mathbf{x}(\mathbf{y})$ solves the modified primal problem,

$$\min z = f(\mathbf{x}) \tag{5.41a}$$

subject to

$$\begin{aligned} g_i(\mathbf{x}) &\leq z_i, \; i = 1, \ldots m, \\ \mathbf{x} &\geq 0 \end{aligned} \tag{5.41b}$$

where

$$z_i = g_i(\mathbf{x}(\mathbf{y})) \quad \text{if } y_i > 0 \tag{5.41c}$$

$$z_i \geq g_i(\mathbf{x}(\mathbf{y})) \quad \text{if } y_i = 0 \tag{5.41d}$$

Moreover, if the Lagrange multipliers have the property, $\mathbf{y}_k^2 > \mathbf{y}_k^1$, $\mathbf{y}_j^2 = \mathbf{y}_j^1$, $j \neq k$ and if $\mathbf{x}(\mathbf{y}^i)$ solves the Lagrangian problem with $\mathbf{y} = \mathbf{y}^i$, then $g_k\{\mathbf{x}(\mathbf{y}^2)\} \leq g_k\{\mathbf{x}(\mathbf{y}^1)\}$.

The dual function of the nonlinear program, described by Equations 5.37 and 5.38, can be expressed as

$$\begin{aligned} \max z &= h(\mathbf{y}) \\ \mathbf{y} &\in D \end{aligned} \tag{5.42a}$$

where the domain, D, and the dual function, $h(\mathbf{y})$, are given as

$$D = \{\mathbf{y} \mid \mathbf{y} \geq 0, \min L(\mathbf{x}, \mathbf{y}) \text{ exists}, \mathbf{x} \geq 0\} \tag{5.42b}$$

$$h(\mathbf{y}) = \min_{\mathbf{x} \geq 0} L(\mathbf{x}, \mathbf{y}) \tag{5.42c}$$

The dual function is concave over any convex subset of D and provides a lower bound on the primal objective.

As in linear programming, the dual programming model can be used to obtain, under certain conditions, the solution of the primal problem. However, the equivalence between the primal and dual problems can only be established if there exists a saddle point for the Lagrangian of the problem. In Chapter 7, we will explore how dual programming algorithms may be used to identify optimal solutions to the conjunctive groundwater supply and quality management problem.

5.9 COMPUTATIONAL METHODS: NONLINEAR PROGRAMMING

5.9.1 Unconstrained Optimization

Computational algorithms in nonlinear programming can be categorized as either unconstrained or constrained methods (Luenberger, 1973). Techniques available for solving the unconstrained problems include the steepest descent (or ascent for maximization) methods, the conjugate direction methods, and quasi-Newton methods. We first consider some common unconstrained optimization algorithms, for solution of the problem.

$$\min_{\mathbf{x}} z = f(\mathbf{x}), \qquad \mathbf{x} \in E^n \tag{5.43}$$

The method of steepest descent is perhaps the best-known unconstrained optimization method. The algorithm, originally developed by Cauchy, is based on the recurrence equation

$$\mathbf{x}_{k+1} = \mathbf{x}_k + \alpha \mathbf{d}_k \tag{5.44}$$

where $\mathbf{x}_k$ is the decision vector at iteration k, $\mathbf{d}_k$ is the direction vector, and α is the step size. The direction vector, or the direction of steepest descent, is related to the gradient vector, or

$$\mathbf{d}_k = -\nabla f(\mathbf{x}_k) \tag{5.45}$$

where ∇f, defined as a column vector, is the gradient of f with respect to $\mathbf{x}$.

The optimal step size, α^*, determined such that $f(\mathbf{x}_{k+1})$ is minimized or

$$\frac{df(\mathbf{x}_k + \alpha \mathbf{d}_k)}{d\alpha} = 0 \tag{5.46}$$

The algorithm can terminate at any type of stationary point ($\nabla f(\mathbf{x}_k) = 0$). However, it is usually necessary to determine if the point is a local minimum or a saddle point. If, for example, the Hessian matrix is positive definite at the point, the solution is a local minimum. If not, nongradient methods may be used to move away from the point, after which the minimization may be continued (see, for example, Himmelblau, 1972).

The Fletcher-Reeves conjugate gradient method is an important unconstrained optimization method. The technique generates a sequence of search directions that are linear combinations of $-\nabla f(\mathbf{x}_k)$, $k = 0, 1, \ldots K$, and weighting factors that are chosen such that the search directions are conjugate. The major elements of the algorithm are:

1. Given an initial trial solution $\mathbf{x}_0$, compute the gradient vector, $\nabla f(\mathbf{x}_0)$, and set the direction vector, $\mathbf{d}_0 = -\nabla f(\mathbf{x}_0)$.
2. Determine the optimum step size, α_k, by minimizing $f(\mathbf{x}_k + \alpha\, \mathbf{d}_k)$ using unidimensional search techniques (e.g., Fibonnaci or Golden Section methods).
3. Set $\mathbf{x}_{k+1} = \mathbf{x}_k + \alpha_k\, d_k$.
4. Evaluate $f(\mathbf{x}_{k+1})$ and $\nabla f(\mathbf{x}_{k+1})$.
5. Determine the new direction vector, $\mathbf{d}_{k+1}$,

$$\mathbf{d}_{k+1} = -\nabla f(\mathbf{x}_{k+1}) + \beta_k \mathbf{d}_k \tag{5.47}$$

where

$$\beta_k = \frac{\nabla f^T(\mathbf{x}_{k+1})\, \nabla f(\mathbf{x}_{k+1})}{\nabla f^T(\mathbf{x}_k)\, \nabla f(\mathbf{x}_k)} \tag{5.48}$$

6. Terminate the algorithm when $\|\mathbf{d}_k\| < \epsilon$, where ϵ is a small constant.

The Fletcher-Powell method is one of the most powerful quasi-Newton or variable metric methods. The technique approximates the Hessian matrix or its inverse using information obtained from the first derivatives of the objective function. The algorithm can be summarized as:

1. Begin with any symmetric positive definite matrix, H_0, and decision vector, $\mathbf{x}_0$.
2. Set $\mathbf{d}_k = -H_k\, \nabla f(\mathbf{x}_k)$.
3. Choose α_k to minimize $f(\mathbf{x}_k + \alpha_k\, \mathbf{d}_k)$ to obtain $\mathbf{x}_{k+1}$, $\mathbf{p}_k = \alpha_k \mathbf{d}_k$, and $\mathbf{q}_k = \nabla f(\mathbf{x}_{k+1}) - \nabla f(\mathbf{x}_k)$.
4. Update the H_{k+1} matrix using the relation:

$$H_{k+1} = H_k + \frac{\mathbf{p}_k \mathbf{p}_k^T}{\mathbf{p}_k^T \mathbf{q}_k} - \frac{H_k \mathbf{q}_k \mathbf{q}_k^T \mathbf{H}_k}{\mathbf{q}_k^T \mathbf{H}_k \mathbf{q}_k} \tag{5.49}$$

Fletcher and Powell (1963) recommend that the optimization be terminated if, after evaluating the search direction or $\alpha_k \mathbf{d}_k$, either of the following occur:

1. Every component of $\mathbf{d}_k$ or $\alpha_k\mathbf{d}_k$ is less than a prescribed value.
2. The predicted length of each of the vectors from the minimum is less than a prescribed value.

We also note that there are a number of unconstrained optimization methods that do not require derivatives. These include Hooke and Jeeves' (1961) pattern search methods, Rosenbrock's method (1960), and Powell's method (1964).

5.9.2 Constrained Optimization Methods

Many groundwater management problems can be represented as the nonlinear programming problem

$$\min z = f(\mathbf{x}) \tag{5.50a}$$

subject to

$$\begin{aligned} \mathbf{g}(\mathbf{x}) \leq \mathbf{b},\ \mathbf{l} \leq \mathbf{x} \leq \mathbf{u} \\ \mathbf{h}(\mathbf{x}) = 0 \end{aligned} \tag{5.50b}$$

where the constraints define the hydraulic or water quality response equations of the aquifer system and water demand requirements. The $\mathbf{b}$ vector contains the boundary conditions and recharge occurring in the groundwater system. The upper and lower bounds constrain the head variations and the magnitude of pumping or artificial recharge. The objective function of the model could reflect the cost or loss in not meeting water targets or the overall net economic benefit. We consider several algorthims for the solution of this class of programming problems. Table 5.1 summarizes the more important constrained optimization algorithm.

Separable convex programming problems can be used when the objective function, $f(\mathbf{x})$, and the constraints, $\mathbf{g}(\mathbf{x})$, are separable, convex functions of the decision variables. A typical program can be expressed as

$$\min_{\mathbf{x}} z = \Sigma f_j(x_j) \tag{5.51a}$$

subject to

$$\begin{aligned} \sum_j g_{ij}(x_j) \leq b_i, \quad i = 1, \ldots m \\ \sum h_{ij}(x_j) = 0, \quad j = 1, \ldots p \\ \mathbf{l} \leq \mathbf{x} \leq \mathbf{u} \end{aligned} \tag{5.51b}$$

TABLE 5.1 Solution Methods for Nonlinear Programming

Solution Methods	*Algorithms*	*Advantages*
Primal Methods	Feasible Direction Gradient Projection Reduced Gradient Projected Lagrangian Methods	Problems with Linear Constraints
Penalty and Barrier Methods	SUMT	Problems with Nonlinear Constraints
Dual Methods	Gradient Algorithm Cutting Plane Method	Convex/Separable Problems
Quasilinearization	Linear, Quadratic Programming	Nonlinear Constraints

Approximate solutions to the separable programming problems can be achieved by replacing each constraint and the objective function by piecewise linear approximations. The resulting optimization model can be solved using linear programming for convex programs and mixed-integer programming for nonconvex programs. The algorithms can identify globally optimal solutions to the programming model (Hadley, 1964).

Penalty function methods are techniques that transform constrained optimization problems into unconstrained optimization problems, or a sequence of unconstrained optimization problems. Much simpler algorithms can then be used for the minimization process. Parametric penalty function methods can be classified as either interior or mixed methods. Exterior methods approximate the solution of the programming problem by solving a sequence of unconstrained optimization problems whose optimal solution approaches the solution of Equation 5.50 from outside the feasible region. The subproblems may be expressed as

$$\min \rho\,(\mathbf{x}, \rho^k) = f(\mathbf{x}) + \frac{1}{\rho^k}\left\{\sum_{i=1}^{m} |\min(0, g_i(\mathbf{x}))|^{\alpha} + \sum_{j=1}^{p} |h_j(\mathbf{x})|^{\beta}\right\} \quad (5.52)$$

where ρ^k is a strictly decreasing sequence of positive numbers and α and β are given constants (usually equal to 1 or 2). The sequence of optimal solutions can be shown to converge to an optimal solution of the original problem provided the feasible region is nonempty and that each subproblem attains its unconstrained minimum in E^n (Avriel, 1976).

Interior or barrier penalty function methods again solve a sequence of unconstrained optimization problems that satisfy the constraint equations. Barrier methods require that the feasible region be nonempty; equality constraints cannot be handled directly by the procedure. The interior penalty function method minimizes the sequence of unconstrained optimization problems given by

$$\min G(\mathbf{x}, \rho^k) = f(\mathbf{x}) + t(\rho^k)q(\mathbf{x}) \tag{5.53}$$

q is a real-valued, continuous function on E^n such that if the sequence of points $\{\mathbf{x}^k\}$ converges to a boundary point, $\mathbf{x}'$, then

$$\lim_{k\to\infty} \{q(\mathbf{x}^k)\} = \infty \tag{5.54a}$$

t is also a real-valued function of $\rho \in R$ such that

$$\rho^1 > \rho^2 \Rightarrow t(\rho^1) > t(\rho^2) > 0 \tag{5.54b}$$

and,

$$\lim_{k\to\infty} \{t(\rho^k)\} = 0 \tag{5.54c}$$

q and t are given functions of the inequality constraint and the penalty parameters. Some common choices for these functions are

$$t_1(\rho) = \rho, \quad t_2(\rho) = (\rho)^2$$

$$q_1(\mathbf{x}) = -\sum_i \log g_i(\mathbf{x}), \quad q_2(\mathbf{x}) = \sum_i \frac{1}{g_i(\mathbf{x})} \tag{5.55}$$

$$q_3(\mathbf{x}) = \sum_i \frac{1}{(g_i(\mathbf{x}))^2}$$

Mixed penalty function methods can be used to solve equality and inequality constrained optimization problems. The method consists of solving a sequence of unconstrained optimization problems described by

$$H(\mathbf{x}, \rho^k, \eta^k) = f(\mathbf{x}) + t(\rho^k)q(\mathbf{x}) + r(\eta^k)s(\mathbf{x}) \tag{5.56}$$

where $s(\mathbf{x}) = \sum_{j=1}^{p} |h_j(\mathbf{x})|^p$, and t, q, r, and s functions are given for the inequality and equality constraints associated with the pure exterior and interior methods $q(x)$ and $s(x)$ are applied only to the inequality and equality constraints, respectively. Under suitable conditions, H converges to $f(\mathbf{x}^*)$ and $\mathbf{x}^{*k}$ converges to a local solution of the optimization model.

The sequential unconstrained minimization technique (SUMT) is perhaps the best-known of the penalty function methods (Fiacco & McCormick, 1968). In this formulation, the penalty function of the programming model (Equation 5.50) can be expressed as (Yeh, 1985)

$$V(\mathbf{x}, r, t) = f(\mathbf{x}) + r \sum_{j=1}^{n} \left\{ -\frac{1}{x_j - u_j} - \frac{1}{l_j - x_j} \right\} + t \sum_{j=1}^{m} [g_j(\mathbf{x})]^2 \quad (5.57)$$

The algorithm can be summarized as:

1. Determine an initial estimate $\mathbf{x}_0$ of the solution.
2. Choose $r > 0$ and $t > 0$ and obtain an unconstrained minimum $\mathbf{x}_1$ of the augmented objective function.
3. Continue with $k = 2, \ldots$ by choosing $r_k < r_{k-1}$ and $t_k > t_{k-1}$ and starting from $\mathbf{x}_{k-1}$, identify the unconstrained minimum $\mathbf{x}_k$ of $V(\mathbf{x}; r, t)$.
4. As $r_k \rightarrow 0$ and $t_k \rightarrow \infty$, if $\|\mathbf{x}_k - \mathbf{x}_{k-1}\|$ and $|f(\mathbf{x}_k) - f(\mathbf{x}_{k-1})|$ are sufficiently small, terminate the process and $\mathbf{x}_k$ is the local solution to the problem.

Conjugate gradient methods or steepest descent methods may be used for the solution of the unconstrained penalty function in step 3 of the algorithm.

In confined or leaky aquifer systems, the response equations are linear functions of the decision variables and the feasible region is a compact, convex set. Optimization models of the groundwater system can be described by the programming problem

$$\min_{\mathbf{x}} z = f(\mathbf{x}) \quad (5.58a)$$

subject to

$$\mathbf{A}\mathbf{x} \leq \mathbf{b}, \; \mathbf{l} \leq \mathbf{x} \leq \mathbf{u} \quad (5.58b)$$

The solution of these types of problems is generally approached using primal methods of nonlinear programming. We first consider a typical example of one of these methods, Rosen's (1960) gradient projection algorithm. The algorithm is motivated by the desire to implement the feasible direction method, while not requiring the solution of a linear program at each iteration. The central idea of the algorithm is that at a feasible point $\mathbf{x}_k$, the active constraints $\mathbf{A}_1^T \mathbf{x}_k = \mathbf{b}_1$ ($\mathbf{A}_2^T \mathbf{x}_k \leq \mathbf{b}_2$) are determined. The negative gradient of the constraint is then projected onto the subspace tangent to the surface determined by these constraints. (A feasible solution to the constraints can be found by application of the Phase I procedure of linear programming.) Assuming that the lower and upper bounds have been included into the linear constraints, $\mathbf{A}\mathbf{x} = \mathbf{b}$, the algorithm can be summarized as

1. Given a feasible point, $\mathbf{x}_k$, determine the projection matrix. If $\mathbf{A}_1$ is vacuous, let $P = I$; otherwise, let the projection matrix $P = I - \mathbf{A}_1^T(\mathbf{A}_1 \mathbf{A}^T)^{-1} \mathbf{A}_1$ and the direction vector, $\mathbf{d}_k = -P \nabla f(\mathbf{x}_k)$.

2. If $\mathbf{d}_k \neq 0$, let α_k be an optimal solution to the following line search problem:

$$\min_{0 \leq \alpha \leq \alpha_{\min}} \quad z = f(\mathbf{x}_k + \alpha \mathbf{d}_k)$$

where $\alpha_{\max} = \max(\alpha\colon \mathbf{x}_k + \alpha \mathbf{d}_k$ is feasible). Update the decomposition and return to step 1.

3. If $d_k = 0$, stop if A_1 is vacuous; otherwise, let $\beta = -(A_1 A_1^T)^{-1} A_1 \nabla f(\mathbf{x}_k)$. If $\beta \leq 0$, stop; $\mathbf{x}_k$ is a Kuhn-Tucker point. Otherwise, delete the row from A_1 with the most negative component of β and return to step 1.

Since the set of active constraints changes by at most one constraint at a time, it is possible to calculate one required projection matrix from the previous one by an updating procedure.

The third class of techniques commonly used in groundwater management and parameter estimation problems is the generalized Newton-Raphson method allied with conventional optimization methods. The optimization algorithm, as presented by Rosen (1966) and Meyer (1970), involves the following steps:

***Step.** 1.* Linearize the constraining equation in function space around an initial feasible decision vector, $\mathbf{x}^n$. The decision vector includes both the state variables of the groundwater system and the pumping or waste injection schedules. The constraints of the optimization problem can then be linearized with a generalized Taylor series as

$$g(\mathbf{x}^{n+1}) \cong g(\mathbf{x}^n) + J_n(\mathbf{x}^n)(\mathbf{x}^{n+1} - \mathbf{x}^n),$$

$$\mathbf{h}(\mathbf{x}^{n+1}) \cong \mathbf{h}(\mathbf{x}^n) + \hat{J}_n(\mathbf{x}^n)(\mathbf{x}^{n+1} - \mathbf{x}^n)$$

where J and $\hat{J}$ are the Jacobian matrixes of the approximation.

***Step** 2.* Incorporate the linearized equation in the optimization model and solve the model using convex programming techniques.

***Step** 3.* Check for convergence of the algorithm within the desired accuracy by examining the relative change in the decision variables over successive iterations, or

$$|\mathbf{x}^{n+1} - \mathbf{x}^n| < \delta$$

where δ is a prescribed small value.

If the algorithm has converged, stop; if not, return to step 1.

As is the case with most nonlinear programming methods, convexity assumptions are needed to prove the global convergence of the algorithm. However, in many cases where these assumptions are not met, the algorithm will converge at least to a local optimum of the objective, rather than an arbitrary stationary point.

Reduced gradient methods are also used for solving optimization problems with linear constraints described by Equation 5.58 where the functions are again continuously differentiable. The method is based on algorithms presented by Wolfe (1959) and the convex-simplex method of Zangwill (1969).

Initially, the matrix A is partitioned into a basis and nonbasis matrix, $A = (B, C)$, where we assume without loss of generality that the first m columns of A correspond to the basic variables $\mathbf{x}_B$. The constraints can then be expressed

$$B\mathbf{x}_B + C\mathbf{x}_N = \mathbf{b} \tag{5.59a}$$

or

$$\mathbf{x}_B = B^{-1}\mathbf{b} - B^{-1}C\mathbf{x}_N \tag{5.59b}$$

where $\mathbf{x}_N$ are the nonbasic variables and we assume that the basic variables are nondegenerate. In the reduced gradient method, $\mathbf{x}_B$ is eliminated from the model; the objective function can then be expressed solely in terms of the nonbasic variables, $\mathbf{x}_N$. The reduced gradient, $\mathbf{r} \in E^{n-m}$, the gradient of the nonbasic variables, can then be expressed as

$$\mathbf{r}(\mathbf{x}_N) = \nabla_{\mathbf{x}N} f(\mathbf{x}_B(\mathbf{x}_N), \mathbf{x}_N) - B^{-1} C \, \nabla_{\mathbf{x}B} f(\mathbf{x}_B(\mathbf{x}_N), \mathbf{x}_N) \tag{5.60}$$

If a small move from the current value of $\mathbf{x}_N$ in the direction of the negative reduced gradient can be made without violating the nonnegativity restrictions, f would decrease. In other words, given a feasible $\mathbf{x}^k$, compute for $i = 1, \ldots n - m$ and define

$$z_i^{N,k+1} = \begin{bmatrix} 0 & \text{if } x_{N,i}^k = 0, \; r_i(\mathbf{x}_N^k) > 0 \\ -r_i(\mathbf{x}_N^k) & \text{otherwise} \end{bmatrix} \tag{5.61}$$

and define $\mathbf{z}^{B,k+1} = -B^{-1} C\mathbf{z}^{N,k+1}$ and $\mathbf{z}^{k+1} = (\mathbf{z}^{N,k+1}, \mathbf{z}^{B,k+1})^T$. The next feasible point is then given by

$$\mathbf{x}^{k+1} = \mathbf{x}^k + \alpha_{k+1}^* \, \mathbf{z}^{k+1} \tag{5.62}$$

where α_{k+1}^* is given from the equations,

$$\alpha_{k+1}^1 = \max \{\alpha_{k+1} : \mathbf{x}_B^k + \alpha_{k+1} \, \mathbf{z}^{B,k+1} \geqslant 0\} \tag{5.63a}$$

and

$$\alpha_{k+1}^2 = \max\{\alpha_{k+1} : \mathbf{x}_N^k + \alpha_{k+1}\, \mathbf{z}^{N,k+1} \geqslant 0\} \tag{5.63b}$$

and

$$f(\mathbf{x}^k + \alpha_{k+1}^* \mathbf{z}^{k+1}) = \min_{\mathbf{x}^{k+1}}\{f(\mathbf{x}^k + \alpha_{k+1}\mathbf{z}^{k+1}) : 0 \leqslant \alpha_{k+1} \leqslant \min(\alpha_{k+1}^1, \alpha_{k+1}^2)\} \tag{5.64}$$

If $\alpha_{k+1}^* < \alpha_{k+1}^1$ then $\mathbf{x}^{k+1}$ is given by Equation 5.62. Otherwise,

$$x_{Bl}^k + \alpha_{k+1}^l\, z_l^{B,k+1} = 0 \tag{5.65}$$

for some l and x_{Bl} is dropped from the vector of basic variables in exchange for the largest positive nonbasic variable. The algorithm terminates if $|\mathbf{z}^{k+1}| < \varepsilon$, where ε is the convergence parameter of the model.

Recently, projected Lagrangian methods have been used in the optimization of nonlinear groundwater management problems (see Noel & Howitt, 1982; Gorelick, et al., 1983). These algorithms recognize that the optimal solution of the programming model is a minimum of the Lagrangian function in the subspace defined by linearization of the active constraints. As a result, given approximations of the Lagrange multipliers, a subproblem can be defined that has only linear constraints. The advantage of the approach is obvious; nonlinear problems with linear constraints are much easier to solve than fully nonlinear problems.

MINOS, a commercially available software package (Murtagh & Saunders, 1981), is designed to solve large-scale nonlinear programming problems. A typical iteration of the algorithm begins with estimates of $\mathbf{x}_k$ and λ_k, the current best approximations of the solution vector and the Lagrange multipliers. At the beginning of each major iteration, the nonlinear constraints are linearized about $\mathbf{x}_k$, or

$$\mathbf{g}(\mathbf{x}) \cong \mathbf{g}(\mathbf{x}_k) + J_k(\mathbf{x}_k)(\mathbf{x} - \mathbf{x}_k)$$

where J_k is the Jacobian matrix evaluated at $\mathbf{x}_k$.

The updated solution, $\mathbf{x}_{k+1}$, is found from the solution of the linearly constrained subproblem,

$$\min z = f(\mathbf{x}) - \boldsymbol{\lambda}_k^T(\mathbf{g} - \mathbf{g}) + 1/2\, \rho(\mathbf{g} - \mathbf{g})^T(\mathbf{g} - \mathbf{g}) \tag{5.66a}$$

subject to

$$\begin{aligned} \tilde{\mathbf{g}}(\mathbf{x}) &\leqslant \mathbf{b} \\ \mathbf{l} \leqslant \mathbf{x} &\leqslant \mathbf{u} \end{aligned} \tag{5.66b}$$

where ρ is a nonnegative scalar penalty parameter.

The objective function, which is the augmented Lagrangian for the problem, contains the approximation of the Langrangian for the current iteration and a modified quadratic penalty function that controls the deviation of the nonlinear constraints from their current linearization. It can be shown, under suitable conditions (Robinson, 1972), that as $\rho \to 0$, the iterates, $\mathbf{x}_k$, converge quadratically, if the Lagrange multiplier vector is taken as $\boldsymbol{\lambda}_{k+1}$ in defining the next subproblem. Broadly speaking, if $\mathbf{x}_k$ is an optimal solution to the kth subproblem, and if it satisfies the nonlinear constraints sufficiently well, then $\mathbf{x}_k$ will probably be an optimal solution to the original nonlinear program. Formally, convergence is assumed to occur if

$\mathbf{x}_k$ is an optimal solution to its subproblem

$\mathbf{x}_k$ satisfies the nonlinear constraints to within a given tolerance

$\boldsymbol{\lambda}_k$ is not substantially different from $\boldsymbol{\lambda}_{k-1}$

$\mathbf{x}_{k+1}$ is an optimal solution to its subproblem, and a basis change does not occur during the solution of the $(k+1)^{\text{st}}$ subproblem.

Reduced gradient methods can be used for solution of the linearized problems.

5.10 MULTI-OBJECTIVE PROGRAMMING

Multi-objective programming models optimize a vector-valued objective function, given the economic, hydraulic, or water quality constraints of the planning problem. The optimal solutions of the model are referred to as the *noninferior solution set*. The central problem in multi-objective programming is to identify these noninferior solutions. Although the noninferior set is infinite, the solutions can be determined by the optimal solutions of a series of scalar objective problems that are equivalent to the multi-objective problem. The optimization process does require, however, information regarding the weighting or preferences associated with each of the system objectives.

Over the past 20 years, various programming methods have been developed for solution of the multi-objective optimization problem. As discussed by Cohon and Marks (1975), Hwang and Yoon (1981), Fandel and Oal (1980), the solution techniques include generating techniques, which completely identify the noninferior set (Gass & Saaty, 1955), and techniques based on preference information articulated by the decision makers prior to the analysis (see, for example, Geoffrion, 1967; Benayoun, et al., 1971: and Monarchi, et al., 1973). Here we consider the two most common multi-objective programming techniques—the weighting and constraint methods.

Formally, the multi-objective programming problem is to maximize a set of

p objectives, or

$$\max z(\mathbf{x}) = [z_1(\mathbf{x}), \ldots, z_p(\mathbf{x})] \tag{5.67}$$

subject to

$$\mathbf{x} \in \overline{\mathbf{X}} \tag{5.68}$$

where Z ($\mathbf{x}$) is the p-dimensional objective function, $\overline{\mathbf{X}}$ is the feasible region, and $\mathbf{x}$ is an n-dimensional vector of decision variables. The noninferior solution set is a subset of the feasible region. Specifically, a noninferior solution is a feasible solution for which there does not exist a solution, $\mathbf{x}' \in \overline{\mathbf{X}}$, such that

$$z_r(\mathbf{x}') > z_r(\mathbf{x}) \text{ for some } r, \qquad r = 1, 2, \ldots, p \tag{5.69}$$

and

$$z_k(\mathbf{x}') \geqslant z_k(\mathbf{x}) \qquad \text{for all } k \neq r \tag{5.70}$$

In the constraint method of multi-objective programming, the vector optimization problem is represented by the programming model

$$\max z = z_r(\mathbf{x}) \tag{5.71a}$$

subject to

$$\mathbf{x} \in \overline{\mathbf{X}}$$

$$z_k(\mathbf{x}) \geqslant L_k \quad \text{for all } k \neq r \tag{5.71b}$$

It can be shown that parametric variation of L_k, the lower bound on the kth objective, generates the noninferior solutions. The dual variables of the model also are the marginal rate of substitution of z_r ($\mathbf{x}$) per unit variation in L_k.

Alternatively, in the weighting method, the programming model becomes

$$\max z = \sum_{k=1}^{P} w_k z_k(\mathbf{x}) \tag{5.72a}$$

$$\mathbf{x} \in \overline{\mathbf{X}} \tag{5.72b}$$

where the given weights w_k are nonnegative. (At least one weight, however, is required to be non-zero.) Varying the weights maps the set of noninferior solutions. However, unless the set of noninferior solutions is convex, the method cannot completely generate the noninferior solution set.

Multi-objective techniques that rely on articulated preference information are designed to substantially reduce the computational burden of identifying the noninferior solutions. This is generally done by prevailing on the decision-makers to state their preferences as to the objectives in some form, either initially or upon the receipt of partial information, and using these preferences to eliminate large portions of the noninferior set. The surrogate worth trade-off method is perhaps the best known of the methods in the second category (see, for examples, Haimes et al., 1975).

5.11 CLOSING COMMENTS

A variety of linear and nonlinear programming techniques have been presented for groundwater systems management. Linear programming, for example, can accommodate relatively high dimensional problems and identify global optimal solutions. Standard computer codes are also readily available. In contrast, DP is capable of handling adaptive, nonlinear, and stochastic variables. However, it requires separability and monotonicity of the constraint and objective functions. Nonlinear programming methods, which exploit the large systems of equality constraints that occur using the response equation method, are based on conjugate gradient and gradient reduction methods.

In Chapters 6 through 8, these optimization algorithms will be used for the solution of groundwater supply and quality management problems.

5.12 REFERENCES

Avriel, M. *Nonlinear Programming: Analysis and Methods*. Englewood Cliffs NJ: Prentice-Hall, 1976.

Beale, E.M.L. "On Quadratic Programming." *Naval Res. Log. Quart.*, 6(3):227–243, 1959.

Bellman, R.E. *Dynamic Programming*. Princeton NJ: Princeton University Press, 1957.

Benayoun, R., de Montgolfier, J., Tergny, J., & Laritchev, O. Linear Programming with Multiple Objective Functions: Step Method (Stem)." *Mathematical Programming*, 1(3):336–375, 1971.

Buras, N. "Dynamic Programming and Water Resources Development." *Advances in Hydroscience*, 3:372–412, 1966.

Cohon, J.L., & Marks, D.M. "A Review and Evaluation of Multiobjective Programming Techniques." *Water Resources Research*, 11(2):208–220, 1975.

Cutler, L., & Pass, D.S. "A Computer Program for Quadratic Mathematical Models to be Used for Aircraft Design and other Applications Involving Linear Constraints." Report R-516-PR, The Rand Corp., 1971.

Dantzig, G.B. "Linear Programming Under Uncertainty." *Management Science*, 1(3) and (4):197–206, 1955.

Dantzig, G.B., & Madansky, A. "On the Solution of Two-stage Linear Programs under Uncertainty." In *Proceedings of the Fourth Berkeley Symposium on Mathematical Statistics and Probability*, University of California Press, Berkeley, California, 1961, pp. 165–176.

Dupacova, J. "Water Resources System Modeling Using Stochastic Programming Models." In *Recent Results in Stochastic Programming*, P. Kall, & A. Prekopa, Eds. New York: Springer-Verlag, 1980.

Everett, H. "Generalized Lagrange Multiplier Methods for Solving Problems of Optimum Allocation of Resources." *Operations Res*, 11:399–417, 1963.

Fandel, O., & Oal, T. "Multiple Criteria Decision-Making-Theory and Application." *Lecture Notes in Economics and Mathematical Systems.* New York: Springer-Verlag, 1980.

Fiacco, A.V., & McCormick, G.P. *Nonlinear Programming: Sequential Unconstrained Minimization Technique*. New York: John Wiley, 1968.

Fletcher, R., & Powell, M.J.D. *J. Computers*, 6:163–168, 1963.

Fletcher, R. *Constrained Optimization*. Chichester: Wiley, 1981.

Gartska, S.J. "The Economic Equivalence of Several Stochastic Programming Models." In *Stochastic Programming*, M.A.H. Dempster, Ed. Academic Press, 1980, based on Proceedings of International Conference, sponsored by the Institute of Mathematics and Applications, Oxford, July 15–17, 1974.

Gass, S., & Saaty, T. "The Computational Algorithm for the Parametric Objective Function." *Naval Research Logistics Quarterly*, 2:39–45, 1955.

Geoffrion, A.M. "Solving Bicriterion Mathematical Programs." *Operations Research*, 15(1):39–54, 1967.

Gill, P.E., Murray, W., & Wright, M.H. *Practical Optimization*. London: Academic Press, 1981.

Gill, P.E., Murray, W., Saunders, M.A., & Wright, M.H. "The Design and Implementation of a Quadratic Programming Algorithm." Dept. of Operations Research, Stanford University, California, 1984.

Gorelick, S.M., Evans, B., & Remson, I. "Identifying Sources of Groundwater Pollution: An Optimization Approach." *Water Resources Research*, 19(3):779–790, 1983.

Hadley, G. *Nonlinear and Dynamic Programming*. Reading MA: Addison-Wesley, 1964.

Haimes, Y.Y., Hall, W.A., & Freedman, H.T. *Multiobjective Optimization in Water Resources System*. New York: Elsevier, 1975.

Hastings, N.A.J. *Dynamic Programming with Management Applications*. New York: Crane, Russel, 1973.

Himmelblau, D.V. *Applied Nonlinear Programming*. New York: McGraw-Hill, 1972.

Hogan, A.J., Morris, J.G., & Thompson, H.E. "Decision Problems Under Risk and Chance Constrained Programming: Dilemmas in the Transition." *Management Science*, 27(6):698–716, 1981.

Hooke, R., & Jeeves, T.A. "Direct Search Solution of Numerical and Statistical Problems." *Journal of the Association Computing Machinery*, 8(2):212–221, 1961.

Howard, R.A. *Dynamic Programming and Markov Processes*. Cambridge MA: MIT Press, 1960.

Hwang, C-L., & Yoon, K. "Multiple Attribute Decision-Making Methods and Applications." *Lecture Notes in Economics and Mathematical Systems*. New York: Springer-Verlag, 1981.

Intriligator, M.D. *Mathematical Optimization and Economic Theory*. Englewood Cliffs NJ: Prentice-Hall, 1971.

Lemke, C.E. "A Method of Solution for Quadratic Programs." *Management Science*, 8(4):442–453, 1962.

Luenberger, D.G. *Introduction to Linear and Nonlinear Programming*. Reading MA: Addison-Wesley, 1973.

Marino, M.A., & Mohammadi, B. "Reservoir Management: A Reliability Programming Approach." *Water Resources Research*, 19(3):613–620, 1983.

Marino, M.A., & Loaiciga, H.A. "Quadratic Model for Reservoir Management: Application to the Central Valley Project." *Water Resources Research*, 21(5):631–641, 1985.

Meyer, R. "The Validity of a Family of Optimization Methods." *J. SIAM Control*, 8(1):41–54, 1970.

Miller, W.L., & Byers, D.M. "Development and Display of Multiobjective Project Impacts." *Water Resources Research*, 9(1):11–20, 1973.

Monarchi, D.E., Kisiel, C.C., & Duckstein, L. "Interactive Multiobjective Programming in Water Resources: A Case Study." *Water Resources Research*, 9(4):837–850, 1973.

Murtagh, B.A., & Saunders, M.A. "A Projected Lagrangian Algorithm and its Implementation for Sparse Nonlinear Constraints." In *Math Prog.*, 16, North Holland, Amsterdam, 1982.

Nemhauser, G.L. *Introduction to Dynamic Programming*. New York: John Wiley, 1966.

Noel, J.E., & Howitt, R.E. "Conjunctive Mullihosin Management: An Optimal Control Approach." *Water Resources Research*, 18(4);753–763, 1982.

Powell, M.J.D. "An Efficient Method for Finding the Minimum of a Function of Several Variables Without Calculating Derivatives." *The Computer Journal*, 7(2):155–162, 1964.

Prekopa, A. "Network Planning Using Two-Stage Programming Under Uncertainty." In *Recent Results in Stochastic Programming*, P. Kall, and A. Prekopa, Eds. New York: Springer-Verlag, 1980.

Robinson, S.M. "A Quadratically Convergent Algorithm for General Nonlinear Programming Problems." *Math. Prog.*, 3:145–156, 1972.

Rosen, J.B. "The Gradient Projection Method for Nonlinear Programming. Part I. Linear Constraints." *J. Soc. Indust. Appl. Math.*, 8(1):181–217, 1960.

Rosen, J.B. "Iterative Solution of Nonlinear Optimal Control Problems." *J. SIAM Control*. 4(1):223–244, 1966.

Rosenbrock, H.H. "An Automatic Method for Finding the Greatest or Least Value of a Function." *The Computer Journal*, 3(3): 175–184, 1960.

Rosenthal, R.E. "The Status of Optimization Models for the Operation of Multi-Reservoirs Systems with Stochastic Inflows and Nonseparable Benefits." Research Report No. 75, Tennessee Water Resources Research Center, May 1980.

Simonovic, S.P., & Marino, M.A. "Reliability Programming in Reservoir Management, 1, Single Multipurpose Reservoir." *Water Resources Research*, 16(5):844–848, 1980.

Walkup, D.W., & Wets, R.J.B. "Stochastic Programs Resource." *SIAM Journal on Applied Mathematics*, 15(5):1299–1314, 1967.

Wets, R.J.B. "Programming under Uncertainty: The Equivlaent Convex Program." *SIAM Journal on Applied Mathematics*, 14(1):89–105, 1966.

Wolfe, P. "The Simplex Method for Quadratic Programming." *Econometrica*, 29:382–398, 1959.

Wolfe, P. "Methods of Nonlinear Programming." In *Nonlinear Programming*, J. Abadie, Ed. Amsterdam: North Holland, 1967.

Yeh, W.W-G. "Reservoir Management and Operative Models: A State-of-the-Art Review." *Water Resources Research*, 21(12):1797–1818, 1985.

Zangwill, W.I. *Nonlinear Programming: A Unified Approach*. Englewood Cliffs NJ: Prentice-Hall, 1969.

6. GROUNDWATER SUPPLY MANAGEMENT MODELS

6.1 INTRODUCTION

The groundwater supply of confined and unconfined aquifers is traditionally developed to satisfy fully, or in part, municipal, industrial, or agricultural water demands. The planning problems associated with the management of the groundwater supply are:

1. The determination of the optimal pumping pattern, i.e., the location of all pumping wells in the system, and the magnitude and duration of pumping necessary to satisfy the given water targets.
2. The timing and staging of well field development, i.e., the capacity expansion problem.
3. The design of surface storage and transport facilities to distribute the groundwater supply to the water demands in the basin.

The optimal management decisions maximize the net discounted benefits from allocating the groundwater supplies over a design or planning horizon while minimizing interference effects between the wells in the system, land subsidence, salt water intrusion, or other water quality problems.

The physical basis of these planning models are the hydrodynamic response equations of the groundwater system, developed in Chapter 4. These response equations, which relate the state and decision variables, are found either through the analytical or numerical solution of the groundwater system's equations.

The response equations have a dual role in groundwater planning because the hydraulic equations can be used for simulation or optimization of the groundwater system. In simulation, the response equations are used to predict (1) the hydraulic or water quality response of the aquifer system to a set of pumping or recharge schedules and (2) the probable hydrologic and environmental impacts associated with groundwater development or conjunctive use. The simulation approach, because it can consider only a limited number of alternatives, generally does not identify the optimal pumping schedules in the context of all the objectives of the planning or design problem.

In contrast, however, optimization models develop optimal planning, design, and operational policies for the groundwater system. And, because the response equations can be incorporated into the management models—the same equations that would normally be used for simulation—the optimal decisions define not only the optimal pumping and recharge schedules but also predict the time and spatial variation in the hydraulic head. As a result, the response equation method combines simulation and optimization in a single management model.

In the following sections and again in Chapter 7, we will examine how this approach can be used to develop planning, design, and an operational model for the management of groundwater supply and groundwater quality. We first consider an example of the simulation approach for analyzing alternative pumping patterns in a semiconfined groundwater system.

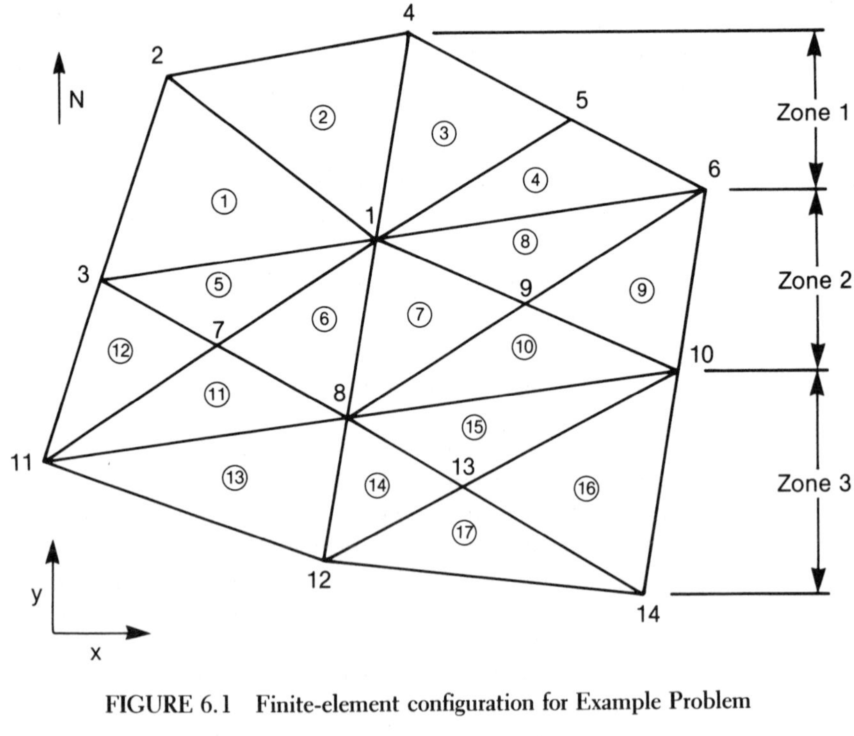

FIGURE 6.1 Finite-element configuration for Example Problem

TABLE 6.1 System Parameters of Example Problem

Zone	K_x (ft/day)	K_y (ft/day)	Angle of Rotation of K Relative to Global Coordinate System (Radius)	Storage Coefficient	Initial Head (ft)
1	114.	23.	0.44	0.01	200.
2	11400.	284.	−0.35	0.01	200.
3	5700.	710.	0.00	0.01	200.

Example Problem 6.1 Finite-Element Groundwater Simulation Model

The semiconfined aquifer shown in Figure 6.1 is to be developed as a source of water supply for agricultural purposes. The anticipated water demand is 5×10^4 ft^3/day. The available groundwater data for the aquifer system is summarized in Table 6.1. There are currently five wells in the system; their location is given in Table 6.2.

A large lake borders the northern edge of the aquifer and the elevation of the water surface, relative to the datum for the groundwater system, is 200 feet. The remaining boundaries are assumed to be impermeable. Develop a simulation model for the groundwater system and investigate alternative pumping schedules for satisfying the water demand.

A Galerkin finite-element simulation model for the groundwater system can be developed using the equations presented in Table 4.4. For the example problem, triangular elements are used to discretize the aquifer system, it is assumed that within any element the head is a linear function of x and y. The basis functions,

TABLE 6.2 Well Location for Example Problem

Well No.	Location x (ft)	Location y (ft)
1	5600.	6000.
2	2800.	4000.
3	8400.	4800.
4	5200.	2800.
5	7200.	1600.

which were developed in Example Problem 4.7, are given as

$$N_i = \{(x_j y_k - x_k y_j) + (y_j - y_k)x + (x_k - x_j)y\}/2A$$

$$N_j = \{(x_k y_i - x_i y_k) + (y_k - y_i)x + (x_i - x_k)y\}/2A$$

$$N_k = \{(x_i y_j - x_j y_i) + (y_i - y_j)x + (x_i - x_j)y\}/2A$$

where A is the area of the element and x_i, x_j, x_k are the nodal coordinates of the finite elements. The discretized aquifer system is shown in Figure 6.1. There are a total of 17 finite elements and 14 nodes in the system.

The Galerkin method transforms the semiconfined flow equation (Equation 2.55),

$$L(h) = \frac{\partial}{\partial x}\left(T_{xx}\frac{\partial h}{\partial x}\right) + \frac{\partial}{\partial y}\left(T_{yy}\frac{\partial h}{\partial y}\right) - \sum_{w\in\pi} Q_w \delta(x - x_w, y - y_w) - S\frac{\partial h}{\partial t} = 0 \tag{2.55}$$

into a system of first-order ordinary differential equations in time. The equations may be expressed as,

$$\mathbf{A}\dot{\mathbf{h}} + \mathbf{B}\mathbf{h} + \mathbf{g} = 0 \tag{6.1a}$$

$$\mathbf{h}(0) = \mathbf{h}_0 \tag{6.1b}$$

where $\mathbf{h}_0$ are the initial conditions of the basin and $\mathbf{h}$ is the vector of nodal values of the hydraulic head. The elemental matrices are defined by the relations,

$$A_e = \iint_{A_e} S N_\alpha N_\beta \, dD_e \tag{6.2a}$$

$$B_e = \iint_{A_e} \left\{ T_{xx}\frac{\partial N_\alpha}{\partial x}\frac{\partial N_\beta}{\partial x} + T_{yy}\frac{\partial N_\alpha}{\partial y}\frac{\partial N_\beta}{\partial y} \right\} dD_e \tag{6.2b}$$

$$g_e = -\iint_{A_e} N_i \sum_{w\in\pi} Q_w \, \delta(x - x_w, y - y_w)\, dD_e$$

$$- \int_{S_e} N_i \sum_j \left[T_{xx}\frac{\partial N_j}{\partial x} d_1 + T_{yy}\frac{\partial N_j}{\partial y} d_2 \right] h_j \, dS_e, \quad \alpha, \beta = x, y \tag{6.2c}$$

where S_e and D_e are the boundary and area of an element.

After elimination of the Dirichlet boundary conditions using the method of deletion of rows and columns (nodes 2, 4, 5, 6), the C and H matrices are 10 × 10 square matrices.

The Crank-Nicolson approximation is used to solve the response equations of the system. The finite-element equations may be written as

$$A\left\{\frac{\mathbf{h}^k - \mathbf{h}^{k-1}}{\Delta t}\right\} + B\left\{\frac{\mathbf{h}^k + \mathbf{h}^{k-1}}{2}\right\} + \mathbf{g} = 0 \qquad (6.3a)$$

where k is an index of the time step, Δt. The recursive or simulation equations that relate the head values over the successive time steps are given as

$$\mathbf{h}^k = \left\{\frac{A}{\Delta t} + \frac{B}{2}\right\}^{-1}\left\{\frac{A}{\Delta t} - \frac{B}{2}\right\}\mathbf{h}^{k-1} - \left\{\frac{A}{\Delta t} + \frac{B}{2}\right\}^{-1}\mathbf{g} \qquad (6.3b)$$

In the example problem, three pumping schedules will be simulated using the finite element numerical model.

Schedule 1 is a uniform pumping schedule: each well in the system is to be pumped at a rate of 10^4 ft^3/day. Schedule 2 is an attempt to exploit the constant head boundary of the system that is assumed to be unaffected by groundwater pumping. All the water demand is withdrawn from the northernmost well in the system, well no. 1. In schedule 3, each well in the system will be pumped for a period of six days; the pumping rate during these periods will be 5×10^4 ft^3/day.

The results of the simulations are presented in Figures 6.2 through 6.4. These illustrations depict the piezometric surface after a one-month operational period.

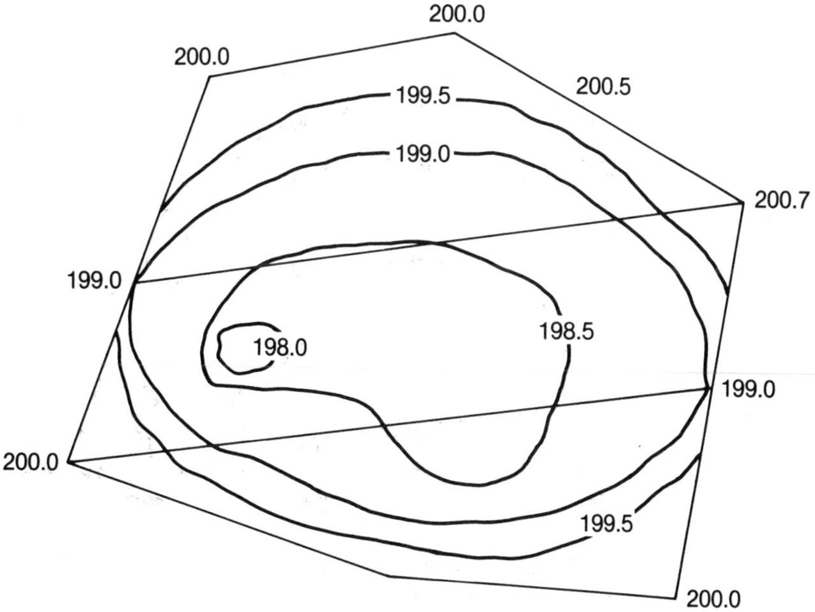

FIGURE 6.2 Simulated piezometric surface: Pumping schedule 1

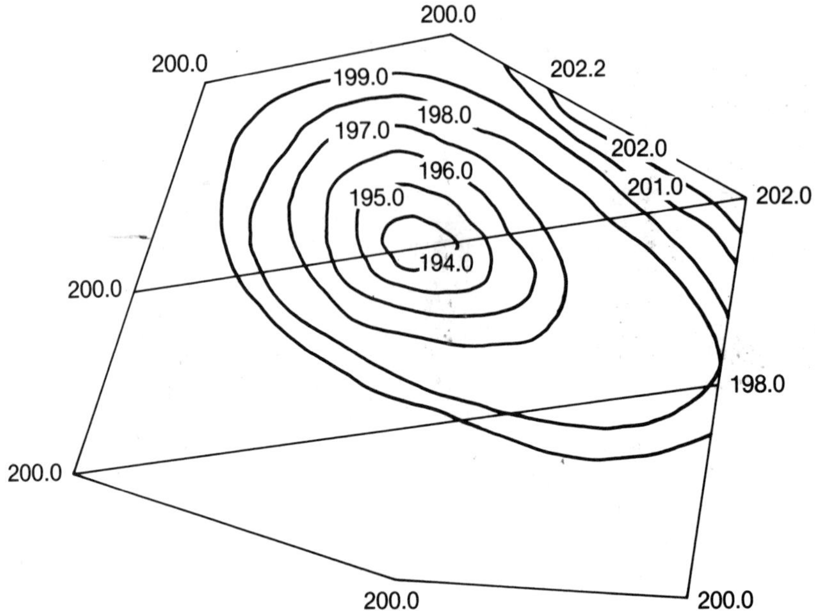

FIGURE 6.3 Simulated piezometric surface: Pumping schedule 2

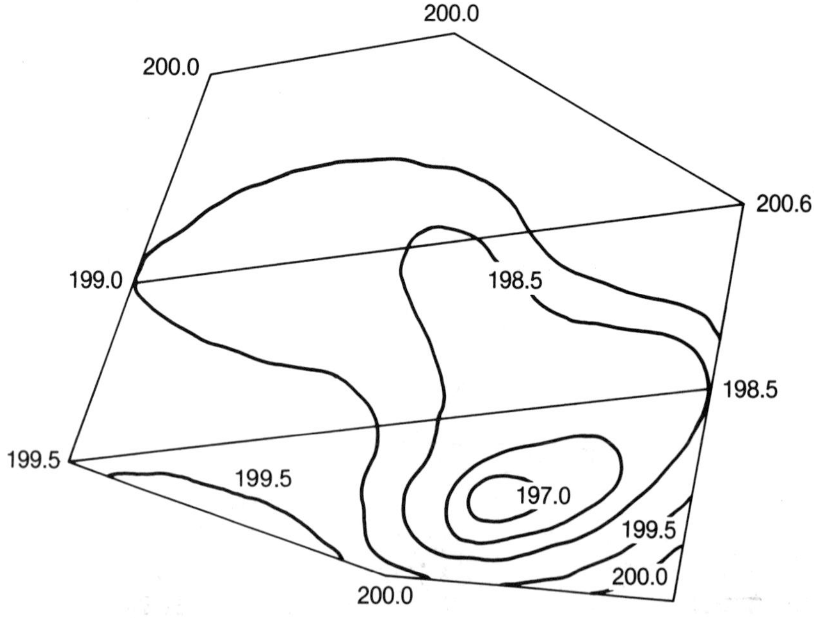

FIGURE 6.4 Simulated piezometric surface: Pumping schedule 3

Schedule 1 produces a relatively uniform distribution of the piezometric head throughout the basin. The maximum drawdown is 2 feet and occurs at well no. 2. In contrast, the maximum drawdown for schedule 2 is 6 feet at well no. 1. Zones 2 and 3 of the aquifer are relatively unaffected by the pumping schedule. In the third simulation, the maximum drawdown occurs at well no. 5, the last well to satisfy the water demand within the one-month planning period. The recovery effects of the other wells in the system minimize the drawdown in other zones of the groundwater basin. Clearly without considering other planning objectives, schedule 2 is the best groundwater management alternative. However, the schedule is totally dependent upon the constant-head boundary condition of the problem.

In the simulation analysis, neither operational costs nor the potential water quality problems that could occur from excessive drawdown conditions have been considered. Rather, the criteria for determining the best alternative were based on the overall response of the basin to a given pumping schedule. Again, it should be stressed that the decision to exploit the aquifer in zone 1 is optimal *only* in relation to the other simulated policies. Because economic and water quality considerations are neglected in the analysis, the results provide only general guidelines for the efficient management of the basin's water resources. A more thorough simulation analysis would consider these environmental and hydrologic objectives, the possible transient variation in the Dirichlet (recharge) boundary condition, and the ultimate or steady-state response of the aquifer system.

6.2 GROUNDWATER ALLOCATION MODEL

From an economic perspective, the management of groundwater is a problem in resource allocation. The allocation, distribution, or development of groundwater is controlled by the economic valuation of the water supply or water quality resource by agricultural, industrial, and municipal demands within the river basin. The optimal allocation of the groundwater supply maximizes the net discounted benefits from developing and extracting the resource while minimizing land subsidence problems, well interference effects, and potential groundwater quality problems. The optimal decisions of the planning model define the expected water targets for the principal agricultural, municipal, and industrial demand centers.

The economic benefit resulting from the allocation of groundwater depends on the demand for the water, as a factor of production, in muncipal, industrial, or agricultural activities. The benefit function, as shown in Figure 6.5, measures the willingness to pay for the water supply (Marglin, 1962). These demand functions relate the price and quantity demanded of the water supply for each water user in the basin. The area beneath the demand function, the economic benefit, represents the sum of the revenue generated by selling the water at a given market price and the consumer surplus provided by developing the water supply. The demand functions can be developed using micro-economic models of farm or industrial production (see, for example, Intriligator, 1971; Gisser, 1970; Sharples, 1969).

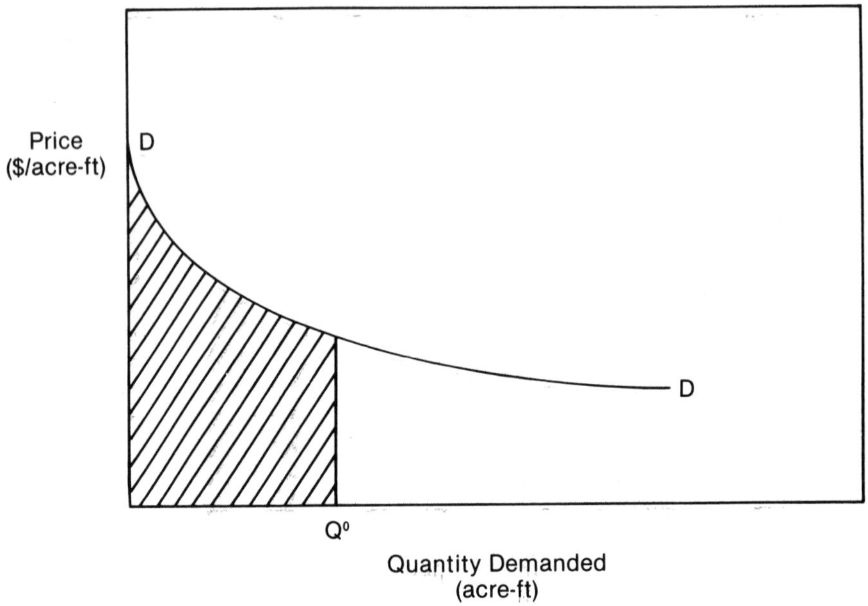

FIGURE 6.5 Aggregate demand function. Shaded area represents willingness to pay for a specified quantity (Q^o) of water

Assuming the demand functions are known or can be inferred from market information, the groundwater allocation model maximizes the net discounted economic benefit associated with developing and providing the water supply. Consider, for example, an extensive river basin consisting of r demand centers or subregions. The objective can be expressed as

$$\max_{Q^k} z = \sum_{k=1}^{m} \left(\frac{1}{1+\rho}\right)^k \sum_r \left\{ \int_0^{Q_r^k} D_r\,(Q_r^k)dQ - \sum_{w \in \pi_r} C(Q_w^k, h_w^k) \right\} \tag{6.4}$$

where D_r is the aggregate demand function for the groundwater in region r, Q_r^k is the groundwater pumped from region r, C is a generalized capital and operational cost function that is dependent on the discharge, Q_w^k, and head, h_w^k, at well site w, for any planning period k. ρ is the discount rate for the planning problem and the planning horizon consists of m planning or operational periods.

The groundwater allocation, Q_r^k, is constrained in each planning period by (1) a set of balance equations relating the allocation for each subregion and the groundwater pumping occurring in the subregion, or

$$Q_r^k = \sum_{w \in \pi_r} Q_w^k \tag{6.5}$$

π_r defines the active well sites in region r. (2) The well capacity constraints for each pumping well,

$$Q_w^k \leq Q_{w,r}^*, \quad w\epsilon\pi_r, \ \forall k \tag{6.6}$$

where Q_w^* is the capacity of the well in region r, period k, and (3) the hydraulic response equations of the groundwater system. For linear, distributed parameter groundwater systems, the response equations can be expressed, using Equation 4.67b, as

$$\mathbf{h}^k - \mathbf{A}_1(T)\mathbf{h}^{k-1} - \mathbf{A}_2(T)\mathbf{g}(Q_r^k) = \mathbf{0} \tag{6.7}$$

where $\mathbf{A}_1$ and $\mathbf{A}_2$ are the coefficient matrices associated with the matrix exponential solution of the finite-difference or finite-element numerical model and T is the length of a planning period. And, (4) the nonnegativity restrictions of the state and decision variables,

$$\mathbf{h}^k, \ Q_{w,r}^k \geq 0, \quad w\epsilon\pi_r, \ \forall k \tag{6.8}$$

The groundwater allocation model determines the optimal groundwater extraction schedule for each subregion of the river basin so as to maximize the overall economic benefit.

The allocation model is a useful planning tool because it can be used to analyze, for example, (1) the feasibility of exporting water from a given region or (2) how the targets can be expected to change for a variety of possible hydrologic and economic conditions. In practice, however, there are several conceptual problems associated with the use of these general allocation models (Howe, 1976). The problems concern:

1. The validity of derived demand schedules. In many instances, market information does not exist or the markets themselves may not exist, e.g., irrigation water. Also, demand functions are related in a complex way not only to the price of the resource but also the income of the resource uses and the prices of substitute commodities. None of these factors is considered in the microeconomic models of agricultural or industrial production.
2. The actual computational problems associated with the solution of a highly nonlinear optimization model. Because the problem is nonlinear and, possibly, nonconvex, there are usually many locally optimal solutions to the planning problem (Mangasarian, 1969). In this case, the trade-off between simulation and optimization is not apparent.

For these reasons, groundwater resources management has been primarily concerned with the development of cost-effective management models. These types of models minimize the operational and capital costs associated with satisfying a water target over a given planning or design horizon. In the following sections, we will examine three cases of these cost-effective models: a general operational or groundwater development model, the capacity expansion problem associated with the timing and staging of well field development, and a conjunctive groundwater, surface water planning model.

Example Problem 6.2 Groundwater Allocation Model

As an example of the groundwater allocation model, consider the problem of determining the optimal groundwater extraction pattern in an agricultural river basin. We assume that a central authority, for example, an irrigation district, has the authority to allocate the groundwater so as to maximize the net return from agricultural production. The district is also responsible for determining the optimal cropping pattern for the region, i.e., the amount of acreage devoted to the various crops in the area.

We also assume that the irrigation water is obtained from an extensive confined aquifer system underlying the irrigation area and that the authority has developed, calibrated, and validated a mathematical model for the aquifer system. The steady-state response equations of the aquifer system may be expressed as

$$\mathbf{B}\mathbf{h} + \mathbf{g}(Q) = \mathbf{0} \tag{6.9}$$

where Q is the total well discharge from the basin.

Before developing the allocation model, the benefit function relating the groundwater allocations and the return from agricultural production will be developed from a simple micro-economic model of farm production (EDI, 1971). Assuming that the irrigation district maximizes the net revenue for the region, the profit function for an individual crop can be defined as

$$(r_j\, Y_j\,(w_j) - pw_j)\, x_j$$

where r_j is the unit revenue of crop j, w_j is amount of groundwater applied per acre to the crop j, p is unit cost of the groundwater per acre, and x_j is the number of acres growing crop j. We assume Y_j, the production function, relates the yield per acre of a crop as a function of the applied water. A typical yield function is shown in Figure 6.6.

The optimal cropping pattern (x_j) and the groundwater allocations (w_j) can be determined by solving the mathematical optimization problem,

$$\max z = \sum_{j=1}^{m} (r_j\, Y_j\,(w_j) - pw_j)\, x_j \tag{6.10a}$$

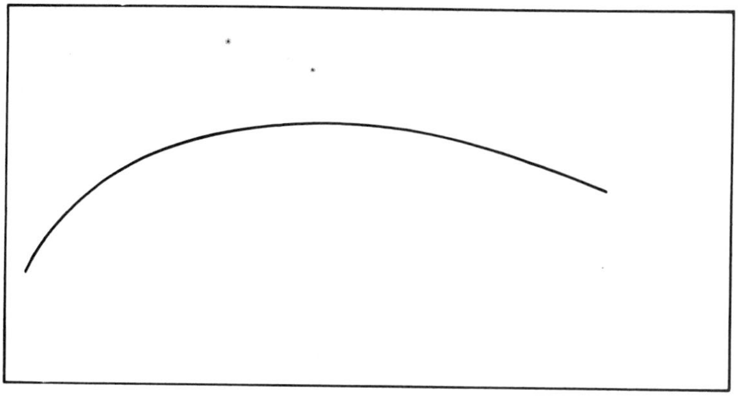

FIGURE 6.6 Typical yield function for crop j

$$Bh + g(Q) = 0 \tag{6.10b}$$

$$Q = \sum_{j=1}^{m} w_j x_j \tag{6.10c}$$

$$Q - \sum_{w \in \pi} Q_w = 0 \tag{6.10d}$$

$$\sum x_j \leq L \tag{6.10e}$$

$$x_j \leq L_j \tag{6.10f}$$

$$Q_w \leq Q_w^*, \ w \in \pi \tag{6.10g}$$

$$h_a \geq h^*, \ a \in \Delta \tag{6.10h}$$

where π is an index set defining the well locations, Q_w^* is the capacity of well w, L is the total available acreage, and L_j is the maximum acreage for crop j. h^* is the minimum permitted head level in the system to minimize well interference effects or possible subsidence problems. The locations of these constraints are defined by the set of points Δ.

The optimal solution of the model (the w_j^* and x_j^*, for all j) determines the total amount of groundwater demanded for all the crops at a given price p, or

$$Q(p) = \sum_j w_j x_j \tag{6.11}$$

This quantity represents one point on the demand schedule. By parametrically varying the price of the water, the demand for the groundwater as a function of the price can be determined. The area beneath the demand function again represents the benefit resulting from the groundwater allocation.

The second problem the authority must address is to determine the optimal groundwater pumping pattern. The groundwater allocations will maximize the willingness to pay for the resource less the operational costs, $K(Q)$, or

$$\max z = \int_0^Q p(Q)dQ - K(Q) \tag{6.12a}$$

The allocation is constrained by

1. The balance equations relating the total groundwater allocation to the individual pumping rates (Q_w)

$$Q = \sum_{w \in \pi} Q_w \tag{6.12b}$$

2. The steady-state response equations,

$$B\,\mathbf{h} + \mathbf{g}\,(Q) = 0 \tag{6.12c}$$

3. The pumping upper bounds,

$$Q_w \leq Q_w^* \quad w \in \pi \tag{6.12d}$$

(Q_w^* is the well capacity limitation).

4. Possible lower bound restriction on the head values, or

$$h_a \geq h_a^*, \; a \in \Delta \tag{6.12e}$$

The mathematical programming problem can be used to determine the optimal groundwater extraction pattern and the economic trade-offs associated with increasing pumping capacity or reducing head lower restriction. The model can also be extended to dynamic water allocation problems and multiple water users.

Example Problem 6.3 Stochastic Groundwater Allocation Model

Develop a stochastic planning model to determine the optimal, temporal allocation of the groundwater supply of a confined aquifer. Assume (1) the dynamics of the groundwater system can be represented by the model

$$S_{t-1} = S_t - P_t + R_t \tag{6.13}$$

where S_t is the quantity of water in groundwater storage at the end of time period t, where t is numbered backwards with respect to real time. P_t and R_t are the groundwater pumping and recharge occurring during period t. And, (2) the groundwater recharge is a random variable described by the probability density function, $h(R_t)dR_t$, which is independently distributed over time.

The stochastic planning problem can be formulated as a dynamic programming problem. The stages of the model correspond to time; the state variable is the groundwater storage. Defining $B_t(S_t, P_t)$ as the net benefit resulting from pumping P_t then $f_t(S)$, the optimal return with t stages remaining, is given by the recursive equation (Burt, 1964)

$$f_t(S) = \max_{P_t \leq S} \left\{ B_t\,(S_t, P_t) + \left(\frac{1}{1 + \rho}\right) E\,[f_{t-1}(S_t + R_t - P_t)] \right\} \qquad (6.14a)$$

or

$$f_t(S) = \max_{P_t \leq S} \left[B_t(S_t, P_t) + \left(\frac{1}{1+\rho}\right) \int_0^{\infty} f_{t-1}(S_t + R_t - P_t) h(R_t) dR_t \right] \qquad (6.14b)$$

where E is the mathematical expectation operator and ρ is the discount rate. The functional equation can be interpreted as the maximization with respect to water consumption at stage t of the immediate net benefit plus the expected return in the $t - 1$ remaining stages, given that an optimal policy will be used during the remaining $t - 1$ stages.

Solution of the dynamic programming model is usually accomplished by discretizing the groundwater storage, recharge, and pumping. The dynamic programming model then becomes a finite Markovian decision process. Defining N discrete levels for groundwater storage ($j = 1, \ldots$ N), m values for R_t, and n values for P_t ($k = 1, \ldots n$), the functional equation can be expressed for a particular groundwater storage level, S_i, as

$$f_t(S_i) = \max_k \left\{ B_t(S_{ti}, P_{tk}) + \frac{1}{(1 + \rho)} \sum_{j=1}^{N} f_{t-1}\,(S_{ti} + R_{tj} - P_{tk}) \cdot P(R_t = R_t^j) \right\}$$

$$P_{tk} \leq S_i$$

For each groundwater storage level, S_{ti}, we can also define a probability, P_{ij}^k, for each P_{tk} as the probability of going to S_{tj}, given that the groundwater storage was S_{ti} at the beginning of the period and P_{tk} is the withdrawal. The recursive equation can then be expressed in terms of these conditional probabilities as

$$f_t(S_i) = \max_k \left\{ B_t^k(S_i) + \frac{1}{(1 + \rho)} \sum_{j=1}^{n} p_{ij}^k\, f_{t-1}\,(S_j) \right\}, \quad t = 0, 1, 2 \qquad (6.15a)$$

where

$$B_t^k(S_i) = B_t(S_{ti}, P_{tk}) \tag{6.15b}$$

An interesting property of the dynamic programming model occurs when the system benefit function is constant over the planning horizon. The solution of the programming model will then converge to an optimal policy that is independent of the stage when the number of stages becomes very large. The optimal policies are found from the solution of the equation

$$(I - \beta P)\,\mathbf{x} = \mathbf{b} \tag{6.16}$$

where I is the identity matrix, $\beta = (1 + \rho)^{-1}$, P is an $n \times n$ matrix containing p_{ij}^k, $\mathbf{b} = [B_t^k(S_i), \ldots .B_t^k(S_i)]^T$, and $\mathbf{x} = [f_1, f_2, \ldots f_t]^T$, the optimal return vector.

6.3 GROUNDWATER OPERATIONAL MODEL

An important class of groundwater planning problems involves the determination of the optimal operational (water extraction) schedules of a groundwater system (Maddock, 1972a). Assuming, for example, that there are k^* (previously developed) well sites in the basin, the planning problem is to determine how the existing well field should be operated over the entire planning horizon to satisfy an exogenous water target. The objective of these types of planning models is to minimize the total discounted operational costs for extraction of the resource.

The operational planning problem can be formulated as a nonlinear optimization model. Defining $C_w\,(h_w, Q_w)$ as the extraction cost function at well site w ($w \epsilon \pi$) and assuming that the planning horizon (T^*) again consists of m planning periods of length T, the objective function of the model is

$$\min z = \sum_{k=1}^{m} \left(\frac{1}{1+\rho}\right)^k \sum_{w \epsilon \pi} C_w\,(h_w^k, Q_w^k) \tag{6.17a}$$

where ρ is the discount rate.

The groundwater extraction policies are constrained in each period by:

1. The water target requirements

$$\sum_{w \epsilon \pi} Q_w^k \geq D^k \quad , \quad \forall k \tag{6.17b}$$

2. The well capacity restriction

$$0 \leq Q_w^k \leq Q_w^* \quad , \quad \forall k,\ w \epsilon \pi \tag{6.17c}$$

where Q_w^* is the capacity of well w.

3. The response equations of the groundwater system. The response equations may be written for each planning period using the matrix exponential solution to the finite-difference or finite-element equations developed in Chapter 4, as

$$\mathbf{h}^k - \mathrm{A}_1(T)\mathbf{h}^{k-1} - \mathrm{A}_2(T)\mathbf{g}(\mathbf{Q}^{k-1}) = 0 \quad , \quad k = 1, \ldots m \qquad (6.17d)$$

and

$$\mathbf{Q}^k = (Q_1^k, Q_2^k, \ldots . Q_w^k)^T, \qquad \forall w \epsilon \pi$$

We have again assumed that, in the event that the aquifer is unconfined, the Boussinesq approximation is valid and that either direct linearization or quasi-linearization has been used to generate a system of linear response equations.

The planning model presented in Equation 6.17 is the basic framework for the analysis of a variety of planning and design problems. The model can incorporate additional restrictions or objectives to reflect differing management strategies or environmental constraints. For example, in the case of salt water intrusion in a coastal aquifer, the head or possibly the hydraulic gradient at the approximate location of the interface may be limited to a specific range of values to prevent the landward migration of the salt water wedge. These constraints can be expressed by the relations

$$h_a^k \geqslant h_a^*, \quad a \epsilon \Delta, \qquad (6.17e)$$

or

$$\frac{\partial h^k}{\partial n} \geqslant \gamma^k, \quad a \epsilon \Delta \qquad (6.17f)$$

where h^* is a lower bound on the head in the aquifer system, n is a unit normal in the direction of the groundwater flow, and γ^k is a prescribed value of the hydraulic gradient.

The state variables of the groundwater system can also be used as surrogates for economic or environmental objectives. Groundwater management problems often involve the determination of the sustainable yield of the groundwater basin or the determination of groundwater dewatering (mining) schedules. In aquifer dewatering problems, the objective maximizes the total yield of the system, or

$$\max z = \sum_{k=1}^{m} \sum_{w \epsilon \pi} Q_w^k$$

The safe yield of an aquifer can be analyzed with the objective,

$$\max z = \sum_{k=1}^{m} \sum_{w \epsilon \pi} h_w^k$$

which is also a surrogate for minimizing operational costs. The objectives are constrained by the hydraulic response equations, well capacity restrictions, and possible head lower bounds. Parametric programming can be used to determine the trade-off between, for example, the total yield and the head lower bounds or the operational costs and the forecasted water demands.

In all of these planning models, it is important to consider the possible impacts of parameter and economic uncertainty on the optimal planning policies. The dual variables associated with the optimal solution of the planning models provide a partial answer to this problem. The dual variables represent the marginal change in the system objectives for a unit relaxation in the constraints of the model. In terms of the water targets of the system, the dual variables or shadow prices indicate how much the system costs are increased (decreased) by increasing (decreasing) the water demand. Similarly, if, for example, the state variables are strictly binding at certain control locations within the aquifer, the shadow prices will determine how the objective will change as these constraints are relaxed. This information is an indication of the economic and hydrologic trade-offs occurring in the basin.

The possible effects of parameter uncertainty on the optimal planning policies is a more difficult problem to assess. As we have discussed, the parameters of the aquifer system are often random variables and, under certain conditions, the direct water demands may form a stochastic process. It it is possible to characterize the underlying distributions of these variables, then Monte Carlo optimization methods or stochastic dynamic programming can be used to determine probabilistic operating rules for the groundwater system (see, for example, Maddock, 1974).

We also note that, in general, the operational model is a nonlinear and, possibly, nonconvex programming program. Local solutions of the planning model can be identified using the constrained nonlinear algorithms presented in Chapter 5 (see, for example, Willis & Newman, 1977; Aguado & Remson, 1980; Rosenwald & Green, 1974).

Example Problem 6.4 Groundwater Operational Problem

Groundwater pumping schedules are to be determined for an extensive confined aquifer system. The confined aquifer is homogeneous and isotropic. Develop a management model to determine the optimal pumping pattern in the basin. Assume, in lieu of cost data, that (1) the objective is to maximize the head at each well site in the basin, (2) the pumping decisions are to be made at the beginning of each planning period, and (3) the operational horizon consists of three planning periods (see Figure 6.7).

Since the aquifer system is homogeneous, isotropic, and essentially infinite, the hydraulic response equations may be developed from the Theis equation (Eq.

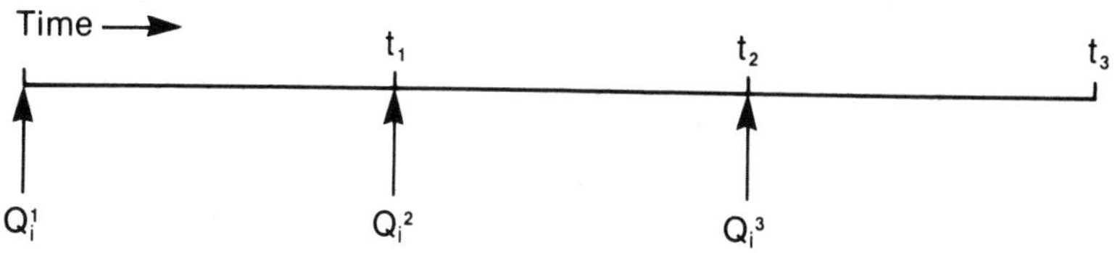

FIGURE 6.7 Planning horizon

2.68). Denoting Q_w^k as the discharge from well site i in period k, the drawdown $s(r, t)$ occurring during the first planning period is

$$s(r,t) = \sum_{w\in\pi} \frac{Q_w^1}{4\pi T} W\left(\frac{\bar{r}_w^2 S}{4Tt}\right) \quad , \quad 0 \leq t \leq t_1 \tag{6.18a}$$

where t_1 is the length of the planning period, π defines the location of the well sites, $\bar{r}_w$ is the distance from the wth well to any point r in the aquifer, and S and T are the storage and transmissivity parameters, respectively. In the second time period, the drawdown is given as,

$$s(r,t) = \sum_{w\in\pi} \frac{Q_w^1}{4\pi T} W\left(\frac{\bar{r}_w^2 S}{4Tt}\right) + \sum_{w\in\pi} \left(\frac{Q_w^2 - Q_w^1}{4\pi T}\right) W\left(\frac{\bar{r}_w^2 S}{4T(t-t_1)}\right), \quad t_1 \leq t \leq t_2 \tag{6.18b}$$

Similarly, for the third period

$$\begin{aligned} s(r,t) = {} & \sum_{w\in\pi} \frac{Q_w^1}{4\pi T} W\left(\frac{\bar{r}_w^2 S}{4Tt}\right) + \sum_{w\in\pi} \left(\frac{Q_w^2 - Q_w^1}{4\pi t}\right) W\left(\frac{\bar{r}_w^2 S}{4T(t - t_1)}\right) \\ & + \sum_{w\in\pi} \left(\frac{Q_w^3 - Q_w^2}{4\pi T}\right) W\left(\frac{\bar{r}_w^2 S}{4T(t - t_2)}\right), \qquad t \geq t_3 \end{aligned} \tag{6.18c}$$

Note that for given r and t values these equations are *linear* functions of the pumping rates.

The management model can then be expressed as

$$\min z = \sum_k \sum_{w\in\pi} s_w^k \tag{6.19a}$$

where s_w^k is the drawdown at well w at the end of period k. The pumping decisions are constrained by (1) the water demand restrictions for each period,

$$\sum_{w\in\pi} Q_w^k \geq D^k, \forall k \tag{6.19b}$$

where D^k is the demand in period k, (2) possible upper bound restrictions on the drawdown at selected control locations in the aquifer,

$$s_a^k \leq s^* \quad , \quad a \epsilon \Delta \tag{6.19c}$$

and (3) the response equations that are written for each individual well site and the control locations ($a \epsilon \Delta$). The management model is a linear optimization problem.

Example Problem 6.5 Groundwater Operational Model—Finite-Difference

The rectangular confined aquifer system shown in Figure 6.8 is to be developed as a municipal water source. The east and west boundaries of the system are Dirichlet conditions, which are assumed to be unaffected by pumping; the remaining boundaries are impermeable. In each of the four zones of the aquifer, the hydraulic conductivity field is homogeneous and isotropic. The aquifer's parameters are summarized in Table 6.3. Assuming steady-state conditions, develop (1) the response equations of the system using finite-difference techniques and (2) an optimal groundwater management model. Specifically investigate (1) how the head levels at the well sites are affected by variations in the water demand and (2) the trade-offs between the aquifer's yield and the minimum head levels in the basin.

For steady-state conditions, the confined flow equation is

$$\frac{\partial}{\partial x}\left(T_{xx}\frac{\partial h}{\partial x}\right) + \frac{\partial}{\partial y}\left(T_{yy}\frac{\partial h}{\partial y}\right) - \sum_{w \epsilon \pi} Q_w \delta\,(x - x_w, y - y_w) = 0 \tag{6.20a}$$

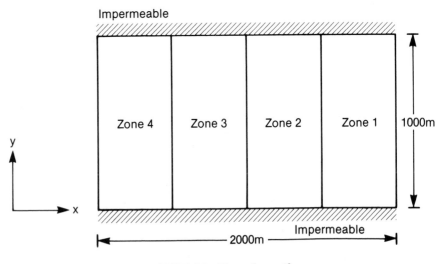

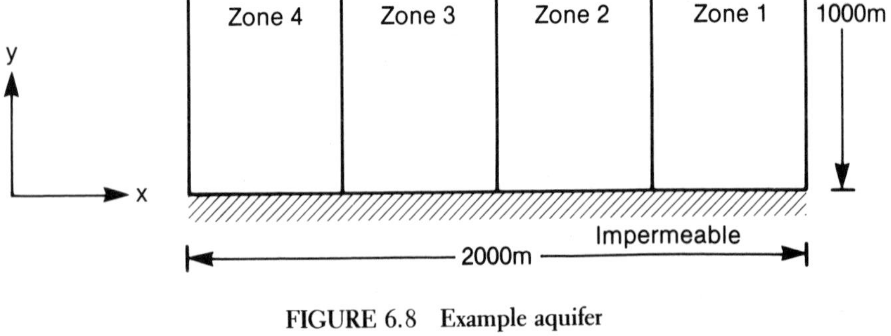

FIGURE 6.8 Example aquifer

TABLE 6.3 Aquifer Parameters

Zone	Transmissivity (m^2/day)	Angle of Rotation
1	5000	0
2	3500	0
3	2000	0
4	1000	0
Dirichlet Boundaries		
h_L(m)	100.	
h_r(m)	70.	

with the boundary conditions

$$h(M_1) = h_1,\ h(M_2) = h_2,\ \frac{\partial h}{\partial n} = 0 \text{ on } M_3 \text{ and } M_4 \tag{6.20b}$$

where n is the unit normal to the impermeable boundaries, M_3 and M_4; $M_1 \cup M_2 \cup M_3 \cup M_4$ is the aquifer's boundary.

We begin by discretizing the aquifer system. The finite-difference grid for the problem is shown in Figure 6.9. For any aquifer zone, the finite-difference

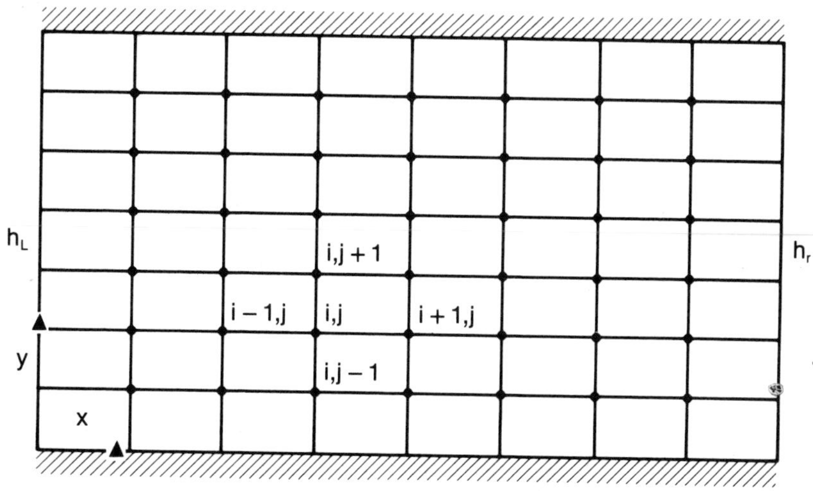

FIGURE 6.9 Finite-difference grid for example aquifer

approximation to the flow equation at a point i, j (x, y) is

$$\frac{T_{xx}}{\Delta x^2}\{h_{i+1,j} - 2h_{i,j} + h_{i-1,j}\} + \frac{T_{yy}}{\Delta y^2}\{h_{i,j+1} - 2h_{i,j} + h_{i,j-1}\} - \frac{Q_{w,i}}{\Delta x \Delta y} = 0 \tag{6.21}$$

where $Q_{w,i}$ is the discharge at node i. The truncation error of the approximation is $0(\Delta x^2 + \Delta y^2)$.

Writing Equation 6.21 for all internal nodal points and taking into account the Neumann and Dirichlet boundary conditions, the response equations may be written as

$$B\mathbf{h} + \mathbf{l} - \frac{1}{\Delta x \, \Delta y}\mathbf{Q} = 0 \tag{6.22}$$

where the B matrix contains the transmissivity parameters, the $\mathbf{l}$ vector contains the Dirichlet boundary conditions, and $\mathbf{Q}$ is the vector of well extraction. (Extractions can occur at any internal nodal point in the aquifer system.)

For the first planning problem, the management model, may be expressed as

$$\max z = \sum_{w \in \pi} h_w \tag{6.23a}$$

$$B\mathbf{h} - \frac{1}{\Delta x \Delta y}\mathbf{Q} = -\mathbf{l} \tag{6.23b}$$

$$\sum_{w \in \pi} Q_w \geq D \tag{6.23c}$$

$$Q_w \leq Q_w^* \tag{6.23d}$$

where Q_w^* is the maximum extraction rate at well site w.

The planning model is a linear programming problem and the response equations are again considered as auxiliary constraints of the model. The overall coefficient matrix of the linear programming problem is shown in Figure 6.10. The model has 43 constraints and 84 decision variables, exclusive of slack, surplus, and upper bounding variables.

The yield of the groundwater system and the attendant affect on the head in the aquifer system can be found from solution of the optimization model,

$$\max z = \sum_{w \in \pi} Q_w \tag{6.24a}$$

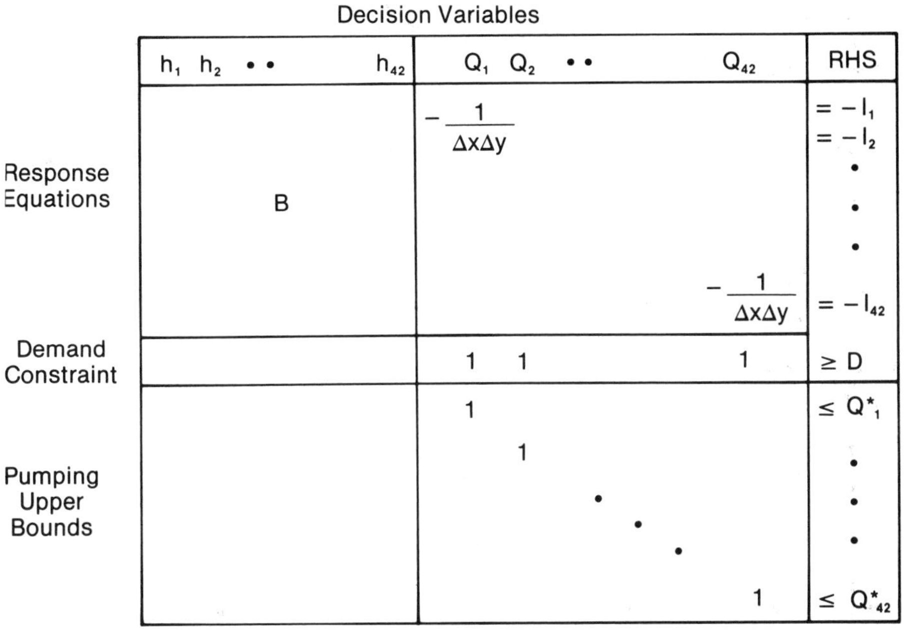

FIGURE 6.10 The constraint matrix of the optimization model

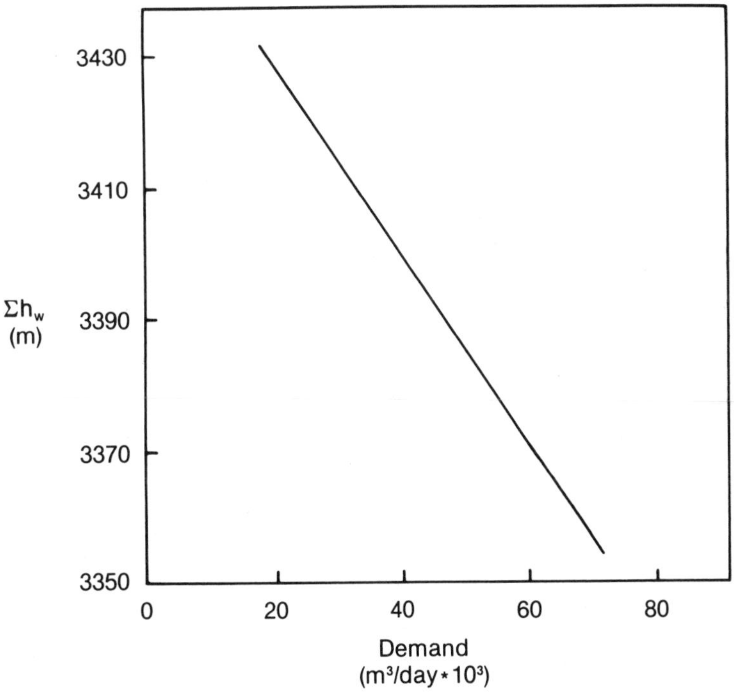

FIGURE 6.11 Head objective versus demand

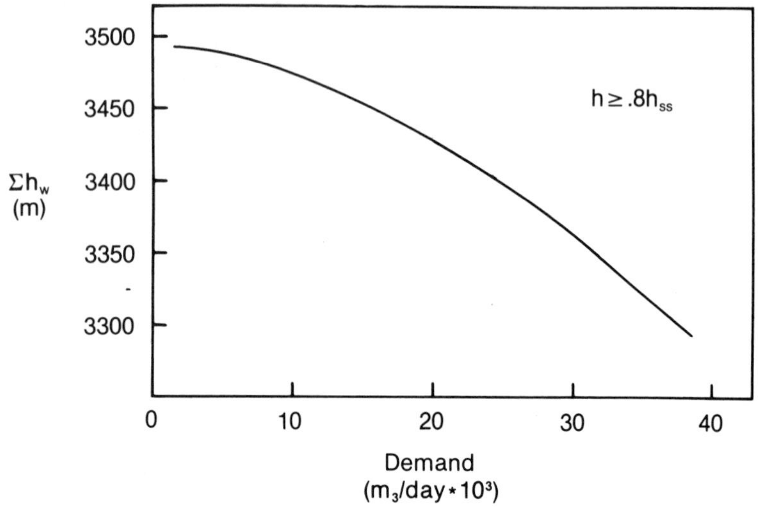

FIGURE 6.12 Head objective versus demand with head lower bounds

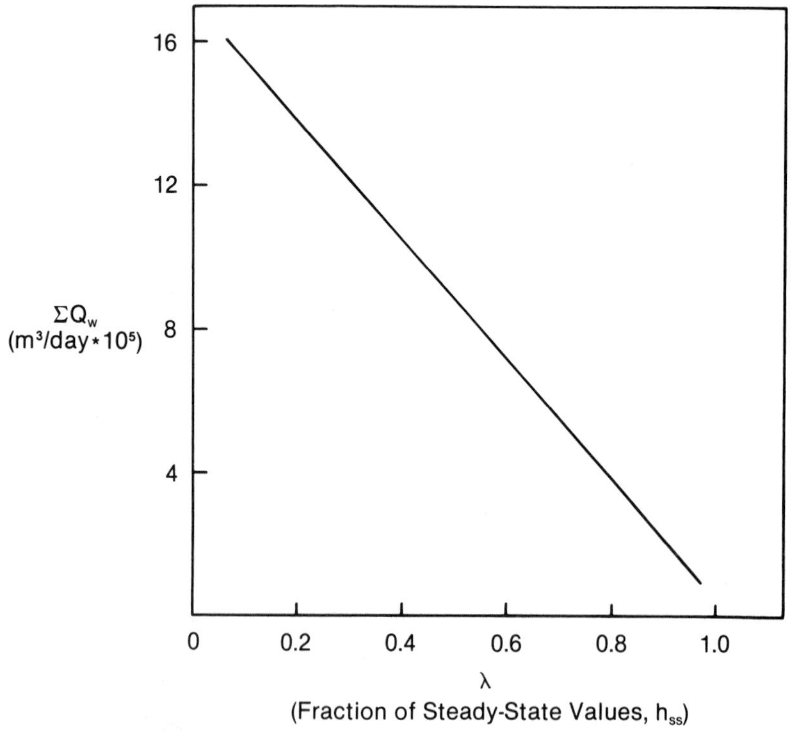

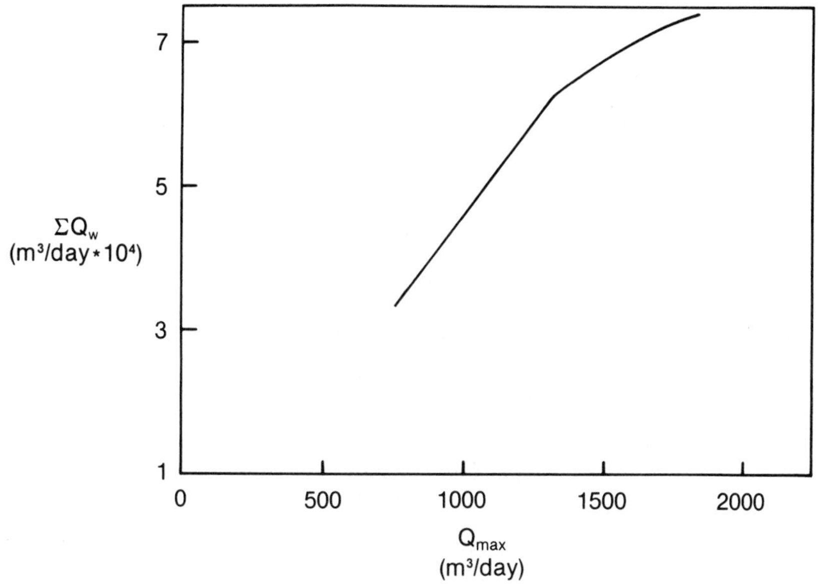

FIGURE 6.14 Yield versus pumping upper bound

$$Bh - \frac{1}{\Delta x \, \Delta y} Q = -1 \qquad (6.24b)$$

$$h_a \geq h_a^*, \; a \epsilon \Delta \qquad (6.24c)$$

$$Q_w \leq Q_w^*, \; w \epsilon \pi \qquad (6.24d)$$

Now, however, lower bound constraints on the head levels (h_a) are added to ensure that excessive depletion of the aquifer does not occur or to minimize potential subsidence or groundwater quality problems.

The results of the optimization analysis, which were obtained using CDC's APEX optimization package, are presented in Figures 6.11 through 6.14. Figure 6.11 shows the relation between the head objective and the demand constraint. As the demand increases, the total heads decrease. The shadow price or a slope of curve is $-.0019$ (m/m^3/day). The optimal pumping pattern is a single well located adjacent to the right boundary.

The effects of lower bound constraints on all the head levels are presented in Figure 6.12. The pumping upper bound is 750 m^3/day. With a lower bound of .8 of the steady-state, undeveloped aquifer head values (h_{ss}), the total head is now a nonlinear, concave function of the demand. Again, increasing the demand decreases the head levels. The pumping pattern reflects the demand variation. As the demand increases, well sites near the left Dirichlet boundary were operational. For low demands, those well sites located near the right boundary were in operation.

Similar trends are apparent in the second model. The relation between the total yield of the system and the head lower bound is presented in Figure 6.13. As the lower bound approaches the steady-state, undeveloped condition, the yield approaches zero. The trade-off between the total yield and the head lower bound is approximately -2×10^5 m^3/day. As the head bounds are increased by 10 percent, the total yield decreases by 100,00 m^3/day. Pumping was again confined to the regions adjacent the right and left boundaries.

The effect of upper bound restrictions on the extraction rates is presented in Figure 6.14. The total yield is a nonlinear, concave function of the pumping upper bound. In the lower portion of the curve, the shadow price is 48; each unit increase in the upper bound produces a 48 m^3/day increase in the total yield. The analysis assumes that the head levels are greater than or equal to 0.8 of the steady-state values.

Example Problem 6.6 Dynamic Operational Model

As an example of the development of time-varying groundwater pumping schedules, consider the rectangular confined groundwater system shown in Figure 6.15. The aquifer system's parameters are given in Table 6.4. Again, Dirichlet boundary conditions are maintained on the east and west boundaries; all other boundaries are impermeable. Assuming that the operational horizon consists of three planning periods, develop the response equations of the system using the Galerkin finite-element method. Investigate the effect of variations in the water targets and head lower bounds on the yield and total head in the system

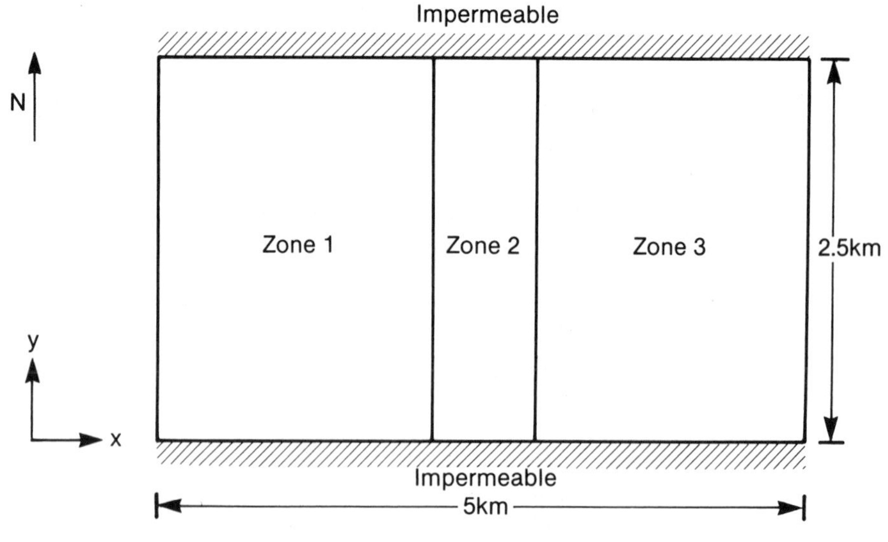

FIGURE 6.15 Example aquifer system

TABLE 6.4 Aquifer Parameters

Zone	T (m^2/day)	Angle of Rotation (radians)	S
1	864.	0.	0.01
2	432.	0.	0.01
3	265.	0.	0.01

The confined flow equation is

$$L(h) = \frac{\partial}{\partial x}\left(T_{xx}\frac{\partial h}{\partial x}\right) + \frac{\partial}{\partial x}\left(T_{yy}\frac{\partial h}{\partial x}\right) - \sum_{w \in \pi} Q_w \delta(x - x_w, y - y_w) - S\frac{\partial h}{\partial t} = 0 \quad (2.55)$$

and the initial and boundary conditions

$$h(\mathbf{x}, 0) = \mathbf{h}_0, \quad h(M_1) = h_1, \quad h(M_2) = h_2, \quad \frac{\partial h}{\partial n} = 0 \quad \text{on } M_3 \text{ and } M_4 \quad (6.25)$$

where $M_1 \cup M_2 \cup M_3 \cup M_4$ is the boundary of the groundwater system.

The finite-element grid for the aquifer system is also shown in Figure 6.16. There are a total of 50 elements and 66 nodes. Within each element, the head is a linear function of x and y, or

$$h = \alpha_1 + \alpha_2 x + \alpha_3 y + \alpha_4 xy \quad (6.26)$$

The basic function, associated with the linear quadrilaterals are, in x, y coordinates,

$$N_i = \frac{1}{4ab}(b - x)(a - y) \quad (6.27a)$$

$$N_j = \frac{1}{4ab}(b + x)(a - y) \quad (6.27b)$$

$$N_k = \frac{1}{4ab}(b + x)(a + y) \quad (6.27c)$$

$$N_l = \frac{1}{4ab}(b - x)(a + y) \quad (6.27d)$$

where a and b are the half-length and width of a quadrilateral element.

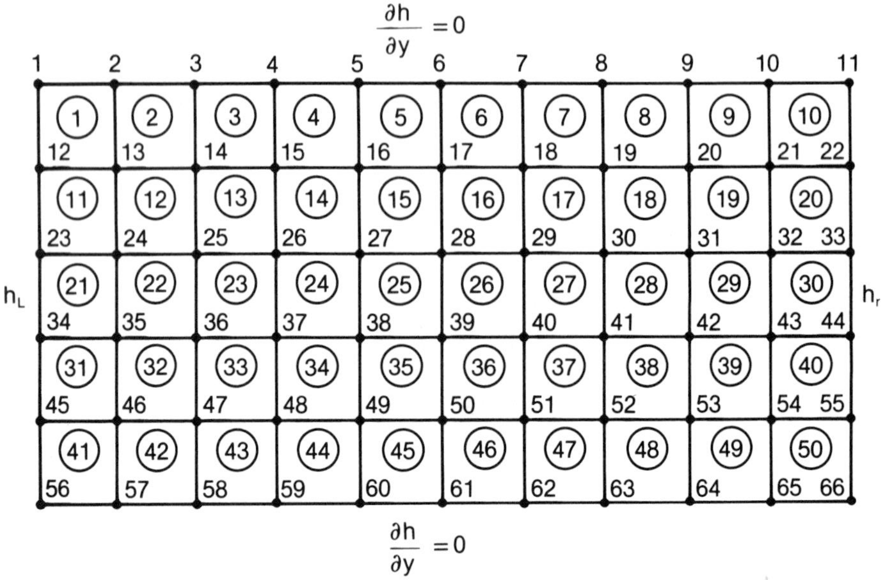

FIGURE 6.16 Finite-element network for example aquifer

Applying the Galerkin procedure, the hydraulic response equations may be expressed as

$$A\dot{\mathbf{h}} + B\mathbf{h} + \mathbf{l} + \mathbf{Q} = 0 \tag{6.28a}$$

$$\mathbf{h}(0) = \mathbf{h}_0 \tag{6.28b}$$

where $\mathbf{h}$ is the vector of the nodal values of the hydraulic head and $\mathbf{h}_0$ are the initial conditions. The conditions are obtained from the steady-state head levels of the basin. The $\mathbf{l}$ vector contains the boundary conditions of the problem. Note also that the elemental matrices have the same form as the equations presented in Example Problem 6.1.

The solution of the dynamic response equations is obtained by approximating the time derivative with a fully implicit finite-difference scheme, or

$$A\left\{\frac{\mathbf{h}^k - \mathbf{h}^{k-1}}{\Delta t}\right\} + B\mathbf{h}^k + \mathbf{l} + \mathbf{Q}^{k-1} = 0 \tag{6.29}$$

It is assumed that the time step, Δt, is equal to the length of each operational period (10 days).

Assuming the objective is to maximize the head at all well sites in the basin, the optimal extraction pattern is found from solution of the programming problem

$$\max z = \sum_{k=1}^{3} \sum_{w \in \pi} h_w^k \tag{6.30a}$$

$$\sum_{w \in \pi} Q_w^k \geqslant D^k \tag{6.30b}$$

$$Q_w^k \leqslant Q_w^*, \quad w \in \pi \tag{6.30c}$$

$$\left\{\frac{A}{\Delta t} + B\right\} \mathbf{h}^k - \frac{A}{\Delta t} \mathbf{h}^{k-1} + Q^{k-1} = -1 \tag{6.30d}$$

where D^k is the demand in period k and Q_w^* is the well capacity restriction.

The linear programming problem for the example problem has 165 constraints and 324 decision variables (excluding the pumping upper bounds and head restrictions) for three operational periods. The constraint matrix of the optimization model is shown in Figure 6.17. In the matrix, I is the identity matrix and Γ is a row

Decision Variables

	h^1	h^2	h^3	Q^1	Q^2	Q^3	
Response Equations	$\left(\frac{A}{\Delta t} + B\right)$			$-\Gamma$			$= \frac{B}{\Delta t} h^0 - I$
	$-\frac{A}{\Delta t}$	$\left(\frac{A}{\Delta t} + B\right)$			$-\Gamma$		$= -I$
		$-\frac{A}{\Delta t}$	$\left(\frac{A}{\Delta t} + B\right)$			$-\Gamma$	$= -I$
Demand Constraints				Γ			$\geq D^1$
					Γ		$\geq D^2$
						Γ	$\geq D^3$
Pumping Upper Bounds				I			$\leq Q_1^*$
					I		$\leq Q_2^*$
						I	$\leq Q_3^*$

FIGURE 6.17 Constraint matrix of the transient optimization model

vector; the elements of Γ are 1 when a well site corresponds to an internal nodal point; otherwise, it is zero.

As in Example Problem 6.4, the potential yield of the aquifer can be developed using a similar optimization model. However, the objective in this case is now to maximize the extractions over all of the planning periods, or

$$\max z = \sum_{k=1}^{3} \sum_{w \in \pi} Q_w^k \tag{6.31a}$$

Pumping in any period is constrained by the well capacity restrictions, the response equation, and the head lower bounds

$$h_a^* \geq h_a^*, \; a \in \Delta \tag{6.31b}$$

where Δ is an index set defining all the internal nodal points of the basin.

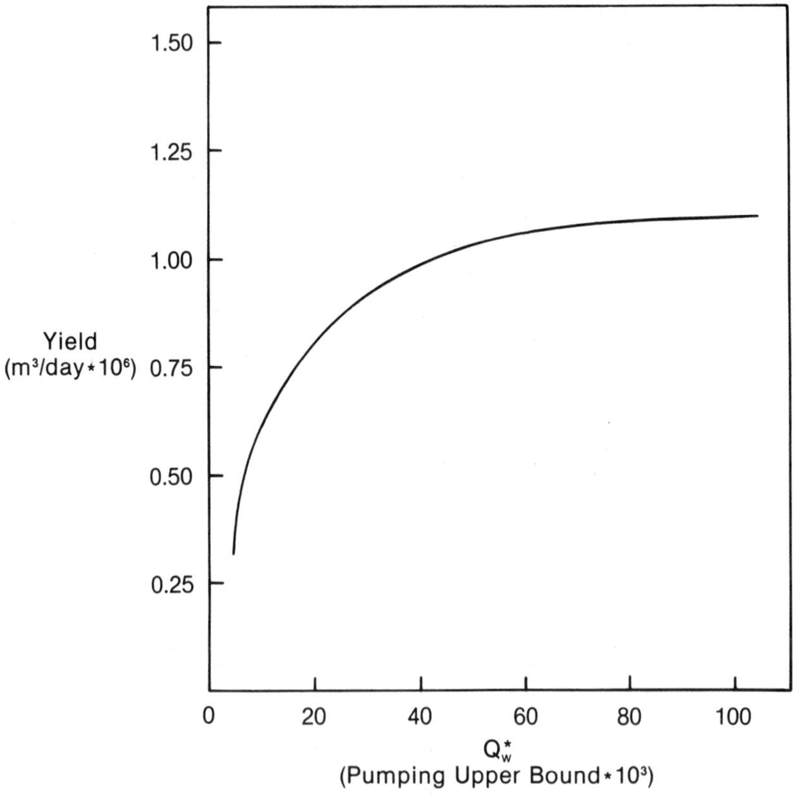

FIGURE 6.18 Yield-pumping upper bound analysis

The linear programming models were again solved using the APEX-III optimization package. The results of the two models are presented in Figures 6.18 through 6.20.

Initially, parametric linear programming was used to relate the total yield of the aquifer system and the pumping well capacity. As shown in Figure 6.18, the yield is a concave (nonlinear) function of the upper bound. The slope of the curve again is the trade-off associated with aquifer's yield and the well capacity. As the capacity approaches 60,000 m^3/day, the trade-off or slope is zero, indicating that the ultimate yield of the system is approximately 1.06×10^6 m^3 (per 30 day period).

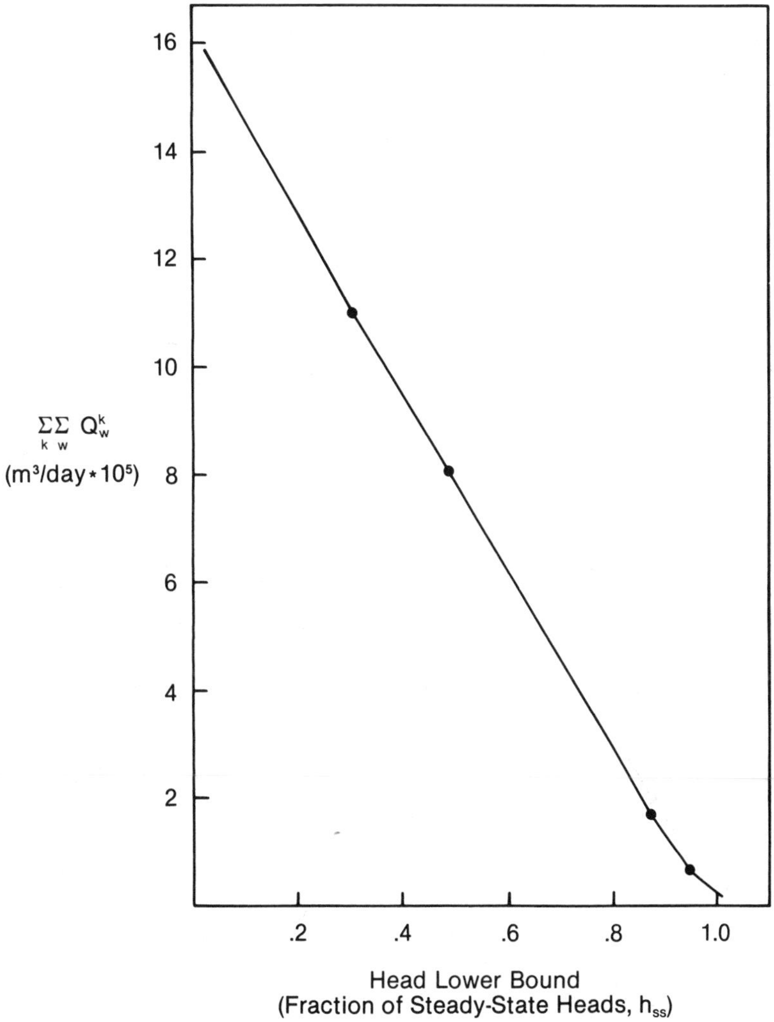

FIGURE 6.19 Yield-head lower bound analysis

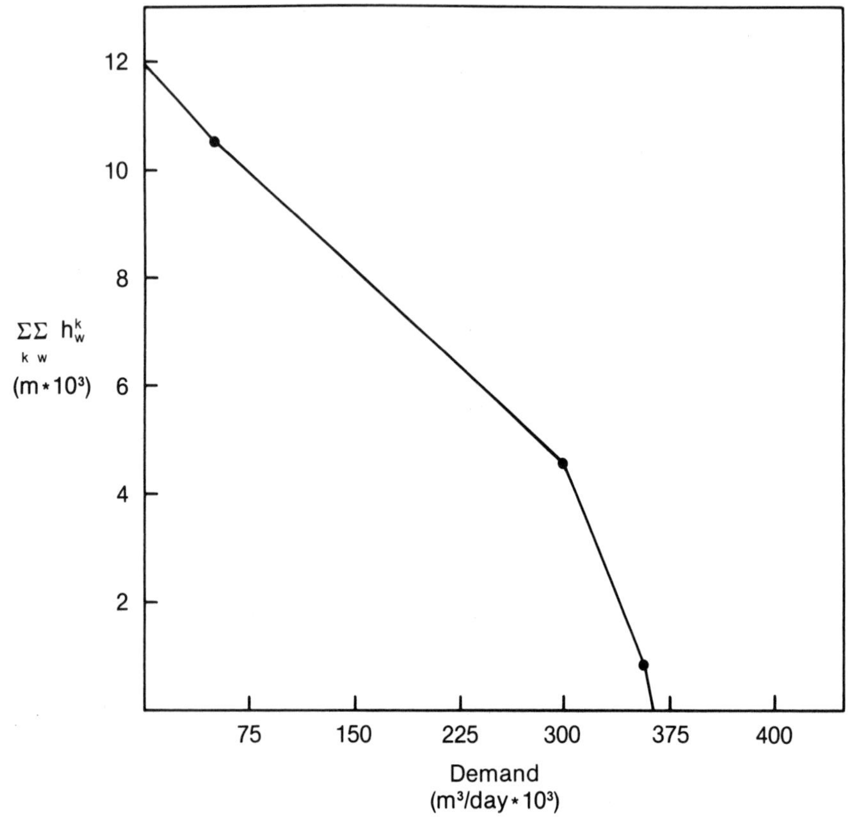

FIGURE 6.20 Head-demand analysis

The optimal pumping patterns for the three operational periods are summarized in Table 6.5. The well capacity constraint is 60,000 m^3/day. For all the demands less than 100,000 m^3/day, the active well sites are adjacent to the system's left boundaries. However, for larger demands, well sites located next to the right boundary of the basin are brought into operation. The optimization model generally selects well sites in the more permeable zones of the aquifer adjacent to the aquifer's boundaries.

The relation between the total well yield and the head levels, expressed as a percentage of the original, steady-state values (h_{ss}), is shown in Figure 6.19. Not surprisingly, decreasing the minimum head constraint increases the overall yield of the system. The relation is approximately linear until the lower constraint approaches the steady-state (original) aquifer condition.

A similar trend is apparent in Figure 6.20, where the objective is to maximize the overall planning periods of the pumping well heads. The dominant constraints of the model are the water targets to be satisfied in each planning period. The objective is a piecewise linear function of the water demand. However, for demands

TABLE 6.5 Optimal Pumping Pattern

Demands[1] (m^3/day)	Well Sites 13	24	35	46	21	32	43	54
10,000		1,2,3[2]	1,3					
50,000		1,2,3	2,3					
100,000		1,2,3	1,2,3					
300,000	1,2,3	1,2,3	1,2,3	1,2,3		1,2,3	1,2,3	

[1] Well capacity 60,000 m^3/day

[2] Planning/operational periods

exceeding 300,000 m^3/day, the head objective decreases significantly, indicating the upper limit for the development of the groundwater supply.

The feasibility of developing the groundwater supply is dependent on the magnitude of the recharge occurring in the basin. Pumping induces recharge through the Dirichlet boundaries. The magnitude of the recharge can be estimated using Darcy's law (Eq. 2.3) or

$$Q_x = -K_x A \frac{\partial h}{\partial x} \tag{6.32a}$$

where K_x is the hydraulic conductivity in the x direction, A is the cross-sectional area, and $\partial h/\partial x$ is the head gradient evaluated at the Dirichlet boundaries. The finite-element approximation of Equation 6.32a is

$$Q_x = - KA\mathbf{N}_x \hat{\mathbf{h}} \tag{6.32b}$$

where $\mathbf{N}_x$ is the derivative of the shape function with respect to x and $\hat{\mathbf{h}}$ is the vector containing the nodal values of the head. Groundwater recharge accounts for 60 percent of the total withdrawal. The balance of the demand is taken directly from groundwater storage. The increase in recharge and any decrease in discharge that results from the groundwater pumping is known as *capture* (Lohman, 1972). Capture may also occur as a decrease in groundwater discharge to streams or lakes or a decrease in evapotranspiration from the unsaturated zone. System equilibrium will occur when the capture is balanced by the water pumped from the groundwater basin.

Example Problem 6.7 Optimization Model for an Unconfined Aquifer System

The unconfined aquifer system shown in Figure 6.21 is to be developed as a source of water supply. The aquifer is homogeneous and isotropic; the permeability is 86.4 m/day. The planning horizon consists of three 50-day operational periods and the

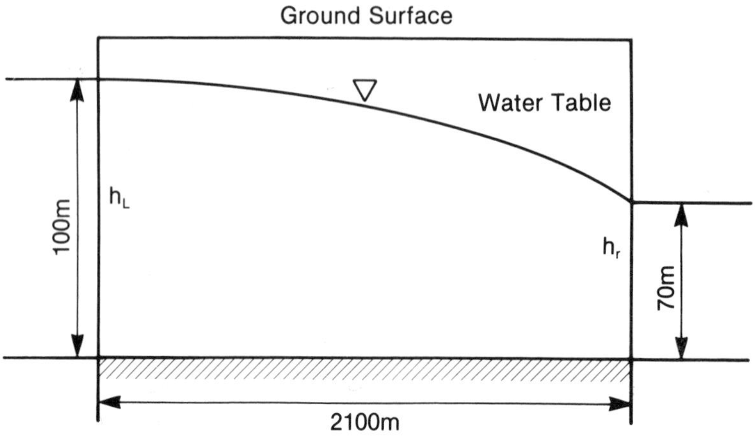

FIGURE 6.21 Unconfined aquifer system

demands for each period are 750, 850, and 950 m^3/day, respectively. Assuming that pumping wells can be located at any point in the aquifer system, determine the optimal pumping schedule and head distribution if the system objective is to maximize the pumping well heads. Determine the optimal pumping policies if the objective is to minimize the operational costs.

The hydraulics of the problem are described by the Boussinesq equation, or

$$K \frac{\partial}{\partial x}\left(h \frac{\partial h}{\partial x}\right) - \sum_{w \in \pi} Q_w \delta(x - x_w) = S_y \frac{\partial h}{\partial t} \tag{2.82}$$

where h is the hydraulic head, K is the hydraulic conductivity, S_y is the specific yield, Q is the pumping rate, and x_w is the coordinate of a well site. The index set, π, again defines the location of all feasible well locations.

The hydraulic response equations can be obtained by developing a finite-difference model of the Boussinesq equation. Introducing a second-order approximation for the spatial derivative, the finite-difference equations may be expressed as

$$S_y \dot{h}_i = \frac{K}{2\Delta x^2}\{h_{i+1}^2 - 2h_i^2 + h_{i-1}^2\} - \frac{Q_w}{\Delta x} \tag{6.33}$$

where Δx is the spatial interval. Writing the equation for all internal nodes in the system and incorporating the boundary conditions, the hydraulic response equations are given as

$$\alpha \dot{\mathbf{h}} = \mathbf{A}\mathbf{h}^2 + \mathbf{r} \tag{6.34}$$

where A is tridiagonal with elements $(1, -2, 1)$, $\mathbf{h}^2 = (h_1^2, h_2^2, \cdots h_n^2)^T$, $\mathbf{r} =$

$(h_1^2 - 2Q_1\Delta x/K, -2Q_2\Delta x/K \ldots\ldots h_r^2 - 2Q_n\Delta x/K)^T$, and $\alpha = 2S_y\Delta x^2/K$. These nonlinear equations relate the heads and pumping rates for each Δt over the entire planning horizon.

The optimization model for the groundwater system is similar to the model presented in Example Problem 6.6. Again, the objective is to maximize the sum of the pumping heads over each operational period, or

$$\max z = \sum_k \sum_{w \in \pi} h_w^k \tag{6.35a}$$

The decisions are constrained in each planning period by the (1) water demands, or

$$\sum_{w \in \pi} Q_w^k \geq D^k, \; \forall t \tag{6.35b}$$

(2) well capacity limitation,

$$Q_w^k \leq Q_{w,\max}, \; w \in \pi, \; \forall t \tag{6.35c}$$

and the nonlinear hydraulic response equations. The planning model is a nonlinear, nonconvex programming problem.

Optimal pumping schedules for the planning model were obtained using MINOS, a projected Lagrangian algorithm discussed in Chapter 5. The programming problem consisted of 300 nonlinear response equation constraints, 300 nonlinear decision variables (the hydraulic head at each nodal point for each time step), and 60 linear variables, the pumping rates. The aquifer system was discretized into 20 nodal points. The system parameters are summarized in Table 6.6. Solution of the model required approximately 1000 CPU seconds on a CDC 170/720 computing system.

The results of the optimization are shown in Figure 6.22. The optimal head profiles are essentially identical to the heads obtained by numerically integrating

TABLE 6.6 System Parameters

Parameter	*Value*
Hydraulic conductivity	86.4 m/day
Length of planning period	50 days
Head lower bound	40 m
Pumping upper bound	100 m^3/day
Specific yield	0.1
Distance step (Δx)	100 m
Time step (Δt)	10 days

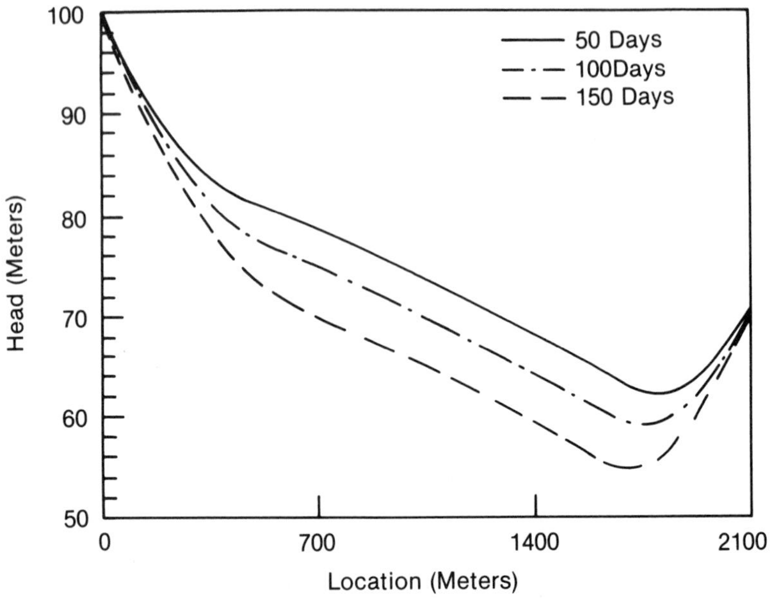

FIGURE 6.22 Optimal head profiles

the nonlinear response equations using a fourth-order Runge-Kutta method. The head profiles are shown for 50, 100, and 150 days. The effects of round-off and truncation error are minimal.

The optimal pumping policies for the example are shown in Table 6.7. The optimal well locations are again located adjacent to the constant head boundaries. As the demand increases over time, more and more wells are operated in the interior of the basin. The wells are always located near the boundary where recharge is the greatest. The optimization model exploits the capture or recharge occurring in the system in determining the optimal well locations.

In the second part of the example problem, the groundwater extraction costs are minimized over the planning horizon. The costs, at first approximation, are proportional to the product of the pumping rate and the lift at each well site (Nelson & Bush, 1967). The cost function for any time period k can be expressed as

$$\sum_{w \in \pi} Q_w^k C_w (L_w - h_w^k) \tag{6.36}$$

where L_w is the depth from the ground surface to the lower boundary of the aquifer at well site w and C_w is the cost per unit lift per unit flowrate at site w. The operational cost is a concave function of the decision variables, Q_w^k and h_w^k.

Again, MINOS was used to obtain solutions to the nonlinear programming problem. The optimal, discounted operational costs are approximately \$117,900 for a two-period operational horizon (100 days). The optimal head distribution at

TABLE 6.7 Optimal Pumping Schedule

Node Number	Planning Period		
	t = 0	*t = 50 days*	*t = 100 days*
1	100	100	100
2	100	100	100
3	100	100	100
4	100	100	100
5	50	100	100
6	0	50	100
7	0	0	50
8	0	0	0
9	0	0	0
10	0	0	0
11	0	0	0
12	0	0	0
13	0	0	0
14	0	0	0
15	0	0	0
16	0	0	0
17	0	0	0
18	100	100	100
19	100	100	100
20	100	100	100

Pumping rate = m^3/day

the end of each planning period is shown in Figure 6.23. The solutions are identical to the Runge-Kutta solution of the response equations. The optimal pumping pattern is summarized in Table 6.8. As in the previous examples, the optimal well pattern places the active wells as close to the boundaries as possible. In contrast to the linear objective problem, however, pumping is non-zero everywhere in the basin. This is caused in part by the nonlinearity of the model and the effect of round-off and discretization errors. Apparently, the more nonlinear the problem, the more pronounced these errors become.

The effects of local optimality were also investigated by varying the initial approximations of the decision variables and the Lagrange multipliers. However, in all cases, changes in the starting points did not alter the optimal head and pumping distribution. The model required approximately 15 minutes of CPU time to identify the optimal planning decisions.

Quasilinearization can also be used for solution of the nonlinear optimization model. As discussed by Willis and Finney (1985), the linearization algorithm involves the iterative solution of a series of convex programming problems. Numerical experiments have demonstrated that the optimal solutions are comparable

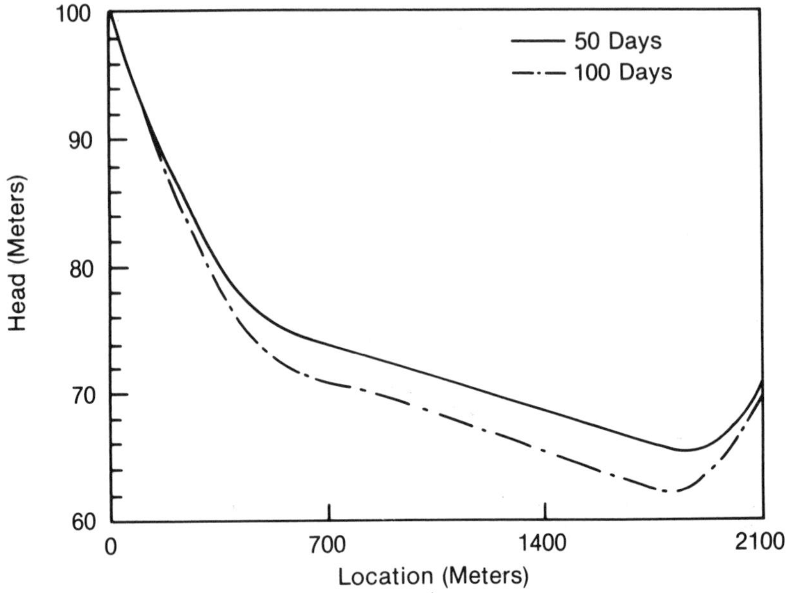

FIGURE 6.23 Optimal head profiles minimizing operational costs

TABLE 6.8 Optimal Pumping Schedule

	Planning Period	
Node	t = *0*	t = *50 days*
1	100	100
2	100	100
3	100	100
4	100	100
5	100	100
6	16.1	52.1
7	0.4	0.5
8	0.4	0.6
9	0.4	0.6
10	0.4	0.6
11	0.4	0.6
12	0.4	0.6
13	0.4	0.6
14	0.4	0.5
15	0.4	0.5
16	0.4	0.5
17	0.4	0.4
18	29.6	92.2
19	100	100
20	100	100

Pumping Rate-m^3/day

to the local solutions identified using MINOS. Computer execution times and storage requirements are substantially less, however.

6.4 THE CAPACITY EXPANSION PROBLEM

The operational and allocation models that were presented in Sections 6.2 and 6.3 were based on the assumption that the well sites of the groundwater system were developed prior to the operational analysis. In other words, the capital costs associated with the development of the well field were neglected. In the capacity expansion problem, both operational and capital investment decisions are considered over the entire planning horizon. Specifically, the capacity expansion problem involves:

1. The selection over the planning horizon of the well sites that are to be developed. This is accomplished in such a way as to always satisfy the water demand.
2. The determination in any time period of the optimal groundwater pumping pattern.

Again, it is assumed that the discounted capital and operational costs are to be minimized. The overall planning problem is usually formulated as a mixed integer programming model (Maddock, 1972b). The integer variables are necessary to define the timing and staging of well-field development and the auxiliary pipeline and storage facilities.

We begin the development of the planning model by considering the capital costs for well-field development and the construction of pipeline and storage facilities. The capital cost for development of the groundwater is dependent on the size of the well field, the formation properties, the well capacity, and the overall depth to the aquifer (Forste, 1973). Assuming the dominant factor influencing the capital cost is the well capacity, Q_w^*, the total discounted capital cost can be expressed as

$$\sum_{k=1}^{T^*} \sum_{w \in \pi} \frac{KC_w(Q_w^*)\, X_{kw}}{(1 + \rho)^k}$$

where we have introduced the zero-one integer variables, X_{kw}, to indicate whether or not well site w is developed in period k, or

$$X_{kw} = \begin{bmatrix} 1 \text{ if site } w \text{ is developed in period } k \\ 0 \text{ otherwise} \end{bmatrix}$$

ρ is the discount factor and T^* is the length of the planning horizon. The index set π defines the set of all possible well sites in the groundwater basin.

Similarly, the capital cost function for the pipeline facilities can also be defined in terms of the integer variable Z_{kw}, or

$$Z_{kw} = \begin{bmatrix} 1 \text{ if pipeline facilities are constructed at well site } w \text{ in period } k \\ 0 \text{ otherwise} \end{bmatrix}$$

The discounted capital cost function for a well site is then

$$\sum_{k=1}^{T^*} \frac{C_p Z_{kw}}{(1 + \rho)^k}$$

where C_p is the capital cost.

Integer variables are also necessary to define the groundwater extraction costs. These variables will determine whether or not a particular well site is in operation during planning period k, or

$$Y_{kw} = \begin{bmatrix} 1 \text{ if site } w \text{ is in operation in period } k \\ 0 \text{ otherwise} \end{bmatrix}$$

The operational costs can then be expressed as

$$\sum_{k=1}^{T^*} \sum_{w \in \pi} C_w(Q_w^k, h_w^k)\, Y_{kw}$$

The overall objective of the capacity expansion problem can now be expressed as the sum of the discounted capital and operational costs, or

$$\min z = \sum_{k=1}^{T^*} \left(\frac{1}{1 + \rho}\right)^k \sum_{w \in \pi} [KC_w\,(Q_w^*)\, X_{kw} + C_w\,(Q_w^k, h_w^k) Y_{kw} + C_p\, Z_{kw}] \tag{6.37a}$$

The optimal decisions will again define the timing and staging of well-field development and the optimal production pattern of the aquifer system.

The policies of the planning model are constrained by:

1. The water target in each planning period, k, D^k

$$\sum_{w \in \pi} Y_{kw}\, Q_w^k \geq D^k \tag{6.37b}$$

2. The well capacity limitation. In the integer programming model, these constraints relate the extraction rates, well capacities, and the operational (integers) variables, or

$$Q^{*k} - Y_{kw}\, Q_w^k \leqslant 0 \tag{6.37c}$$

3. The response equations,

$$\mathbf{h}^k - A_1(T)\mathbf{h}^{k-1} - A_2\,(T)\,\mathbf{g}\,(\mathbf{Q}^{k-1}) = \mathbf{0} \tag{6.37d}$$

4. Possible state variable restrictions,

$$h_a^k \leqslant h_a^*,\ a \epsilon \Delta \tag{6.37e}$$

5. and the non-negativity of the decision variables

$$\mathbf{h}^k, Q_w^k \geqslant 0, \quad w \epsilon \pi, \qquad \forall k$$

The integer variables are also required to be 0, 1 variables, or

$$X_{kw}, Y_{kw}\, Z_{kw} = [0, 1] \tag{6.37f}$$

We also insure that, over the entire planning horizon, a maximum of one well and one pipeline facility can be developed at each site, or

$$\sum_{k=1}^{T^*} X_{kw} \leqslant 1,\ w \epsilon \pi \tag{6.37g}$$

$$\sum_{k=1}^{T^*} Z_{kw} \leqslant 1,\ w \epsilon \pi \tag{6.37h}$$

There also may be a limitation on the availability of capital in any planning period, or

$$\sum_{w \epsilon \pi} \{C_p\, Z_{kw} + KC_w\, X_{kw}\} \leqslant M_k \tag{6.37i}$$

where M_k is the upper bound on capital investment in period k.

Finally, we have the constraints reflecting the time lag between well and pipeline construction and the actual operation of the wells. Assuming a one-period time lag between construction and eventual operation of the well site, these constraints can be expressed as

$$Y_{kw} - \sum_{j=1}^{k-1} X_{jw} = 0, \qquad \forall w \epsilon \pi \qquad (6.37j)$$

$$Y_{kw} - \sum_{j=1}^{k-1} Z_{kw} = 0, \qquad \forall w \epsilon \pi \qquad (6.37k)$$

Equations 6.37 constitutes the mixed integer capacity expansion model.

The capacity expansion model is most useful for long-term groundwater planning. The optimal policies of the model, which will have to be determined using mixed integer programming algorithms, will define whether or not the basin can satisfy the imposed water targets of the problem and the potential role of groundwater in regional water resources management.

Because of the uncertainties associated with demand projection and changing cost functions, the optimal policies of the capacity expansion model should be viewed as guidelines for the optimal development of the groundwater supply of the basin. In reality, as more and better information becomes available, the model may be resolved and the optimal capacity expansion policies updated. This iterative approach allows the planner to more accurately assess the optimality of the current development pattern of the basin as the cost functions and demands of the model are updated with current economic and demographic data.

Example Problem 6.8 Mixed Integer Programming Model

The groundwater system shown in Figure 6.24 is to be developed as a source of water supply. The system parameters are summarized in Table 6.9. Develop a steady-state integer programming model to determine the optimal well sites and extraction rates. Assume that the response equation may be expressed as

$$B\mathbf{h} + \mathbf{Q} = -\mathbf{l} \qquad (6.38)$$

where $\mathbf{l}$ contains the Dirichlet boundary conditions of the problem.

We first consider the objective function of the planning model. In lieu of economic data, the capital and operational costs will be represented by surrogate objectives and constraints. Assuming that the operational objective is to maximize the yield of the system, then the objective may be expressed as

$$\max z = \sum_{w \epsilon \pi} Q_w Y_w \qquad (6.39a)$$

where π again defines the set of potential well sites and Y_w is a zero-one integer

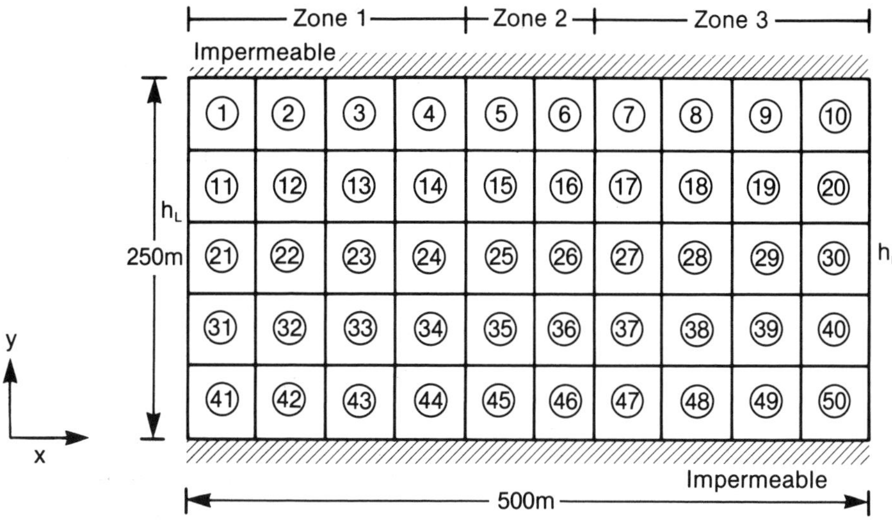

FIGURE 6.24 Example aquifer and finite-element grid

variable defining whether or not well site w is in operation. However, the objective necessitates that the heads be constrained by lower bounds, or

$$h_a \geq h_a^*, \; a \epsilon \Delta \tag{6.39b}$$

where Δ is the set of all internal nodal points.

The capital cost function is defined in terms of the integer variables Y_w. As a surrogate for minimizing the capital costs, we introduce a constraint limiting the

TABLE 6.9 Example Aquifer Parameters

Zone	*Transmissivity* (m^2/day)	*Angle of rotation* (rad)
1	864	0
2	432	0
3	216	0
Dirichlet Boundaries		
h_L(m)	100	
h_r(m)	50	

maximum number (W) of potential well sites, or

$$\sum_{w} Y_w \leq W \tag{6.39c}$$

Additional constraints of the planning model also include (1) the well capacity restriction,

$$Q_w \leq Q_w^*, \quad w \in \pi \tag{6.39d}$$

and, (2) an upper bound on development at well site w, (D_w), or

$$Q_w^* \geq Y_w D_w, \quad w \in \pi \tag{6.39e}$$

Typical results of the planning model are presented in Figures 6.25 and 6.26. The aquifer yield for both capacity limitations is a linear function of the head lower bound. The trade-off or slope of the curve for the 10-well case is 50,000 m³/day

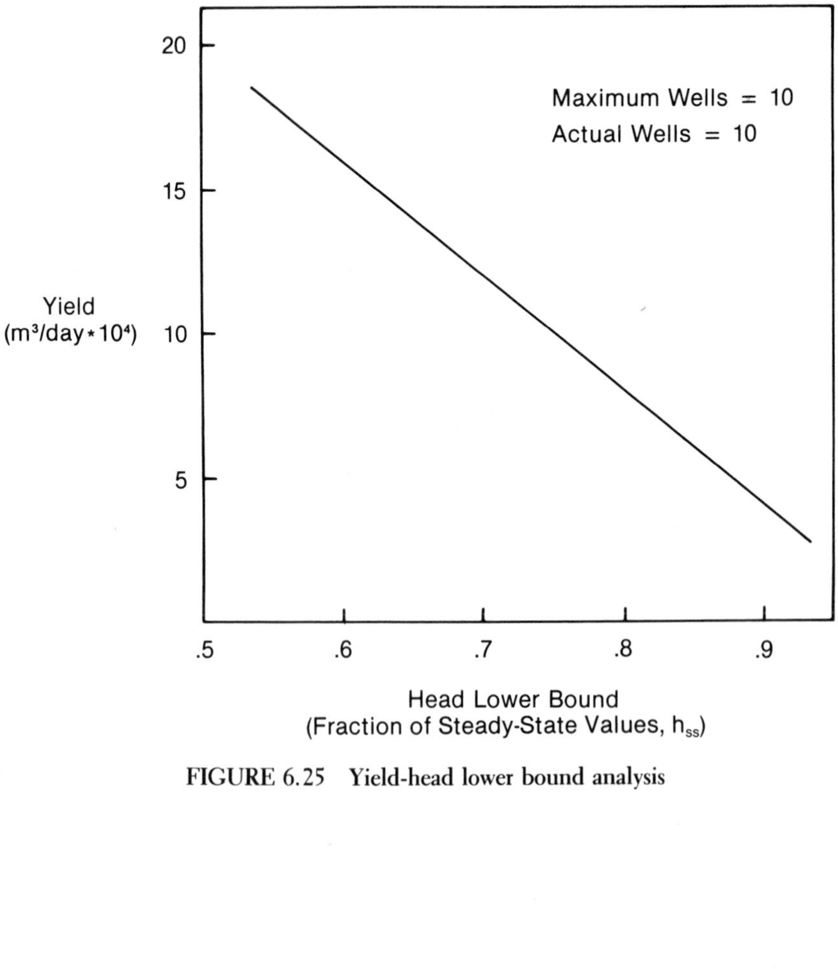

FIGURE 6.25 Yield-head lower bound analysis

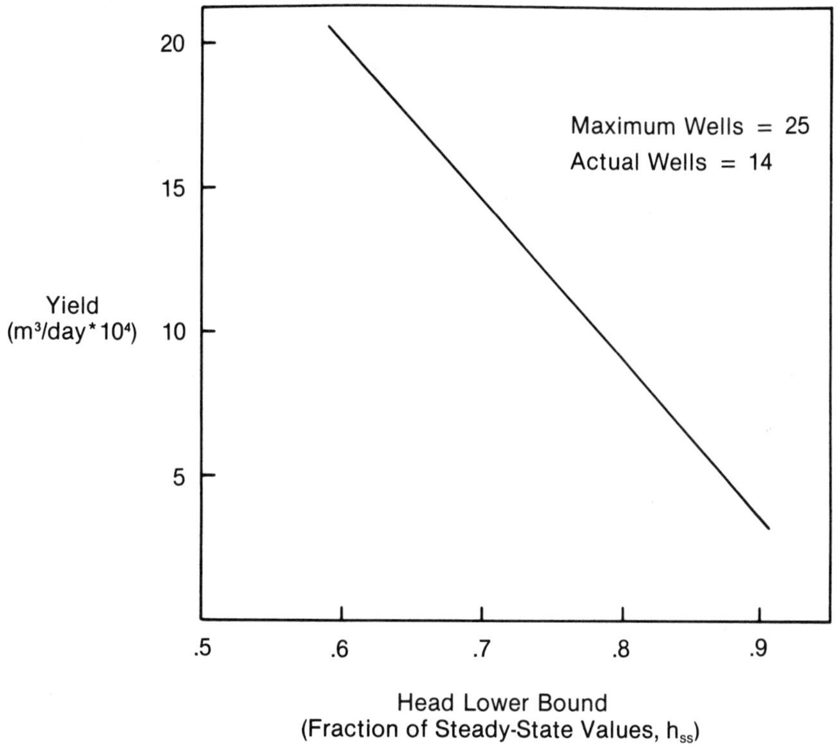

FIGURE 6.26 Yield-head lower bound analysis

for a 10 percent reduction in the lower bounds. As the number of wells increases, the slope of the yield function also increases to 60,000 m^3/day. As the yield increases, there is a relatively larger decrease in the overall head in the groundwater system. The yield of the aquifer also increases as the upper bound on the maximum number of wells increases. If, for example, the head lower bounds are 80 percent, then the change in the yield per unit increase in the maximum number of well sites is approximately 13,000 m^3/day.

The optimal pumping pattern is again similar to the other example problems. Wells were generally located adjacent to the Dirichlet boundaries and in the more permeable regions of the aquifer.

The effect of a $\pm$ 10 percent variation in the hydraulic conductivity is presented in Table 6.10. A 10 percent decrease in K and a head lower bound of 90 percent of the initial, undeveloped conditions (h_{ss}) increases the number of active well sites by 50 percent. The yield also increased by 12.7 percent. A 10 percent increase in K increases the yield of the system slightly (6.4 percent). However, the number of active well sites remains unchanged.

TABLE 6.10 Transmissivity Sensitivity Analysis

T	*Number of Wells*	*% Change*	ΣQ_i *(m³/day)*	*% Change*
Original	10	—	77532	—
−10%	10	0	67507	−12.7
+10%	10	0	82351	6.4

Example Problem 6.9 Capacity Expansion Problem

The confined aquifer shown in Figure 6.27 is to be developed as a source of water supply. Develop an integer programming model to identify the optimal well sites. The water demands for a four-year planning horizon are shown in Table 6.11. The problem information is summarized in Table 6.12. Assume a one-period time lag between well construction and operation and that each operational period is four months in duration.

Neglecting the pipeline design problem, the capacity expansion model is described by Equations 6.37a, d, and e. In the example problem, however, the operational costs will be replaced by an objective maximizing, over time, the sum of the pumping well heads. The objective function of the planning model can then

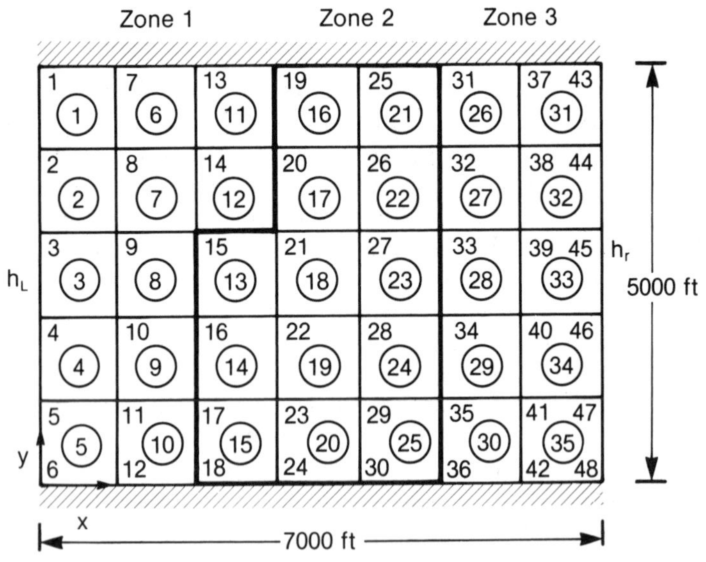

FIGURE 6.27 Example aquifer

TABLE 6.11 Demand Schedules

Period	*Schedule A*[1]	*Schedule B*[1]
1	500	5000
2	1000	100000
3	2000	200000
4	500	50000
5	3000	300000
6	4000	400000
7	500	50000
8	5000	500000
9	6000	600000
10	500	50000
11	7000	700000
12	8000	800000

[1] m^3/day

be expressed as

$$\max z = \sum_k \left\{ \sum_{w \in \pi} h_w^k - \lambda \sum_{w \in \pi} \frac{KC_w X_w^k}{(1+\rho)^k} \right\} \tag{6.40}$$

where λ is a weighting factor representing the relative weighting of the pumping well heads, a surrogate for operational costs, and the capital costs.

The constraints of the planning model include well capacity constraints, head lower bounds, the well construction constraints, and the hydraulic response equations.

The response equations were developed in Example Problem 4.3 using the Galerkin finite-element method. The finite-element grid, which is also shown in Figure 6.26, consists of 35 quadrilateral elements and 48 nodes. The response equations were solved analytically in time using the matrix calculus. The solutions

TABLE 6.12 System Parameters

Parameter	*Zone 1*	*Zone 2*	*Zone 3*
Storage Coefficient	0.075	0.060	0.050
Transmissivity (ft^2/day)	30	24	20
Thickness	250	250	250
Q_{max} (ft^3/day)	70000	70000	70000
h_{min} (ft)	300	300	300

TABLE 6.13 Well Development Schedule[1]

Demand Schedule	*Nodes*[2]											
	8	9	10	11	14	15	16	17	20	21	22	23
A[3]				1		7		4				11
B[3]	7			1			4					10
B[4]	10	4	1	1								
B[5]						7	10	4				1

[1] Beginning of period
[2] Feasibile nodes
[3] Head ≤ 80%
[4] Head ≤ 50%
[5] Head ≤ 90%

were then used to relate the head values and pumping rates over each four-month operational period. The optimization model had a total of 432 linear response equations and 864 head and pumping decision variables. There were also 72 [0, 1] integer variables for each planning period.

The capacity expansion model was solved using CDC's mixed integer programming package. Computer runs averaged 2600 seconds on a CYBER 170/720 computing system.

The optimal development pattern of the groundwater basin is shown in Table 6.13. Development schedules are presented for various demand schedules and head constraints. The optimal solutions are constrained by well capacity limitations of 70,000 ft^3/day and head bounds that are constrained to be at least greater than 80 percent of the initial levels in the basin. Four wells are sufficient to meet the water demands for both demand schedules. The active well sites are again located in the western, more permeable regions of the groundwater basin.

The effect of the lower bound constraints on the optimal well development sites is also illustrated in Table 6.13. If, for example, the head bounds are decreased to 50 percent, all the active well sites are in zone 1, adjacent to the recharge boundary. Increasing the bounds shifts the well locations to the most permeable regions of the aquifer system.

6.5 CONJUNCTIVE GROUNDWATER AND SURFACE WATER PLANNING MODEL

Conjunctive groundwater and surface water management models are resource allocation models of the water resources system. These types of models, as we saw in Section 6.2, optimally distribute, over a planning or design horizon, the water resources of a river basin to competing water demands or water uses. By controlling the total water resources of a region, conjunctive use planning can increase the efficiency, reliability, and cost-effectiveness of water use, particularly in river basins with spatial or temporal imbalances in water demands and natural supplies.

Rarely do regions of high rainfall and run-off coincide with those of extensive water development and water demand. Rather, periods of lowest streamflow and recharge usually coincide with the largest demand, or vice versa. Conjunctive water management can reduce these deficiencies by using groundwater to supplement scarce surface water supplies during the drier seasons. During periods of medium or high run-off, surface water can then be used to satisfy the water demands and to recharge the groundwater systems in spreading basins, abandoned stream channels, and wells. In hydraulically coupled stream-aquifer systems, as occur, for example, in alluvial formations, conjunctive use can also affect the magnitude and timing of irrigation return flows and capture intended for downstream water use.

The conjunctive use, planning, or operational problem can again be formulated as an optimization model of the water resources system. The decision or control variables of the model are the groundwater and surface water allocations in each planning period. The optimal decisions maximize the objectives of the water resource system while satisfying the hydraulic response equations of the surface and groundwater systems, and any constraints limiting the head variations and the surface water availability.

In the optimization analysis, we will assume that the water resources system consists of M groundwater sources ($i = 1, \ldots M$), N surface water supplies ($j = 1, \ldots N$) and L water demands ($k = 1, \ldots L$) (see Figure 6.28). Considering first the objective function of the planning model, we will assume, as in Section 6.2, that the water demands are known or can be inferred from market information. The appropriate economic benefit resulting from the allocating of groundwater and surface water is the sum of the consumer surplus and the revenue generated through the sale of the water. Equivalently, the benefit function is the willingness to pay for the resource or the area beneath the demand functions for each water user. Defining GW^t_{jk} as the groundwater allocated from groundwater basin i to demand k, SW^t_{jk} as the surface water from source j used to satisfy demand k in period t, $p_k(Q^t_k)$ is the demand function for each water use, the economic benefit function for any time period can be expressed as

$$\sum_k \int_0^{Q^t_k} p_k(Q^t_k)\, dQ$$

where Q^t_k is the total water allocation for demand k in period t, or

$$Q^t_k = \sum_i GW^t_{ik} + \sum_j SW^t_{jk} \tag{6.41}$$

The system objective will also include the capital, operation, and maintenance

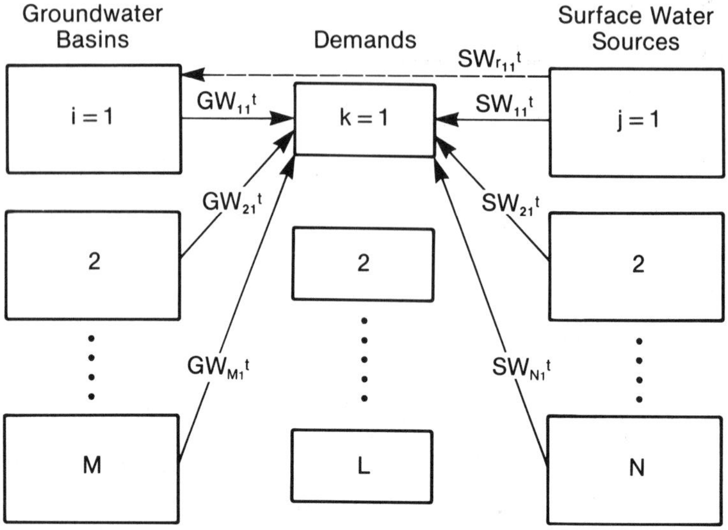

FIGURE 6.28 Conjunctive use model

costs. We assume that these costs, for any time period and demand, are functions of the groundwater and surface water allocations. The capital, operation, and maintenance costs for any time period are then,

$$\sum_k \sum_i f_{ik}^t (GW_{ik}^t) + \sum_k \sum_j \bar{f}_{jk}^t (SW_{jk}^t) + \sum_i \sum_j f_{ji}^t (SWR_{ji}^t) \tag{6.42}$$

where f_{ik}^t, $\bar{f}_{jk}^t$ and f_{ji}^t are the costs associated with allocating groundwater from basin i to demand k, surface water from source j to demand k, and the surface water for the artificial recharge of basin i. SWR_{ji}^t is the volume of water for artificial recharge in basin i, from source j, in period t.

The objective function of the planning model is to maximize the net discounted economic benefits

$$\max z = \sum_t \left(\frac{1}{1+\rho}\right)^t \left\{ \sum_k \left[\int_0^{Q_k^t} p_k (Q_k^t) dQ - \sum_i f_{ik}^t (GW_{ik}^t) - \sum_j \bar{f}_{jk}^t (SW_{jk}^t) \right] - \sum_i \sum_j f_{ji}^t (SWR_{ji}^t) \right\} \tag{6.43}$$

The groundwater and surface water allocations are constrained within any planning period by the hydraulic response equations, balance equations for pumping and recharge, and possible limitations on surface water availability and head variations in each groundwater system. The groundwater recharge constraints prescribe that the recharge target for each groundwater system is satisfied for any planning period. The recharge balance equation is

$$\sum_j SWR_{ji}^t = GWR_i^t, \quad \forall i,t \tag{6.44}$$

where GWR_i^t is the target recharge rate for basin i, period t.

The hydraulics of the surface and groundwater systems are again described by the system's equations. Assuming no hydraulic interaction between the surface and groundwater systems, the surface water system's equations may be expressed as

$$\overline{SW}_j^t = \overline{SW}_j^{t-1} + R_j^t - \sum_k SW_{jk}^t - \sum_i SWR_{ji}^t \tag{6.45}$$

where $\overline{SW}_j^t$ is the volume of water in surface storage, source j, at the end of time period t, and R_j^t is the inflow or additions to the surface water supplies during period t. The variables are generally random, exogenous variables of the planning model.

The surface water storage volumes are also constrained in any period by the relations

$$\overline{SW}_j^t \leq SW_j^*, \quad \forall j,t \tag{6.46}$$

where SW_j^* is the capacity of surface source j.

The response equations of the groundwater system can be developed from the models presented in Chapters 2 and 4. For a linear groundwater system, the response equations can be expressed functionally as

$$\mathbf{h}_i^t = \mathrm{A}_1^i \mathbf{h}_i^{t-1} + \mathrm{A}_2^i \mathbf{g}\left(\sum_i \mathrm{GW}_{ik}^{t-1}, \quad \sum_j \mathrm{SWR}_{ji}^{t-1}\right) \tag{6.47}$$

where $\mathbf{h}_i^t$ is the vector of heads in groundwater system i at the end of period t and A_1^i and A_2^i are the response matrices of each basin.

The balance equations relating the groundwater allocations and the pumping schedules in each basin can be expressed as

$$\sum_k \mathrm{GW}_{ik} = \sum_{w \in \pi i} Q_{w,i}^t, \quad \forall i \tag{6.48}$$

where π_i defines the active well sites in groundwater basin i. We assume that the artificial recharge of the groundwater system can occur in injection wells or surface water spreading operations, or

$$\sum_j SWR_{ji}^t = \sum_{r \in \psi i} Q_r^t, \ \forall i$$

or (6.49)

$$= \sum_{s \in w i} R_{s,i}$$

where ψ_i is an index set defining the injection well locations, Q_r^t is the injection rate, and R_s is recharge rate at spreading basin s. The location of the recharge facilities in basin i is defined by the set, ω_i.

Finally, to prevent excessive groundwater overdraft or water logging of the agricultural areas of the basin, constraints are also introduced to limit the head variation in each aquifer system, or

$$h_{i,a}^t \leq h_{i,a}^*, \; a \epsilon \Delta_i \tag{6.50}$$

where h^* are the permitted head levels and Δ_i is an index set defining the control locations in groundwater basin i.

Equations 6.41 through 6.50 constitute the optimal conjunction use planning model. In general, the model is a nonlinear programming problem.

The example problems illustrate the same common uses of the conjunctive use model in water resources planning.

Example Problem 6.10

Develop a conjunctive use model to determine the optimal allocation of surface and groundwater in the river basin shown in Figure 6.29. Assume that the groundwater system is a confined aquifer and the estimated demand for water use in any period t is given as

$$p(Q^t) = \alpha(Q^t)^\beta \tag{6.51}$$

where Q^t is the total surface and groundwater allocation in period t, p is the unit price, and α and β are parameters.

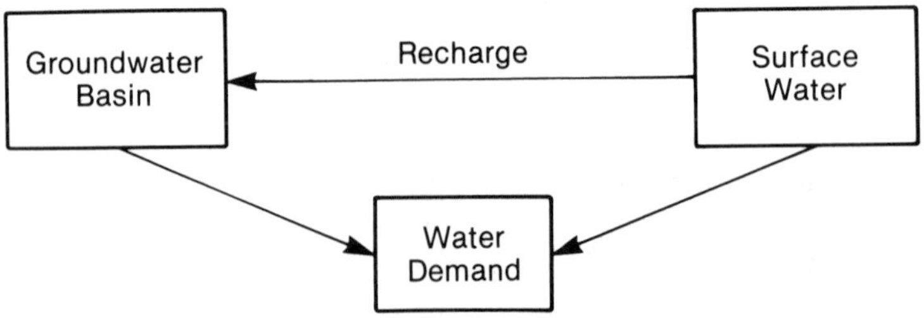

FIGURE 6.29 Conjunctive use model for Example Problem 6.10

The planning model is similar to the model presented in Section 6.5. The example problem has, however, a single water demand and two water sources—a surface reservoir and a groundwater basin. The groundwater and surface allocations in each period satisfy the balance equations

$$Q^t = GW^t + SW^t \tag{6.52}$$

where GW^t and SW^t are the groundwater and surface allocations in period t.

The objective function of the planning model again includes the cost of groundwater pumping, water distribution costs, and the economic benefit that is measured by the willingness to pay for the water supply. The cost of groundwater can include the cost of pumping and the cost of boosting the water to a distribution reservoir. For any time period, we assume that these costs can be expressed as

$$\sum_{w \in \pi} \left\{ C_B Q_w^t + Q_w^t C_p (L_w - h_w^t) \right\}$$

where C_B is the unit cost to boost the water from each well site to the surface distribution reservoir, and C_p is the unit operational costs for each well site. Q_w^t is the well discharge at site w in period t, and L_w is the distance from the ground surface to the base of the aquifer at site w. Q_w^t is related to the total groundwater allocations in any period by the equation,

$$GW^t = \sum_{w \in \pi} Q_w^t$$

The cost of surface water usually includes the cost of source development and the annual costs associated with the operation and maintenance of the system. Assuming that these costs can be represented by a fixed-unit cost, C_s, the cost in any time period can be expressed as

$$C_s SW^t$$

Similarly, water distribution costs that include annual costs for the depreciation of fixed assets such as distribution reservoirs and pipeline networks, will also be expressed as

$$C_d \left(SW^t + \sum_{w \in \pi} Q_w^t\right)$$

where C_d is the unit annual operation and maintenance cost.

We will also assume that the costs of artificial recharge in any period can be expressed in terms of the unit recharge cost, C_r, as

$$C_r SWR^t$$

where SWR^t is the surface water allocated to groundwater recharge in period t.

The system objective is again to maximize the net discounted benefits from operating the system over the planning horizon, or

$$\max z = \sum_t \left(\frac{1}{1+\rho}\right)^t \left\{\int_0^{Q^t} \alpha(Q^t)^\beta dQ - \sum_{w\in\pi} \left\{C_B Q_w^t + Q_w^t C_p(L_w - h_w^t)\right\}\right.$$

$$\left. - (C_s + Cd)SW^t - C_d \sum_{w\in\pi} Q_w^t - C_r SWR^t\right\} \quad (6.53a)$$

where ρ is the discount rate. The constraints of the planning model include:

1. The balance equations for the surface water system,

$$\overline{SW^t} = \overline{SW^{t-1}} - SW^t + R^t - SWR^t \quad (6.53b)$$

   where $\overline{SW^t}$ is the surface storage at the end of period t and R^t is the addition or inflow in period t to the surface water system.
2. The capacity limitations of the surface water system.

$$\overline{SW^t} \leq \overline{SW^*},\ \forall t \quad (6.53c)$$

3. The groundwater system's equations,

$$\mathbf{h}^t = A_1\mathbf{h}^{t-1} + A_2\mathbf{g}\left\{\sum_{w\in\pi} Q_w^{t-1},\ SWR^{t-1}\right\},\ \forall t \quad (6.53d)$$

4. Constraints limiting the maximum and minimum groundwater levels

$$h_{a,\min} \leq h_a^t \leq h_{a,\max},\quad a\in\Delta,\ \forall t \quad (6.53e)$$

Again, the planning model is a nonlinear programming problem.

The conjunctive use planning model was solved using the sequential unconstrained minimization technique (SUMT) discussed in Chapter 5, using demand and parameter information presented by Mobasheri and Grant (1971). The planning horizon consisted of 25 yearly operational periods.

The results of the model are presented in Figure 6.30 and 6.31. The variation in the water supplies over the planning horizon is shown in Figure 6.30. The temporal changes in the pumping lifts are summarized in Figure 6.31. Initially, the market price for the water is \$135/acre-feet. This, however, increases to \$148/acre-feet at the end of the planning horizon, reflecting changing demand conditions.

Figure 6.30 also presents the variation in the water supply if the pricing

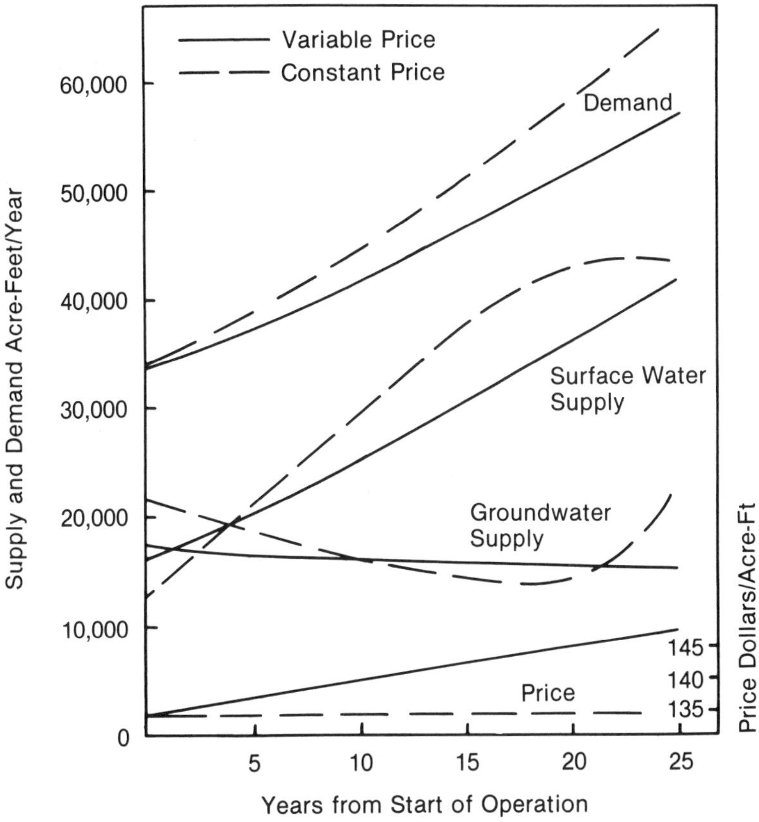

FIGURE 6.30 Demand, supply, and water price for optimum operation of the conjunctive system

structure of the allocation problem is ignored. The benefit function in this case is the revenue generated by supplying the surface and groundwater at a given market price. The results demonstrate that, for fixed-market prices, water supplies tend to decrease over time and pumping lift costs become significantly higher than in the variable pricing case.

Example Problem 6.11 Conjunctive Water Use

Groundwater pumping and canal diversion schedules are to be determined for the water resources system shown in Figure 6.32 (Longenbaugh, 1970). The groundwater basin is an unconfined aquifer; seepage and percolation of precipitation and irrigation water account for the majority of groundwater recharge. Develop an optimization model to determine the optimal groundwater and surface water al-

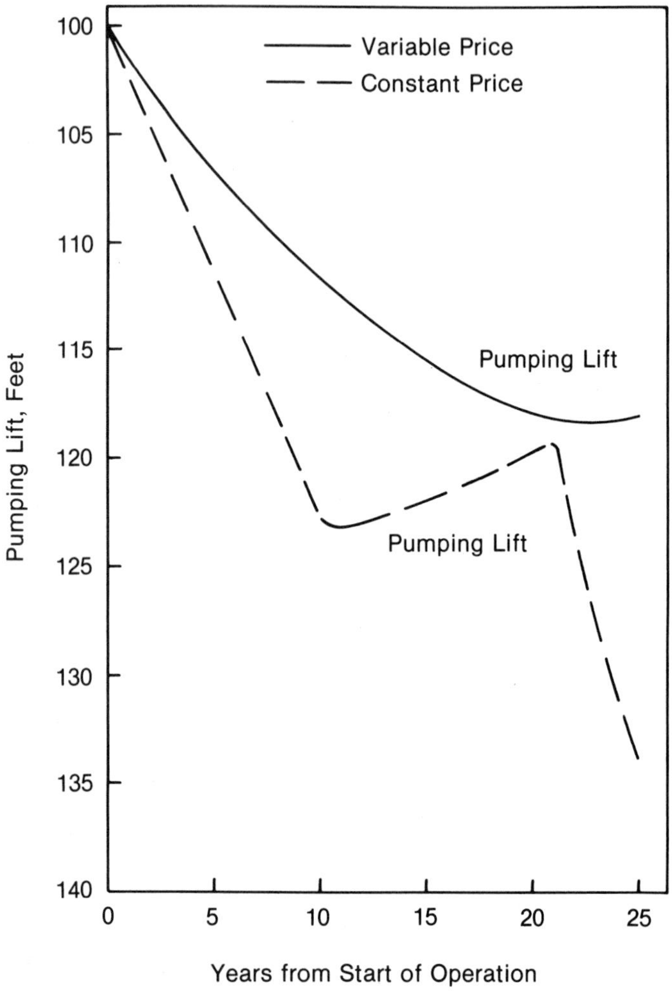

FIGURE 6.31 Pumping lift change over planning time horizon

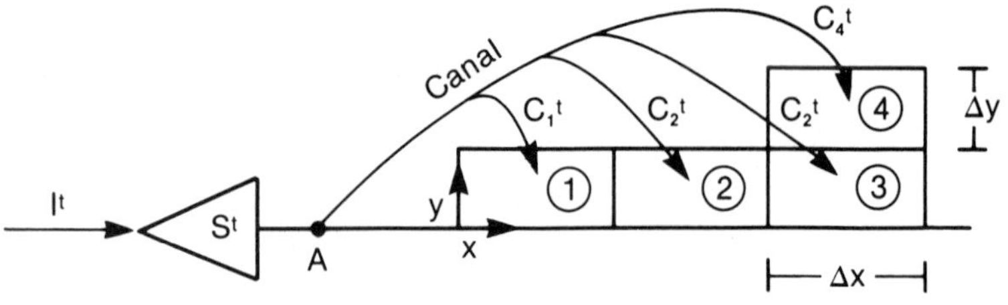

FIGURE 6.32 Water resources system

locations. Assume that the system objective is to minimize the total cost of meeting the forecasted agricultural water demand and that the stream can be represented by a constant head (Dirichlet) boundary condition.

Assuming the drawdowns are small in comparison with the saturated thickness of the aquifer, the hydraulics of the groundwater systems can be described by the confined flow equation,

$$\frac{\partial}{\partial x}\left(T\frac{\partial h}{\partial x}\right) + \frac{\partial}{\partial y}\left(T\frac{\partial h}{\partial y}\right) + R(x,y,t) - P(x,y,t) = S_y\frac{\partial h}{\partial t} \qquad (2.55)$$

where h is the head, T is the transmissivity, P is the groundwater pumping, and R is the groundwater recharge. S_y is again the specific yield. A second-order finite-difference model was developed in Example Problem 6.5 for this type of groundwater flow model. The hydraulic response equations are again obtained by writing the difference equation for each computational element and solving the resulting ordinary differential equations. For the four-element systems shown in Figure 6.32, the response equations are

$$\mathbf{h}^t = A_1\mathbf{h}^{t-1} + A_2\mathbf{g}(R_E^{t-1}, P_E^{t-1}) \qquad (6.54)$$

where $\mathbf{h}$ is the vector of heads at the end of planning period t, A_1 and A_2 are the response coefficient matrices, and R^t and P^t are the recharge and pumping occurring in zone or element E during time period t. Note, the $\mathbf{g}$ vector also contains the Dirichlet boundary conditions.

The groundwater recharge that occurs in any zone or element is assumed to be a linear function of the canal diversion in that zone, C_E^t and the precipitation, P_E^t, or

$$R_E^t = R_E^t(C_E^t, P_E^t) \qquad (6.55)$$

The hydraulics of the surface water system are again described by the continuity equation. Defining S^t as the surface water storage at the end of period, the surface reservoir equation is

$$S^t = S^{t-1} + I^t - E^t - SR^t \qquad (6.56)$$

where I^t is the flow in period t, E^t is the evaporation loss, and SR^t is the reservoir release. The surface water storages are also constrained by the relations

$$S_{min} \le S^t \le S_{max}, \ \forall t \qquad (6.57)$$

where S_{max} and S_{min} are the maximum and minimum storage volumes.

The balance constraint at diversion point A can be expressed as

$$\sum C_E^t \le SR^t, \ \forall t \qquad (6.58)$$

The constraint ensures that the total diversion in any time period cannot exceed the reservoir release.

Streamflow downstream of the diversion point is also governed by the continuity equation. For any reach or finite-difference cell, the continuity or mass balance equation may be expressed as

$$SF_k^t = SF_k^{t-1} - ET_k^t \pm RFLO_k^t + DR_k^t \tag{6.59}$$

where SF_k^t is the streamflow at the head of reach k, period t, ET is the evapotranspiration loss, $RFLO_k^t$ is the inflow to or outflow from the aquifer system, and DR_k^t is the distributed run-off entering reach k. The inflow (outflow) to reach k is given by Darcy's law, averaged over the planning period, T, or

$$RFLO_k^{T-1} = -\frac{KA}{T}\int_0^T \left(\frac{\partial h}{\partial y}\right)_k\Bigg|_{y=0} dt \tag{6.60}$$

where K is the hydraulic conductivity, A is the cross-sectional area of the reach, and $\left(\frac{\partial h}{\partial y}\right)_k$ the hydraulic gradient, is evaluated at the stream-aquifer boundary of reach k ($y = 0$).

The water demand constraint for each element or zone during each planning period can be expressed as

$$C_E^t + P_E^t + DEF_E^t \geq D_E^t, \forall t \tag{6.61}$$

where D_E^t is the water demand in zone E and DEF_E^t is the water deficit in zone E.

The objective function of the planning model minimizes the total cost of meeting the water demand. These costs include the pumping well costs, the canal delivery costs, and the costs associated with a water deficit in any planning period. Assuming unit costs for each of these costs, the objective may be expressed as

$$\min z = \sum_t \sum_E (C_d DEF_E^t + C_p P_E^t + C_c C_E^t) \tag{6.62}$$

where C_p, C_c, and C_d are the unit costs associated with pumping, canal diversions, and any water deficits. The conjunctive use model is a linear programming problem.

The optimization model was solved using the simplex algorithm for a three-operational period (30 days) and four finite-difference cells. The model consisted of 49 constraints, 24 pumping and canal decision variables, and 30 state variables characterizing the hydraulic heads in the system. The data for the problem is representative of the Arkansas River Valley in Colorado. The optimal water allocations are summarized in Table 6.14.

TABLE 6.14 Optimal Water Allocations (acre-feet/month)

	Operational Period		
	1	2	3
Canal Diversions			
C_1	139.2	373.8	0.
C_2	426.7	711.1	59.4
C_3	0.	81.4	237.3
C_4	426.7	711.1	533.3
Groundwater Pumping			
P_1	101.4	216.8	342.0
P_2	0.	0.	304.7
P_3	274.3	404.8	190.3
P_4	0.	0.	0.

From R.A. Longenbaugh, "Determining Optimum Operational Policies For Conjunctive Use of Ground and Surface Water Using Linear Programming." *ASCE, 18th Annual Specialty Conference*, Minneapolis, Minnesota, 1970. Reprinted with permission from the American Society of Civil Engineers.

6.6 CLOSING COMMENTS

The planning problems associated with the management of the groundwater supply have been based on the hydraulic response equations of the aquifer system. As we have seen, these equations may be developed for regional distributed parameter groundwater systems using finite-difference or finite-element methods. The hydraulic response equations, because they relate the state, hydraulic head, and decision variables and the pumping or artificial recharge schedules, can be incorporated as a set of auxiliary constraints in the management or planning problem. The optimal solution of the operational or capacity expansion model identifies, not only the optimal planning or design policies, but simultaneously predicts the hydraulic head distribution in the aquifer system. As a result, the response equation approach combines simulation and optimization in a single management model. In Chapter 7, we will examine how this approach can be used to develop optimal planning models for the control of groundwater quality.

6.7 REFERENCES

Aguado, E., & Remson, I. "Groundwater Management with Fixed Charges." ASCE, *J. of Water Resources Planning and Management Division*, 106(WR2):375–382, 1980.

Aron, G., & Scott, V.H. "Dynamic Programming for Conjunctive Water Use." ASCE, *J. Hydraulics Division*, 97(HY5):705–720.

Burt, O.R. "The Economics of Conjunctive Use of Ground and Surface Water." *Hilgardia*. University of California, Division of Agricultural Sciences, Davis, 1964.

EDI. *An Economic Analysis of Water Uses within the Truckee-Carson System*. Bureau of Outdoor Recreation, U.S. Dept. of the Interior, 1971.

Forste, R.H. "Verification of Groundwater Capital Costs." Technical Completion Report, A-023-HN, Water Resources Research Center, University of New Hampshire, Durham, 1973.

Gisser, M. "Linear Programming Models for Estimating the Agricultural Demand Function for Imported Water in the Pecos River Basin." *Water Resources Research*, 6:1025–1032, 1970.

Howe, C.W. "Economic Models." In *Systems Approach to Water Management*, A.K. Biswas, Ed. New York: McGraw-Hill, 1976.

Intriligator, M.D. *Mathematical Optimization and Economic Theory*. Englewood Cliffs NJ: Prentice-Hall, 1971.

Lohman, S.W. "Groundwater Hydraulics." *U.S.G.S. Prof. Paper 708*, 1972.

Longenbaugh, R.A. "Determining Optimum Operational Policies for Conjunctive Use of Ground and Surface Water Using Linear Programming." *ASCE, 18th Annual Speciality Conference*, Minneapolis, Minnesota, 1970.

Maddock, T., III. "Algebraic Technological Function from a Simulation Model." *Water Resources Research*, 8(1):129–134, 1972a.

Maddock, T., III. "A Groundwater Planning Model—A Basis for Data Collection Network." *Proceedings of the International Symposium on Uncertainties in Hydrologic and Water Resource Systems*, University of Arizona, Tucson, 1972b.

Maddock, T., III. "The Operation of a Stream-Aquifer System under Stochastic Demands." *Water Resources Research*, 10(1):1–10, 1974.

Mangasarian, O.L. *Nonlinear Programming*. New York: McGraw-Hill, 1969.

Marglin, S. "The Objectives of Water-Resource Development, A General Statement." In *Design of Water Resource Systems*, A. Maass et al., Eds. Cambridge MA: Harvard University Press, 1982.

Moench, A.F., & Visocky, A.P. "A Preliminary Least Cost Study of Future Groundwater Development in Northeastern Illinois." *Circular 102*, Illinois State Water Bureau, Urbana, 1971.

Nelson, A.G., & Busch, C.D. "Costs of Pumping Water in Central Arizona." *Technical Bulletin 182*, Arizona Agricultural Experiment Station, 1967.

Noel, J.E., & Howitt, R.E. "Conjunctive Multibasin Management: An Optimal Control Approach." *Water Resources Research*, 18(4):758–763, 1982.

Rosenwald, G.W., & Green, D.W. "A Method for Determining the Optimum Location of Wells in a Reservoir Using Mixed-Integer Programming." *J. of Petroleum Engineering*, 44–54, 1974.

Schwarz, J. "Linear Models for Groundwater Management." *J. of Hydrology*, 28:377–392, 1976.

Sharples, J.A. "The Representative Farm Approach to Estimation of Supply Response." *Am. J. Agric. Econ.*, 51–D353-361, 1969.

Willis, R., & Newman, B.A. "Management Model for Groundwater Development." *ASCE, J. Water Resource Planning and Management Division*, 103(WR1):159–171, 1977.

Willis, R., & Finney, B.A. "Optimal Control of Nonlinear Groundwater Hydraulics: Theoretical Development and Numerical Experiments." *Water Resources Research*, 21(10):1476–1482, 1985.

7. GROUNDWATER QUALITY MANAGEMENT MODELS

7.1 INTRODUCTION

Mathematical models of the groundwater quality system are based on the hydraulic and mass transport equations that were developed in Chapters 2 and 3. In simulation or optimization modeling, these equations are used to describe how the groundwater quality system responds to, or is affected by, waste contaminants. These substances may be introduced to the aquifer system either through hydraulic interaction with polluted surface waters or directly through injection wells and/or spreading basins or seepage from waste disposal sites.

Well injection of wastewaters is the most common form of waste disposal. In the United States, there are over 300 injection wells disposing of oil field brines and other noxious and radioactive substances in deep, saline aquifers. Although the economics of deep-well injection are attractive in comparison with other forms of conventional wastewater treatment, the normally high injection pressures of the disposal operations create the potential for serious environmental problems. Water quality problems have resulted from the migration of contaminated groundwater through faults, abandoned wells, or leakage into freshwater aquifers. Increasing evidence also suggests the correlation between injection and increased seismic activity, possibly because of the lubrication or pressurization of stressed faults.

The surface disposal of domestic and industrial wastes in sanitary landfills, lagoons, septic tanks, and land disposal systems are other common forms of waste disposal. In land disposal systems, the wastewaters are discharged as spray irrigation or overland flow, or in spreading basins such as abandoned stream channels. The assimilative waste capacity of the soil and groundwater systems provides additional wastewater treatment and containment of the waste residuals. This *water quality resource* or assimilative capacity is the ability of the aquifer to attenuate certain wastewater constituents through ion-exchange, adsorption, chemical and biochemical reactions, and the natural filtering capacity of the porous media. For example, coliforms are usually attenuated within the first few feet of the ground surface. Suspended solids and biochemical oxygen demand (BOD) removal efficiencies have also been reported as high as 90 percent. Phosphorous removal by adsorption can range from 70 to 95 percent while overall nitrogen removal can vary from 10 to 80 percent of the initial nitrogen load. In return, however, these disposal operations return nutrients and trace minerals to the soil and recharge groundwater reservoirs.

Subsurface waste disposal invariably poses problems of groundwater contamination. The Environmental Protection Agency (EPA) has identified, for example, over 15,800 potentially hazardous waste sites in the United States and has targeted 418 of these sites for clean-up in conjunction with the Superfund. Importantly, in 70 percent of these sites, contaminants present some threat to groundwater supplies.

Aquifer reclamation or restoration involves implementing remedial control measures for the rehabilitation of contaminated groundwater supplies. These options include physical containment, in situ rehabilitation, and withdrawal followed by treatment and use (Lehr & Nielsen, 1982). Physical containment systems prevent the flow of contaminated groundwater by controlling the flow field via slurry trenches, cut-off walls, or grout curtains or by altering the circulation pattern of the aquifer system through pumping or injection. Typically, aquifer rehabilitation involves injection and recharge systems that are augmented by chemical treatment. A neutralizing or biochemical degradation agent may also be injected into the contaminated zone. Withdrawal and treatment does not, however, exploit or utilize the aquifer's assimilative waste capacity; it simply removes the contaminated water from the groundwater system. Chemical treatment processes for aquifer decontamination are described by Landon and Sylvester (1982), Yaniga (1982) for hydrocarbon contamination; Stover (1982), McBride (1982), and Ohneck and Gardner (1982) for removal of toxic organic chemicals; and Molsather and Barr (1982) and Giddings (1982) for landfill leachate containment.

From a planning perspective, the central issues in groundwater quality management are

1. To ensure that surface or subsurface waste disposal is accompanied by minimal impact on the groundwater environment.
2. To optimize the waste treatment capacity of the soil and groundwater systems.

The feasibility of subsurface waste disposal or waste treatment is predicated on the hydraulic, water quality, and environmental response of the groundwater system. In this chapter, we examine how the response equations of the aquifer system may be used to develop simulation and optimization groundwater quality management models. Specifically, planning models will be developed to:

1. Simulate or predict the behavior of waste materials in aquifer systems.
2. Conjunctively manage the water supply and water quality of regional groundwater systems.

We begin by considering the simulation or predictive approach to groundwater quality management.

7.2 GROUNDWATER QUALITY SIMULATION MODELS

The simulation or prediction of groundwater quality can be described by two classes of groundwater quality models. In the first category are the density-dependent mass transport models. The physical processes of these type of models are described by:

1. The continuity equation of the aquifer system,

$$-\nabla \cdot \rho \mathbf{q} = \rho(\alpha + n\beta)\frac{\partial p}{\partial t} \tag{2.45}$$

2. The motion equation or momentum balance for the system,

$$\mathbf{q} = -\frac{k\rho g}{\mu}\nabla\left(\frac{p}{\rho g} + z\right) \tag{2.5}$$

3. The mass transport equation for each constituent,

$$\nabla \cdot \rho \mathbf{D}\, \nabla(c_s/\rho) - \nabla \cdot c_s\mathbf{q} = nc_s\alpha\frac{\partial p}{\partial t} + \frac{\partial}{\partial t}(nc_s), \quad s\epsilon\chi \tag{3.11}$$

   where χ is an index set defining the constituents in the groundwater basin.
4. The equations defining the (a) solute concentration,

$$\rho = \sum_{s\epsilon\chi} c_s \tag{3.2}$$

   (b) the variation in the solute density

$$\rho = f(c_s; p) \tag{3.12a}$$

   and (c) the viscosity changes

$$\mu = f_2(c_s; p) \tag{3.12b}$$

This coupled set of nonlinear partial differential equations can be used to describe salt water intrusion in coastal aquifers or to evaluate contamination problems where the solution density of the groundwater system is significantly affected by the waste materials.

For many contaminant transport problems, however, these equations can be simplified to a linear system of partial differential equations. This presumes that the overall solution density is relatively constant and that there are negligible temporal pressure variations. This second class of groundwater quality models is described by

1. The continuity equation, which now can be expressed in terms of the hydraulic head

$$-\nabla \cdot \mathbf{q} = S_s \frac{\partial h}{\partial t} \tag{2.47}$$

2. The motion equation (Darcy's law)

$$\mathbf{q} = -\mathbf{K}\nabla h \tag{2.5}$$

3. and the mass transport equation for each constituent, or

$$\nabla \cdot \mathbf{D}\nabla c_s - \nabla \cdot c_s\mathbf{q} = \frac{\partial}{\partial t}(nc_s), \; s \epsilon \chi \tag{3.14}$$

The complete mathematical description of the groundwater quality problem again requires a set of boundary and initial conditions for the aquifer system. As discussed in Chapter 3, the density-dependent transport models require that $\mathbf{q}, \rho$, and c_s or their derivatives be specified on the aquifer's boundaries. For the linear groundwater model, the hydraulic head and the mass concentrations (or their normal derivatives) must also be given on the boundary of the groundwater system.

In the next section, we explore several methods for the solution of the groundwater quality problem.

7.3 SOLUTION METHODS FOR THE GROUNDWATER QUALITY PREDICTION PROBLEM

The numerical solution of the mass transport problem has been approached using finite-difference, particle tracking (the method of characteristics), and finite-element techniques. The wide range in solution methods is primarily the result of the failure of conventional techniques to minimize the stability problems and the oscillations and numerical diffusion that are associated with the solution of the convection-dispersion equation. We first show how the problems arise in the finite-difference solution of the dispersion model and then illustrate two alternative solution approaches—the method of characteristics and the multiple-cell balance method.

7.3.1 Finite-Difference Methods

Early numerical models of the convection-dispersion equation often relied on conventional second-order finite-difference approximations. However, the truncation error of these approximations introduces oscillations in the numerical solution of the mass transport model, especially in those regions with sharp concentration gradients where convection dominates dispersion (Price, et al., 1968). The oscillations can be controlled at the expense of a smearing of the concentration front provided that an artificial dispersion coefficient is introduced in the numerical model. Several studies have shown, for example, that the smearing could be reduced without generating oscillatory numerical solutions (Chaudhari, 1971; Bresler, 1973; Lantz, 1971).

To illustrate the development of a finite-difference model and the attendant numerical problems, consider the one-dimensional convection-dispersion equation,

$$\frac{\partial(nc)}{\partial\tau} = D\frac{\partial^2 c}{\partial\eta^2} - q_\eta \frac{\partial c}{\partial\eta} \tag{7.1}$$

where τ is time, η is the spatial variable, c is the solute concentration, D is the dispersion coefficient, which is assumed constant, q_η is the Darcy velocity, and n is the porosity. The initial and boundary conditions for the problem are described by

$$\begin{aligned} c(\eta, 0) &= 0,\ \eta > 0 \\ c(0, t) &= C_0,\ t \geq 0 \\ c(\infty, t) &\quad \text{finite} \end{aligned} \tag{7.2}$$

To facilitate the analysis of the transport equation, we introduce the dimensionless time and spatial variables

$$t = \tau/\Delta\tau, \qquad x = \eta/\Delta\eta$$

where $\Delta\tau$ and $\Delta\eta$ are uniform finite-difference intervals (with $\Delta\tau = \Delta\eta$). Equation 7.1 can then be expressed as

$$\frac{\partial c}{\partial t} = D^*\frac{\partial^2 c}{\partial x^2} - v_x\frac{\partial c}{\partial x} \tag{7.3}$$

where $v_x = [q_\eta\Delta\tau/\Delta\eta n]$, $D^* = [D\Delta\tau/n(\Delta\eta)]^2$, and v_x is the pore velocity.

As was described in Chapter 4, finite-difference approximations of the model can be developed by developing Taylor series approximations of the time and spatial derivatives. Defining i and k as indices of the space and time discretization steps, the first-order spatial derivatives can be represented by the series,

$$\frac{c_{i,k} - c_{i-1,k}}{\Delta x} = \frac{\partial c}{\partial x} - \frac{\Delta x}{2}\frac{\partial^2 c}{\partial x^2} + \frac{(\Delta x)^2}{3!}\frac{\partial^3 c}{\partial x^3} + \cdots \tag{7.4}$$

The second-order spatial derivatives are given as

$$\frac{c_{i+1,k} - 2c_{i,k} + c_{i-1,k}}{(\Delta x)^2} = \frac{\partial^2 c}{\partial x^2} + \frac{(\Delta x)^2}{12!}\frac{\partial^4 c}{\partial x^4} + \cdots \tag{7.5}$$

Introducing a backward approximation for the first-order spatial derivative, an explicit finite-difference model can be expressed as

$$\frac{c_{i,k+1} - c_{i,k}}{\Delta t} = \frac{D^*}{(\Delta x)^2}\{c_{i+1,k} - 2c_{i,k} + c_{i-1,k}\} - v_x\left\{\frac{c_{i,k} - c_{i-1,k}}{\Delta x}\right\} \tag{7.6}$$

Comparing the difference equations with the original governing equation, the actual differential equation that is being approximated is

$$\frac{\partial c}{\partial t} + \frac{\Delta t}{2}\frac{\partial^2 c}{\partial t^2} = D^*\frac{\partial^2 c}{\partial x^2} - v_x\frac{\partial c}{\partial x} + \frac{v_x \Delta x}{2}\frac{\partial^2 c}{\partial x^2} \tag{7.7}$$

where only the second-order terms of the expansion have been retained. Apparently, the difference approximation introduces an additional dispersional or diffusional term that is equal to $v_x\Delta x/2$. The time derivative is similarly affected. The truncation error is proportional to $\Delta t/2$. Equivalence between the time and spatial derivatives can be obtained by differentiation of Equation 7.1 with respect to time and space. The second-order time derivative can then be expressed as

$$\frac{\partial^2 c}{\partial t^2} = -v_x D^*\frac{\partial^3 c}{\partial x^3} + v_x^2\frac{\partial c}{\partial x} + \frac{v_x \Delta x}{2}\frac{\partial^2 c}{\partial x^2} \tag{7.8}$$

The total dispersion coefficient or diffusivity is the sum of the actual physical dispersion coefficient and the numerical diffusivity, or

$$D^* + \frac{v_x \Delta x}{2} + \frac{v_x^2 \Delta t}{2}$$

A more general method for developing higher-order approximations to the convection-dispersion equation is given by Van Genuchten and Gray (1978). In the approach, the difference approximation to Equation 7.3 can be expressed as

$$\frac{\Delta c}{\Delta t} = \left[-a_1 v_x \frac{\Delta c}{\Delta x} + a_2 D^* \frac{\Delta^2 c}{\Delta x^2}\right]^{k+1} + \left[-b_1 v_x \frac{\Delta c}{\Delta x} + b_2 D^* \frac{\Delta^2 c}{\Delta x^2}\right]^{k} \tag{7.9}$$

where $\Delta c/\Delta t$ is given by

$$\frac{\Delta c}{\Delta t} \cong \frac{c^{k+1} - c^k}{\Delta t} \tag{7.10}$$

The coefficients a, b are determined such that the approximation becomes a higher-order accurate approximation to the mass transport equation.

If we consider time corrections only, the spatial derivatives of the approximation can be replaced by the appropriate time derivatives. This equivalence can again be obtained by progressively solving the transport equation for the convective transport term, or

$$\begin{aligned} v_x \frac{\partial c}{\partial x} &= -\frac{\partial c}{\partial t} + D^* \frac{\partial c}{\partial x^2} \\ v_x^2 \frac{\partial c}{\partial x} &= -v_x \frac{\partial c}{\partial t} + D^* \frac{\partial}{\partial x}\left(v_x \frac{\partial c}{\partial x}\right) \\ v_x^2 \frac{\partial c}{\partial x} &= -v_x \frac{\partial c}{\partial t} - D^* \frac{\partial}{\partial x}\left(\frac{\partial c}{\partial t}\right) + D^{*2} \frac{\partial^3 c}{\partial x^3} \end{aligned} \tag{7.11}$$

Substituting these equations in the difference approximation, we obtain

$$\begin{aligned} \frac{\partial c}{\partial t} = &\left[a_1 \frac{\partial c}{\partial t} + (a_2 - a_1)\frac{D^*}{v_x^2}\frac{\partial^2 c}{\partial t^2} + \frac{2D^{*2}}{v_x^4}(a_1 - a_2)\frac{\partial^3 c}{\partial t^3} + \frac{5D^{*3}}{v_x^6}(a_2 - a_1)\frac{\partial^4 c}{\partial x^4}\right]^{k+1} \\ &+ \left[b_1 \frac{\partial c}{\partial t} + (b_2 - b_1)\frac{D^*}{v_x^2}\frac{\partial^2 c}{\partial t^2} + \frac{2D^{*2}}{v_x^4}(b_1 - b_2)\frac{\partial^3 c}{\partial t^3} + \frac{5D^{*3}}{v_x^6}(b_2 - b_1)\frac{\partial^4 c}{\partial x^4}\right]^{k} \end{aligned} \tag{7.12}$$

A second-order correct implicit difference scheme can be easily developed from Equation 7.12. In the implicit method, the right-hand side of the equation is zero (the time level k is not considered). b_1 and b_2 are then zero. Optimal values for the two independent parameters a_1, a_2 can be determined by requiring that Equation 7.12 be exact for $c = t^2$. Substituting this relation into Equation 7.12 and recognizing that $\Delta\tau = 1$, we obtain

$$(t + 1)^2 - t^2 = 2a_1(t + 1) + \frac{2(a_2 - a_1)D^*}{v_x^2} \tag{7.13}$$

Equating the coefficients of like powers of t on both sides of the equation and solving the equations yields

$$a_1 = 1 \tag{7.14a}$$

$$a_2 = 1 - v_x^2/2D^* \tag{7.14b}$$

Substituting these parameters into the difference model, we obtain the second-order, time corrected finite-difference scheme,

$$\frac{\Delta c}{\Delta t} = \left[- v_x \frac{\Delta c}{\Delta x} + \left(D^* - \frac{v_x^2}{2} \right) \frac{\Delta^2 c}{\Delta x^2} \right]^{k+1} \tag{7.15}$$

TABLE 7.1 Corrected Finite-Difference Approximations to the Convection-Dispersion Equations

Second-Order, Time Corrected
Implicit

$$\frac{\Delta c}{\Delta t} = \left[- v_x \frac{\Delta c}{\Delta x} + \left(D^* - \frac{v_x^2}{2} \right) \frac{\Delta^2 c}{\Delta x^2} \right]^{k+1}$$

Second-Order, Time Corrected
Explicit

$$\frac{\Delta c}{\Delta t} = \left[- v_x \frac{\Delta c}{\Delta x} + \left(D^* + \frac{v_x^2}{2} \right) \frac{\Delta^2 c}{\Delta x^2} \right]^{k}$$

Third-Order, Time Corrected
Crank-Nicolson

$$\frac{\Delta c}{\Delta t} = \frac{1}{2}\left[- v_x \frac{\Delta c}{\Delta x} + \left(D^* - \frac{1}{6} v_x^2 \right) \frac{\Delta^2 c}{\Delta x^2} \right]^{k+1} + \frac{1}{2}\left[- v_x \frac{\Delta c}{\Delta x} + \left(D^* + \frac{1}{6} v_x^2 \right) \frac{\Delta^2 c}{\Delta x^2} \right]^{k}$$

Fourth-Order, Time-Space Corrected
Scheme

$$\frac{\Delta c}{\Delta t} = \theta \left[- v_x \frac{\Delta c}{\Delta x} + \left(D^* - \frac{\gamma_1 v_x^2}{6} \right) \frac{\Delta^2 c}{\Delta x^2} \right]^{k+1} + (1 - \theta) \left[- v_x \frac{\Delta c}{\Delta x} + \left(D^* + \frac{\gamma_2 v_x^2}{6} \right) \frac{\Delta c^2}{\Delta x^2} \right]^{k}$$

$$\theta = \frac{1}{2} + \frac{D^*(2v_x^2 - 1)}{v_x^4 + 12D^* - v_x^2}$$

$$\gamma_1 = \frac{60D^{*2} - 6D^* + 12D^* v_x^2 - 2 + v_x^2 + v_x^4}{12D^{*2} - 2D^* + 4D^* v_x^2 - v_x^2 + v_x^4}$$

$$\gamma_2 = \frac{60D^{*2} + 6D^* - 12D v_x^2 - 2 + v_x^2 + v_x^4}{12D^{*2} + 2D^* - 4D^* v_x^2 - v_x^2 + v_x^4}$$

From M. T. Van Genuchten and W. G. Gray, "Analysis of Some Dispersion Corrected Numerical Schemes for Solution of the Transport Equation." *International Journal for Numerical Methods in Engineering*, 12:387-404, 1978. Copyright 1978 by John Wiley & Sons, Ltd. Reprinted by permission.

Similar approximations can be developed for higher-order spatial derivative approximations.

Table 7.1 summarizes the higher-order approximations to the convention-dispersion equations. The fourth-order correct schemes have been given by Van Genuchten and Gray (1978).

The effects of the truncation error of the numerical solution of the one-dimensional convection dispersion equation are shown in Figures 7.1 and 7.2. The analytical solution, shown in the illustrations, is given by Lapidus and Amundson (1952) for the initial and boundary conditions $c(x, 0) = 0, x > 0$ $c(0, t) = 1, t \geq$

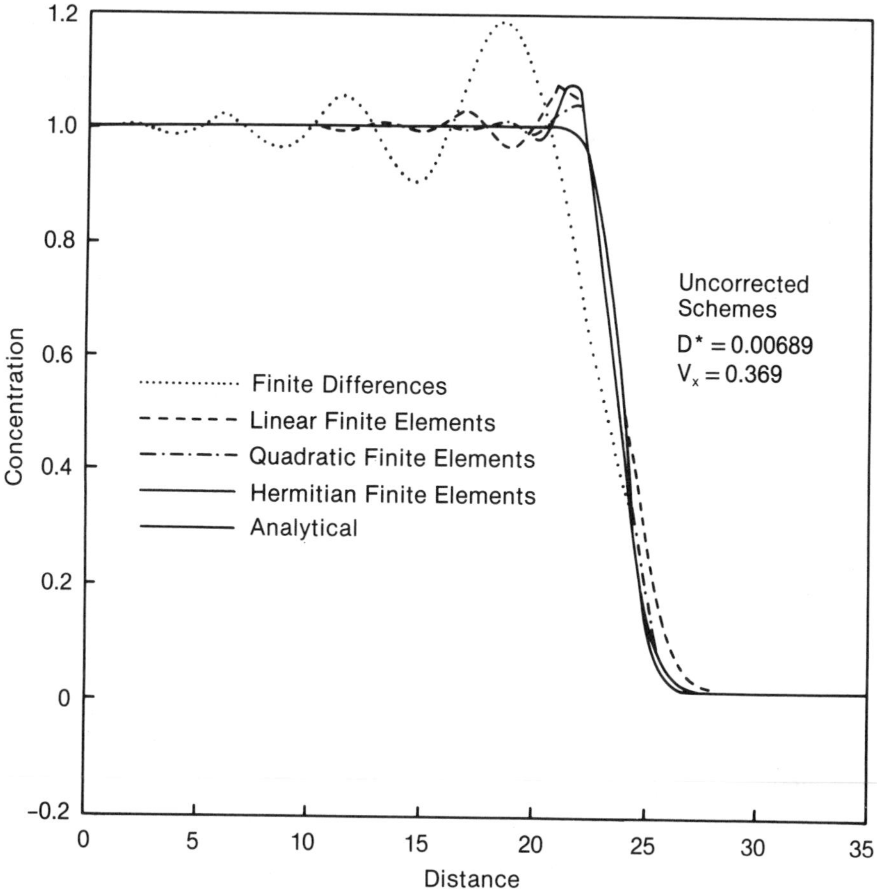

FIGURE 7.1 Concentration profiles obtained with various uncorrected finite-difference and finite-element schemes. (From M. Th. Van Genuchten and W. G. Gray, "Analysis of Some Dispersion Corrected Numerical Schemes for Solution of the Transport Equation." *International Journal for Numerical Methods in Engineering,* Vol. 12, 1978, p. 395. Copyright © 1978 by John Wiley & Sons, Ltd. Reprinted with permission)

0, and $c(\infty, t)$ finite, as

$$c = \frac{1}{2}\,\mathrm{erfc}\left\{\frac{x - v_x t}{\sqrt{4D^* t}}\right\} + \frac{1}{2}\,\exp\left\{\frac{v_x x}{D^*}\right\}\,\mathrm{erfc}\left\{\frac{x + v_x t}{\sqrt{4D^* t}}\right\} \tag{7.16}$$

The concentration profiles for both finite-difference and various finite-elements are shown in Figure 7.1. All the results are uncorrected for numerical diffusion; all the schemes are also time-centered. Note that the finite-difference approximation performs poorly and none of the methods can remove the oscillations upstream of

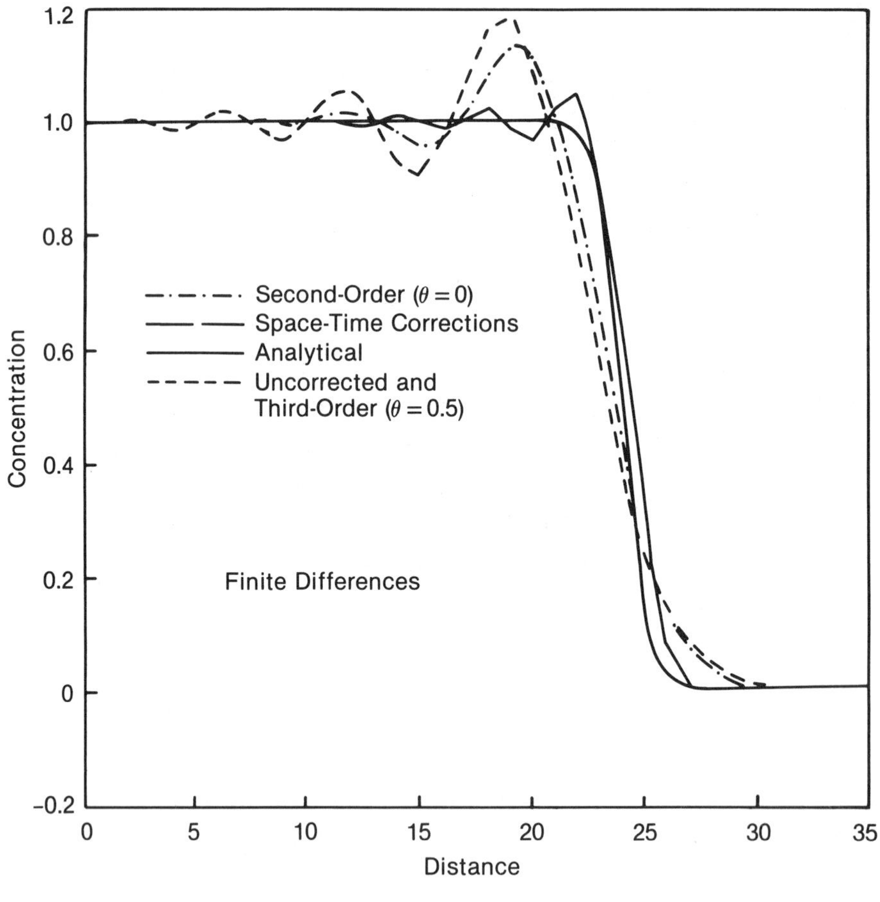

FIGURE 7.2 Concentration profiles obtained with several dispersion corrected finite-difference schemes. (From M. Th. Van Genuchten and W. G. Gray, "Analysis of Some Dispersion Corrected Numerical Schemes for Solution of the Transport Equation." *International Journal for Numerical Methods in Engineering*, Vol. 12, 1978, p. 397. Copyright © 1978 by John Wiley & Sons, Ltd. Reprinted with permission)

the concentration front. In contrast, the results for the dispersion corrected model shown in Figure 7.2 perform better. The results confirm that the space-time corrected scheme is superior to second-order and third-order correct time approximations. Also, the second-order correct implicit scheme is unconditionally unstable.

Recently, the Galerkin finite-element method has been used to approximate the solution of the convection-dispersion equation. Similar oscillation and smearing problems occur in many finite-element models. However, numerical dispersion, especially in the case of sharp concentration profiles, can be minimized by using

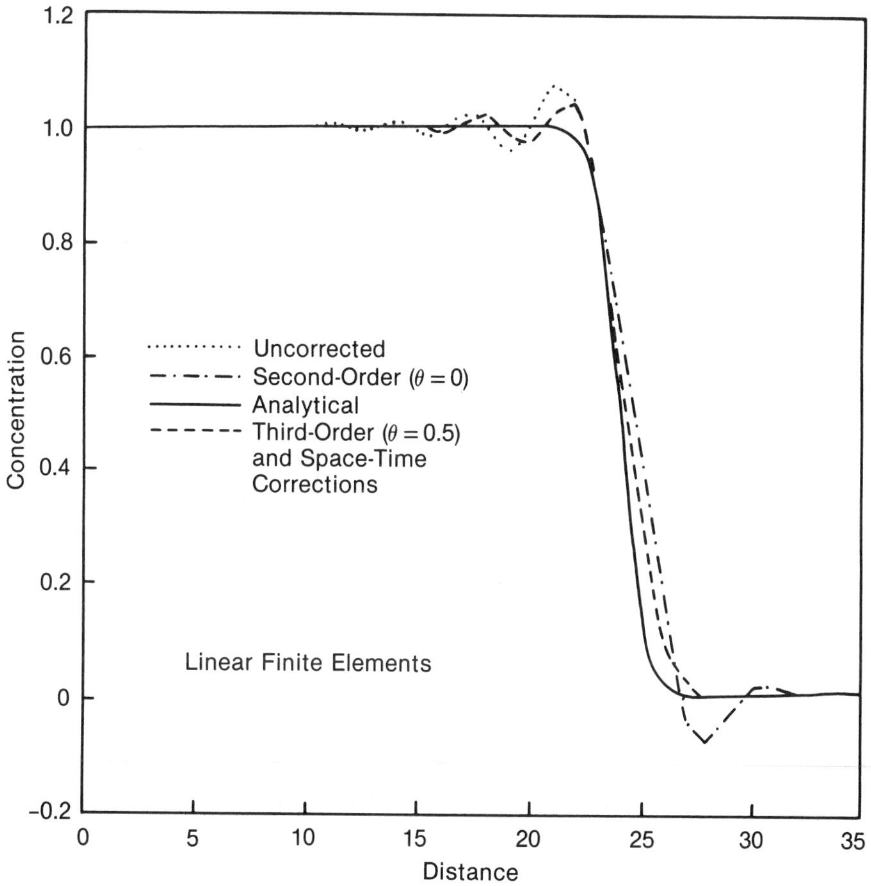

FIGURE 7.3 Concentration profiles obtained with several dispersion corrected linear finite-element schemes. (From M. Th. Van Genuchten and W. G. Gray, "Analysis of Some Dispersion Corrected Numerical Schemes for Solution of the Transport Equation." *International Journal for Numerical Methods in Engineering*, Vol. 12, 1978, p. 399. Copyright © 1978 by John Wiley & Sons, Ltd. Reprinted with permission)

higher-order basis functions (Van Genuchten, 1977), nonsymmetric weighting functions (Heinrich, et al., 1977; Huyakorn & Nilkiha, 1976; Huyakorn, 1976) or orthogonal collocation (Pinder & Shapiro, 1979; Herbst, 1981). All these variations of the finite-element method have been shown to be effective in minimizing numerical dispersion and instability, at the expense of increased computational effort.

Typical finite-element results for the solution of the convection-dispersion equation (Eq. 7.1) are shown in Figures 7.3 through 7.5. Figure 7.3 illustrates the concentration profiles obtained using linear finite elements. Again, oscillations are produced for the uncorrected approximations. However, the space-time corrected

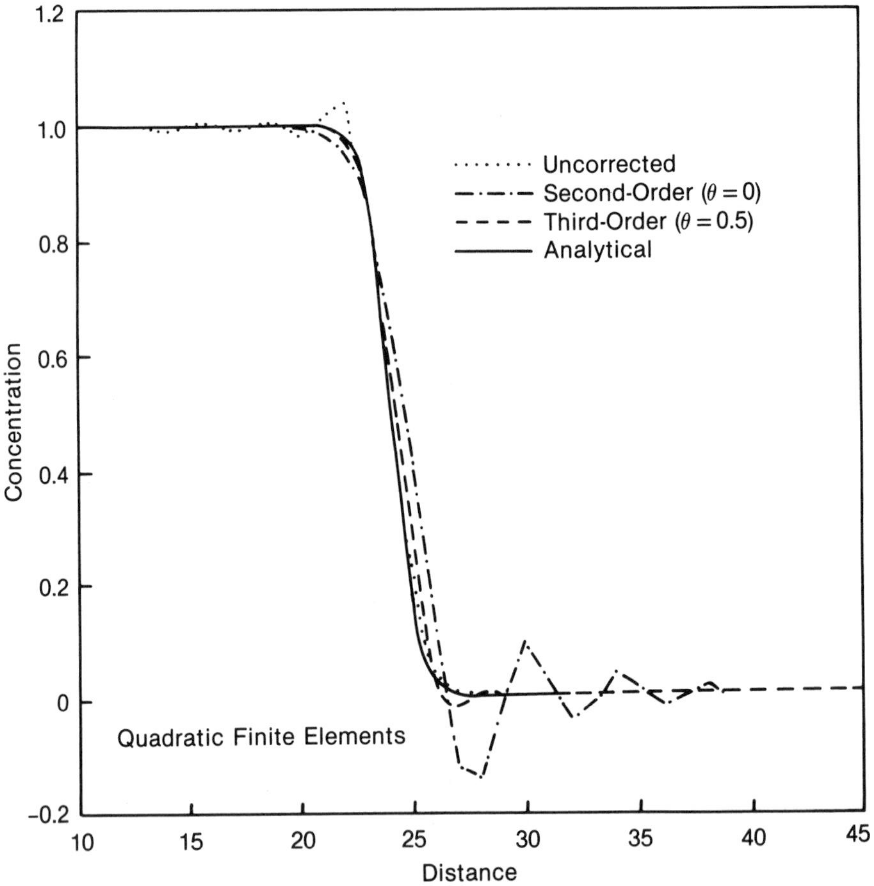

FIGURE 7.4 Concentration profiles obtained with several dispersion corrected quadratic finite-element schemes. (From M. Th. Van Genuchten and W. G. Gray. "Analysis of Some Dispersion Corrected Numerical Schemes for Solution of the Transport Equation." *International Journal for Numerical Methods in Engineering*, Vol. 12, 1978, p. 401. Copyright © 1978 by John Wiley & Sons, Ltd. Reprinted with permission)

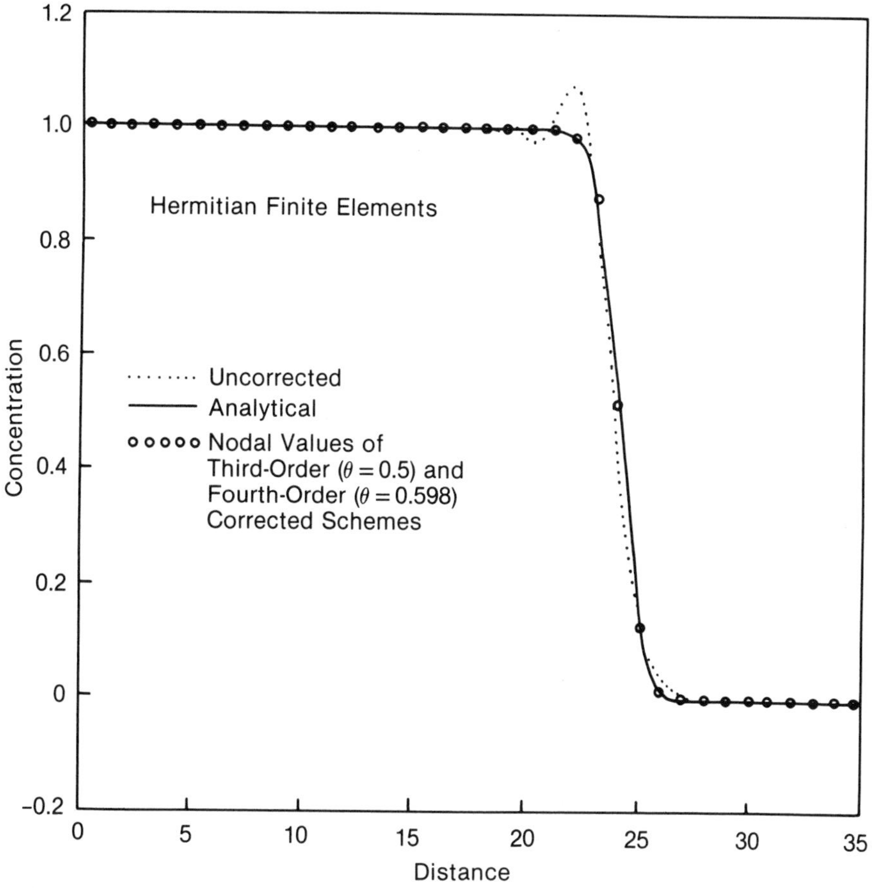

FIGURE 7.5 Concentration profiles obtained with several dispersion corrected Hermitian finite-element schemes. (From M. Th. Van Genuchten and W. G. Gray, "Analysis of Some Dispersion Corrected Numerical Schemes for Solution of the Transport Equation." *International Journal for Numerical Methods in Engineering*, Vol. 12, 1978, p. 403. Copyright © 1978 by John Wiley & Sons, Ltd. Reprinted with permission)

solutions and the third-order time approximation produce approximately the same results. The second-order explicit scheme, however, smears the concentration front and, as a result, is inferior to the other finite-element schemes.

Similar results are presented in Figure 7.4. Here, however, quadratic finite elements are used for discretization of the spatial domain. The third-order correct in time scheme provides some improvement over the uncorrected scheme. Note also that the third-order in time corrected method reduces the oscillations in comparison to the uncorrected scheme by distributing the oscillations between the upstream and downstream sides of the concentration front. In the second-order explicit scheme, the oscillations appear almost exclusively on the downstream side of the concentration front.

Figure 7.5 illustrates the concentration profiles for Hermitian finite elements. The third- and fourth-order corrected schemes are in very close agreement with the analytical solution. In contrast, the uncorrected scheme exhibits severe upstream overshoot of the concentration front. The higher-order, corrected finite-element approach obviously produces superior results for the one-dimensional simulation of the convection-dispersion equation.

The development of Galerkin finite-element models of the convection-dispersion equation are illustrated in the following example problems.

Example Problem 7.1 Soil Water Quality Simulation

A spreading basin is used for the disposal of wastewaters from a secondary wastewater treatment plant. The disposal site overlies a shallow unconfined aquifer and soil surveys indicate that the soil is homogeneous and isotropic (see Figure 7.6). Assuming steady-state flow conditions, simulate the behavior of the constituent in the soil system.

The water quality simulation model for the problem has been developed using the Galerkin finite-element method in Example Problem 4.5, where it was assumed that the soil system was completely saturated. Here we make the less restrictive assumption that the flow regime is steady-state, implying that the soil is partially saturated. The flow problem is then described by the continuity equation (see Example Problem 2.5),

$$\frac{\partial}{\partial z}(q_z) = 0, \; z \leqslant 0 \tag{7.17}$$

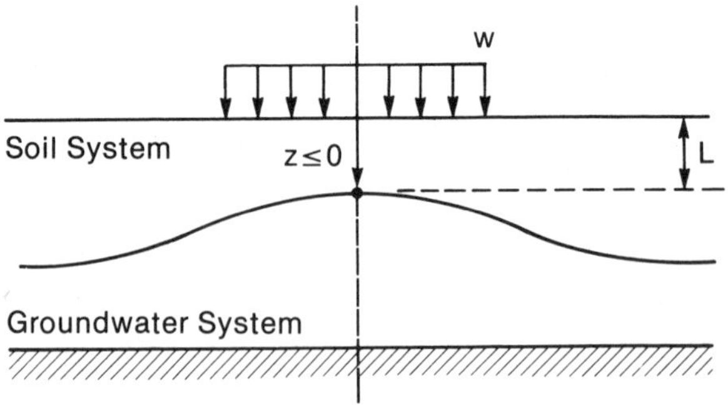

FIGURE 7.6 The land disposal system

The Darcy velocity is given as

$$\begin{aligned} q_z &= -K_z(\theta)\, \nabla\, (\psi + z) \\ &= -D(\theta)\frac{\partial\theta}{\partial z} - K_z(\theta) \end{aligned} \tag{7.18}$$

where we have introduced the soil moisture diffusivity, $D(\theta)$ (Section 2.4) and assumed that $D(\theta)$ and $K(\theta)$ are single-valued functions of the moisture content (hysteresis effects are neglected). Substituting Darcy's law into the continuity equation produces a nonlinear differential equation for the volumetric water content,

$$\frac{d}{dz}\left[-D(\theta)\frac{d\theta}{dz} - K_z(\theta)\right] = 0 \tag{7.19}$$

The integration of Equation 7.19 yields

$$\frac{d\theta}{dz} = \frac{C_1 - K_z(\theta)}{D(\theta)} \tag{7.20}$$

where C_1 is the constant of integration. The boundary condition at the ground surface, $z = 0$, can be used to evaluate the constant of integration, $C_1 = W$.

The differential equation describing the moisture variation is

$$\frac{d\theta}{dz} = \frac{W - K_z(\theta)}{D(\theta)}, \qquad z \leqslant 0 \tag{7.21}$$

The boundary condition for the problem is

$$\theta(-L) = n$$

where L is the approximate distance from the groundwater surface to the water table and n is the porosity.

The solution of Equation 7.21 can easily be obtained using quasilinearization and finite differences. For example, defining

$$f(\theta) = \frac{W - K_z(\theta)}{D(\theta)}$$

and assuming a solution, $\theta^n(z)$, $0 \leq z \leq -L$, the linearized flow equation can be expressed as

$$\left.\frac{d\theta}{dz}\right|_{n+1} = f(\theta)\Big|_n + \left.\frac{\partial f}{\partial\theta}\right|_n (\theta^{n+1} - \theta^n)$$

where n denotes the iteration. The differential equation can be solved with a forward finite-difference scheme, beginning with the boundary condition at the water table. A typical steady-state water content profile is shown in Figure 2.8. Experimental conductivity and diffusivity functions have been used for the numerical solution of the flow equation.

We now return to the mass transport problem. Referring to Example Problems 4.5 and 4.9, the reduced pore velocity is

$$\bar{v}_z = \frac{W}{\theta(1 + eR)}$$

The velocity, and consequently, the dispersion coefficient will then vary from element to element in the finite-element equations. For the example problem, linear shape (Chapeau) functions have been used to discretize the soil system. The response equations can then be expressed as

$$P\dot{\mathbf{c}} + R\mathbf{c} + \mathbf{f} = 0 \tag{7.22}$$

where $\mathbf{f}$ contains the input mass concentration (c^*), which, in the simulation analysis, is assumed to be known.

Using an implicit approximation for the time derivative, the Galerkin model was used to simulate the variation in the concentration in the soil system, assuming the concentration of the contaminant in the spreading basin is c^*. Figure 7.7 shows the dimensionless concentration (c/c^*) plotted as a function of the distance to the water table, for various times within an eight-hour disposal cycle.

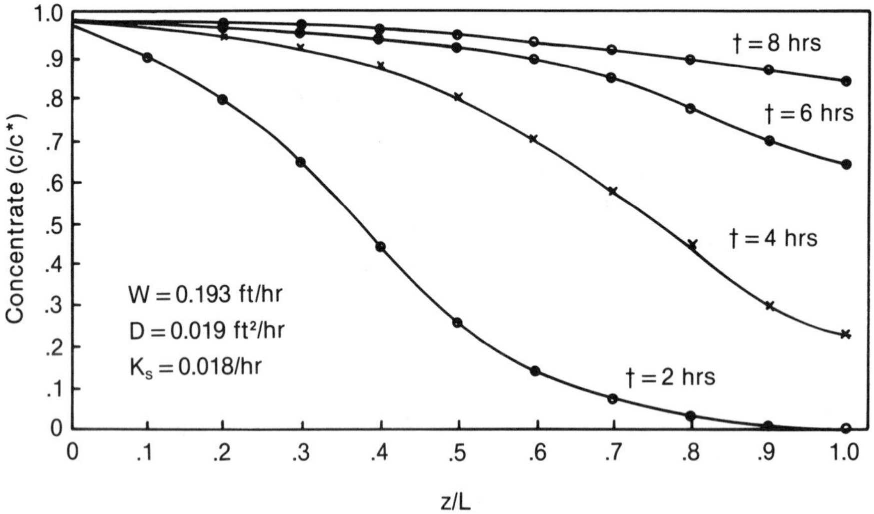

FIGURE 7.7 Simulated concentration profiles in the soil system

Several things are apparent from the simulation results. First, after two hours there is little contamination reaching the groundwater system. However, after four and six hours, the concentration of the constituent at the lower boundary of the system has increased to 0.26 and 0.65 of the surface concentration. Evidently, the maximum exit concentration occurs at the end of the disposal cycle, the critical time for maintaining any groundwater quality standards.

The simulation results are based on several, limiting modeling assumptions. At the lower boundary of the soil system, dispersive mass transport is neglected. The model also does not consider the redistribution and evaporation of soil water following the cessation of waste disposal. Increased oxygen transfer could then occur in the system, possibly increasing the rate of biochemical reactions. In addition, the temporal variations in the volumetric water content are also neglected, which can affect the velocity field and, consequently, the soil water quality. A more detailed analysis of the environmental implications of the waste disposal operation would address these limitations as well as the possibility of alternating wetting and drying periods of the disposal cycle (see, for example, Bresler, 1973).

Example Problem 7.2 Density-Dependent Mass Transport Model

Develop a Galerkin finite-element model to simulate the salt water intrusion occurring in the coastal aquifer shown in Figure 7.8. Assume that (1) the confined aquifer is homogeneous and isotropic, (2) the viscosity of the fluid is constant, and (3) the pressure-compressibility effects in the mass transport equation are negligible.

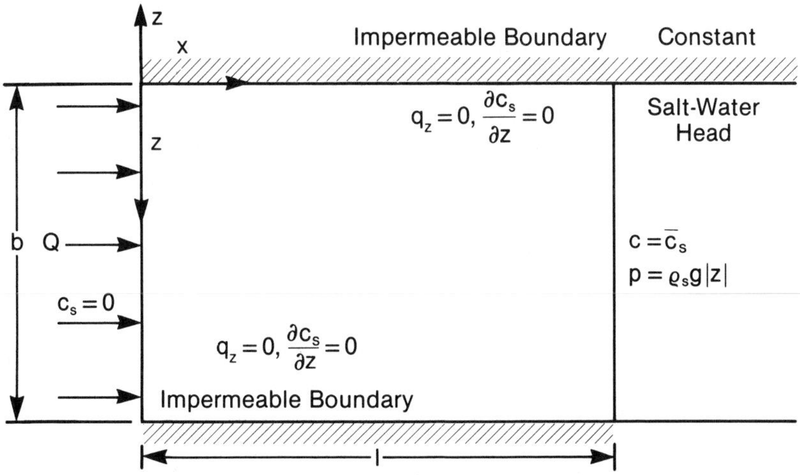

FIGURE 7.8 Salt water intrusion problem (From H. R. Henry, "Transitory Movements of the Saltwater Front in an Extensive Artesian Aquifer." *U.S. Geological Survey Professional Paper 450-B*, 1962)

We begin the development of the model by analyzing the governing equations of the aquifer system. The fluid in the aquifer consists of fresh water (mass concentration, c_f) and salt water (mass concentration c_s). The concentration of each constituent is described by Equation 3.11, or

$$\nabla \cdot \rho \mathbf{D} \nabla (c_s/\rho) - \nabla \cdot c_s \mathbf{q} = \frac{\partial}{\partial t}(nc_s) \tag{7.23}$$

and

$$\nabla \cdot \rho \mathbf{D}(c_f/\rho) - \nabla \cdot c_f \mathbf{q} = \frac{\partial}{\partial t}(nc_f) \tag{7.24}$$

where we have assumed that the temporal pressure changes are small in relation to the concentration variation. ρ is the overall solution density and $\mathbf{q}$ is the Darcy velocity, the mass average velocity of the solution. The Darcy velocity is

$$\mathbf{q} = -\frac{k\rho g}{\mu} \nabla \left(\frac{p}{\rho g} + z \right) \tag{7.25}$$

where k is the intrinsic permeability. Neglecting the grain velocity, the continuity equation for the system is given by Equation 2.45, or

$$-\nabla \cdot \rho \mathbf{q} = \rho(\alpha + n\beta)\frac{\partial p}{\partial t} \tag{7.26}$$

These four nonlinear partial differential equations contain five unknowns: the solution density (ρ), the mass concentrations for the salt water and fresh water (c_s, c_f), the fluid pressure (p), and the Darcy velocity ($\mathbf{q}$). To close the system of equations, we introduce an empirical relation developed for salt solutions relating ρ, c_s, and c_f as (Baxter & Wallace, 1916)

$$\rho = c_f + (1 - E)c_s \tag{7.27}$$

E is a constant and, for sea water, has a value of 0.3.

The boundary conditions for the flow problem can be expressed as

$$\begin{aligned} q_x(0, z, t) &= Q/b\,, & 0 \leqslant z \leqslant -b \\ q_z(x, 0, t) &= 0\,, & 0 \leqslant x \leqslant l \\ q_z(x, -b, t) &= 0\,, & 0 \leqslant x \leqslant l \\ p(l, z, t) &= \rho_s g|z|\,, & 0 \leqslant z \leqslant -b \end{aligned} \tag{7.28}$$

The boundary and initial conditions for the salt water prediction model are

$$
\begin{aligned}
& c_s(0, z, t) = 0 \qquad , 0 \leqslant z \leqslant -b \\
& c_s(l, z, t) = \bar{c}_s \qquad , 0 \leqslant z \leqslant -b \\
& \frac{\partial c}{\partial z}(x, 0, t) = 0, \qquad \frac{\partial c}{\partial z}(x, -b, t) = 0 \; , \; 0 \leqslant x \leqslant l \\
& c(x, z, 0) = \bar{c}_s H(x - l)
\end{aligned}
\tag{7.29}
$$

where ρ_s is the density of sea water, $\bar{c}_s$ is the concentration of salt in sea water, and H is the Heaviside step function. The Dirichlet boundary condition cannot, however, be modeled by the Galerkin method because of the apparent conflict between the fluid concentration as it approaches the exit boundary and the given boundary condition. Instead, a mixed boundary condition is used, or

$$\frac{\partial c_s}{\partial z}(l, z, t) = 0, \quad z \in \Gamma_+$$

$$c_s(l, z, t) = 0, \quad z \in (\Gamma - \Gamma_+)$$

where Γ_+ is the segment of the boundary, S, where the fluid leaves the system.

The set of continuity, transport, and pressure equations is highly nonlinear. In the mass transport equation, for example, the solution density appears nonlinearly in the dispersion terms, while the convective terms contain the products of the Darcy velocity and the solute concentrations. The Darcy equation also contains nonlinear density and pressure-state variables; multiplicative density and pressure terms also occur in the continuity equation.

Solution algorithms for density-dependent models usually involve the iterative solution of the equations using quasilinearization or localized linearization of the state variables. In the linearization approach, the overall solution density is assumed constant over the solution interval or time step. As a result, the continuity and Darcy equations will update the pressure and velocity fields at the end of the solution interval.

The solution of the mass transport equation, in turn, yields the concentrations of fresh and salt water and the overall solution density at the end of the time interval. At this stage of the algorithm, a comparison is made between the updated solution density and the original estimate. If the difference is greater than a prescribed tolerance, the algorithm resolves the continuity, Darcy, and mass transport equations with the updated solution density. If, however, the density difference is small (within the tolerance level), the algorithm proceeds to the next time step.

The iterative algorithm is based on the numerical transformation of the system's equations. The Galerkin finite-element method will be used to approximate the solution of these equations. Assuming a constant density field over the time interval, (Δt), we first introduce the differential operators,

$$L_1(\rho, \mathbf{q}) = \frac{\partial}{\partial x}(\rho q_x) + \frac{\partial}{\partial z}(\rho q_z) - \rho(\alpha + n\beta)\frac{\partial p}{\partial t} = 0 \tag{7.30}$$

$$L_2(\rho, q_x) = q_x + \frac{k}{\mu}\frac{\partial p}{\partial x} = 0 \tag{7.31}$$

$$L_3(\rho, q_z) = q_z + \frac{k}{\mu}\left(\frac{\partial p}{\partial z} + \rho g\right) = 0 \tag{7.32}$$

Next, we assume the trial solutions to these equations may be represented by the finite series

$$q_x \cong \hat{q}_x = \sum_{i=1}^{n} N_i(x, z)\,\tilde{q}_{x,i} \tag{7.33}$$

$$q_z \cong \hat{q}_z = \sum_{i=1}^{n} N_i(x, z)\,\tilde{q}_{z,i} \tag{7.34}$$

$$p \cong \hat{p} = \sum_{i=1}^{n} (x, z)\,\tilde{p}_i \tag{7.35}$$

where N_i is the set of basis function for the problem and $\tilde{q}_{x,i}, \tilde{q}_{z,i}$ and $\tilde{p}$ are the nodal values of the x, z velocities and pressure. The orthogonality conditions of the Galerkin method require

$$\iint N_i L_1(\rho, \mathbf{q})dA = 0, \quad \iint N_i L_2(\rho, q_x)dA = 0, \quad \iint N_i L_3(\rho, q_z)dA = 0, \forall_i \tag{7.36}$$

The first set of conditions produces the matrix system of equations, or

$$\mathrm{A}\tilde{\mathbf{q}}_x + B\tilde{\mathbf{q}}_z + D\frac{\partial \tilde{\mathbf{p}}}{\partial t} = 0 \tag{7.37}$$

where $\tilde{\mathbf{q}}_x$, $\tilde{\mathbf{q}}_z$ and $\tilde{\mathbf{p}}$ are the vectors of nodal values of the x and z velocities and the fluid pressure. Typical elements of the elemental matrices are

$$A_{ij}^e = \rho\iint N_i \frac{\partial N_j}{\partial x}\,dA_e, \qquad B_{ij}^e = \rho\iint N_i \frac{\partial N_j}{\partial z}\,dA_e$$

$$D_{ij}^e = \rho(\alpha + n\beta)\iint N_i N_j dA_e$$

(A_e is the elemental area.) Transformation of the Darcy equations produces the system of equations

$$E\tilde{q}_x + F\,\tilde{p} = 0 \tag{7.38}$$

$$E\tilde{q}_z + G\tilde{p} + z = 0 \tag{7.39}$$

where the elemental matrices are given as

$$E^e_{ij} = \iint N_i N_j dA_e, \qquad E^e_{ij} = \frac{k}{\mu}\iint N_i \frac{\partial N_j}{\partial z} dA_e$$

$$G^e_{ij} = \frac{k}{\mu}\iint N_i \frac{\partial N_j}{\partial z}\, dA_e, \qquad z^e_i = \frac{k\rho g}{\mu}\iint N_i dA_e$$

Finite-difference methods are generally used for the solution of these linearized equations over a time step. For example, using an implicit scheme for the pressure equation, we have

$$A\tilde{q}_x^{k+1} + B\tilde{q}_z^{k+1} + D\left(\frac{\tilde{p}^{k+1} - \tilde{p}^k}{\Delta t}\right) = 0 \tag{7.40}$$

$$E\tilde{q}_x^{k+1} + F\tilde{p}^{k+1} = 0 \tag{7.41}$$

$$E\tilde{q}_z^{k+1} + G\tilde{p}^{k+1} + z = 0 \tag{7.42}$$

or, in vector-matrix form,

$$\begin{bmatrix} D/\Delta t & A & B \\ F & E & 0 \\ G & 0 & E \end{bmatrix} \begin{bmatrix} \tilde{p}^{k+1} \\ \tilde{q}_x^{k+1} \\ \tilde{q}_z^{k+1} \end{bmatrix} = \begin{bmatrix} D\tilde{p}^k/\Delta t \\ 0 \\ -z \end{bmatrix} \tag{7.43}$$

where 0 is the zero matrix. The coefficient matrix is nonsymmetric and contains zero elements along the main diagonal. This can hamper the use of conventional Gaussian elimination techniques that pivot on the diagonal elements.

Next, the mass transport equations can be solved using the pressure and velocity field information obtained from Equation 7.43. Applying the Galerkin procedure to the mass transport equations, we have, for either constituent

$$P\dot{c} + Rc + f = 0 \tag{7.44}$$

where

$$P^e_{ij} = n \iint N_i N_j dA_e$$

$$R^e_{ij} = \iint N_i \left\{ N_{jx} \left(\sum_k N_k q_{kx} \right) + N_j \left(\sum_k N_{kx} q_{ix} \right) \right\} dA_e$$

$$+ \iint N_i \left\{ N_{jz} \left(\sum_k N_k q_{kz} \right) + N_j \left(\sum_k N_{kz} q_{kz} \right) \right\} dA_e$$

$$- \iint \left\{ D_{xx} \frac{\partial N_i}{\partial x} \frac{\partial N_j}{\partial x} + D_{zx} \frac{\partial N_i}{\partial x} \frac{\partial N_j}{\partial z} + D_{zz} \frac{\partial N_i}{\partial z} \frac{\partial N_j}{\partial z} \right\} dA_e$$

and

$$f_i = - \int_\Gamma \mathbf{q}_n \cdot \mathbf{n}\, N_i dS$$

where $\mathbf{n}$ is the unit normal to the boundary Γ and $\mathbf{q}_n$ is the normal velocity.

Although we have assumed that the dispersion parameters are constant over an element, we could also introduce functional representations for these parameters. This improves the accuracy of the approximations and ensures a continuous dispersion field over the solution domain. In the approach, we assume that D can be expressed as

$$D_{\alpha\beta} \cong \hat{D}_{\alpha\beta} = \sum_{k=1}^{n} N_k(x, z)\, \tilde{D}_{k,\alpha\beta}, \qquad \alpha, \beta = x, z \tag{7.45}$$

where $\tilde{D}_{k,\alpha\beta}$ are the nodal values of the dispersion coefficients. The dispersion transport terms in the elemental equations can then be expressed as

$$\iint \left\{ N_{i,x} \sum_k N_k \tilde{D}_{k,xx} \frac{\partial N_i \partial N_j}{\partial x\, \partial x} + \sum_k N_k \tilde{D}_{k,zx} \frac{\partial N_i \partial N_j}{\partial x\, \partial z} + \sum_k N_k \tilde{D}_{k,zz} \frac{\partial N_i \partial N_j}{\partial z\, \partial z} \right\} dA_e$$

The solution of the mass transport equation for either constituent can again be obtained using finite-difference methods. However, the time step for solution of the equations is usually much smaller than for the flow and pressure equations.

Simulation results for the groundwater quality model will be presented in Example Problem 7.4.

7.3.2 The Upstream Weight Multiple-Cell Balance Method

An alternative to the Galerkin finite-element method in the simulation of groundwater quality is the upstream weight multiple-cell balance method. The technique is based on the upstream weighting of the basis or shape functions of the finite-element method and the integrated form of the convection-dispersion equation. The weighting coefficients are determined by the local Peclet number (the ratio of convection/dispersion) of the flow domain. In comparison with the Galerkin method, the method has been shown to be effective in eliminating numerical oscillations with only a nominal increase in computational effort. (Sun, 1981; Sun & Yeh, 1983).

The governing equation of the multiple-cell balance method is obtained by integrating the convection-dispersion equation over a given region, A, using Green's theorem. Assuming a constant porosity, the equations may be expressed as

$$-\int_S m\mathrm{D}\,\nabla\, c \cdot n\; dS + \int_S \mathrm{v}mc \cdot n\; dS = \iint_A \left[\frac{\partial(mc)}{\partial t} + M\right] dx\, dy \qquad (7.46)$$

where m is the aquifer thickness and M is a source term.

In a two-dimensional Cartesian coordinate system, the scalar form of the equation is

$$\int_S m \left[\left(\mathrm{D}_{xx}\frac{\partial c}{\partial x} + \mathrm{D}_{xy}\frac{\partial c}{\partial y}\right)dy - \left(\mathrm{D}_{xy}\frac{\partial c}{\partial x} + \mathrm{D}_{yy}\frac{\partial c}{\partial y}\right)dx\right] + \int_S m\; c(v_y dx - v_x dy) = \iint_A \left[\frac{\partial(mc)}{\partial t} + M\right] dx\, dy \qquad (7.47)$$

The finite-element representation of the equation is based on linear triangular elements (see Figure 7.9). The mass concentration at nodes i, j, and k is related to the concentration at node m, the center of the element, by the relation,

$$c_m = \frac{1}{w}(w_i c_i + w_j c_j + w_k c_k), \qquad w = w_i + w_j + w_k \qquad (7.48)$$

where w_i, w_j, and w_k are the upstream weights corresponding to nodes i, j, and k. By linking the center node m and the nodes i, j, and k, any element can be discretized into three subtriangular elements. The concentration can be represented approximately by a linear function for each subelement. For example, for subelement Δijm, the concentration is given as

$$c(x, y, t) = \mathrm{N}_{ki}c_i + \mathrm{N}_{kj}c_j + \mathrm{N}_{km}c_m, \qquad (x, y) \in \Delta ijm \qquad (7.49)$$

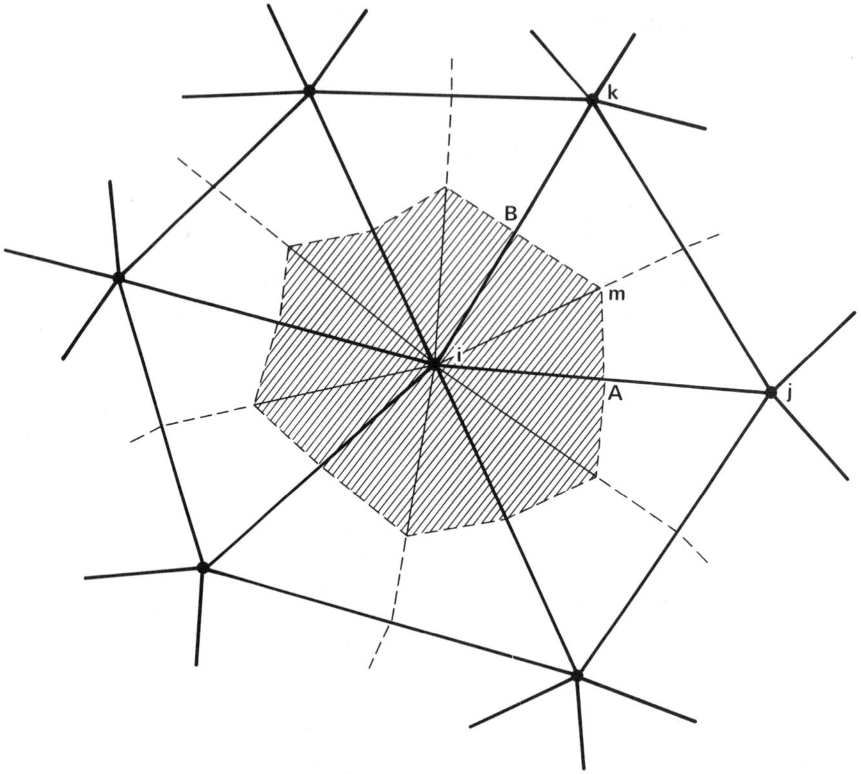

FIGURE 7.9 Multiple cell balance mesh and the exclusive subdomain of node i (From N.Z. Sun and W.W-G. Yeh, "A Proposed Upstream Weight Numerical Method for Simulating Pollutant Transport in Groundwater." *Water Resources Research*, 19(6):1983, p. 1492. Copyright © 1983 by the American Geophysical Union)

where N_{ki}, N_{kj}, and N_{km} are the general linear basis functions in Δijm for nodes i, j, and m, respectively. The basis functions are given as

$$N_{kl} = \frac{3}{2\Delta_e}(a_{kl} + b_{kl}x + c_{kl}y), \qquad l = i, j, m \tag{7.50}$$

where Δ_e is the area of the triangular element and,

$$a_{ki} = x_j y_m - x_m y_j \qquad b_{ki} = y_j - y_m \qquad c_{ki} = x_m - x_j$$

$$a_{kj} = x_m y_i - x_i y_m \qquad b_{kj} = y_m - y_i \qquad c_{kj} = x_i - x_m$$

$$a_{km} = x_i y_j - x_j y_i \qquad b_{km} = y_i - y_j \qquad c_{km} = x_j - x_i$$

Substituting the equations into Equation 7.49 yields

$$c(x, y, t) = \overline{N}_{ki}c_i + \overline{N}_{kj}c_j + \overline{N}_{kk}c_k, \qquad (x, y) \in \Delta ijm \tag{7.51}$$

where the upstream weight basis functions are

$$\begin{aligned} \overline{N}_{ki} &= N_{ki} + \frac{w_i}{w} N_{km} \\ \overline{N}_{kj} &= N_{kj} + \frac{w_j}{w} N_{km} \\ \overline{N}_{kk} &= \frac{w_k}{w} N_{km} \end{aligned} \tag{7.52}$$

Similar expressions can be developed for each subelement.

The partial derivatives of the convection-dispersion equation within any subelement can then be expressed as

$$\frac{\partial c}{\partial x} = \frac{3}{2\Delta_e}(\overline{b}_{ki}\, c_i + \overline{b}_{kj}c_j + \overline{b}_{kk}\, c_k), \qquad (x, y) \in \Delta ijm \tag{7.53a}$$

$$\frac{\partial c}{\partial y} = \frac{3}{2\Delta_e}(\overline{c}_{ki}\, c_i + \overline{c}_{kj}\, c_j + \overline{c}_{kk}\, c_k), \qquad (x, y) \in \Delta ijm \tag{7.53b}$$

$$\frac{\partial c}{\partial t} = \overline{N}_{ki}\frac{\partial c_i}{\partial t} + \overline{N}_{kj}\frac{\partial c_j}{\partial t} + \overline{N}_{kk}\frac{\partial c_k}{\partial t}, \qquad (x, y) \in \Delta ijm \tag{7.53c}$$

and

$$\frac{\partial c}{\partial x} = \frac{3}{2\Delta_e}(\overline{b}_{ji}\, c_i + \overline{b}_{jj}\, c_j + \overline{b}_{jk}\, c_k), \qquad (x, y) \in \Delta imk \tag{7.54a}$$

$$\frac{\partial c}{\partial y} = \frac{3}{2\Delta_e}(\overline{c}_{ji}\, c_i + \overline{c}_{jj}\, c_j + \overline{c}_{jk}\, c_k), \qquad (x, y) \in \Delta imk \tag{7.54b}$$

$$\frac{\partial c}{\partial t} = \overline{N}_{ji}\frac{\partial c_i}{\partial t} + \overline{N}_{jj}\frac{\partial c_j}{\partial t} + \overline{N}_{jk}\frac{\partial c_k}{\partial t}, \qquad (x, y) \in \Delta imk \tag{7.54c}$$

The parameters are defined as

$$
\begin{aligned}
\bar{b}_{ki} &= b_{ki} + \frac{w_i}{w} b_{km} & \bar{b}_{ji} &= b_{ji} + \frac{w_i}{w} b_{jm} \\
\bar{b}_{kj} &= b_{kj} + \frac{w_j}{w} b_{km} & \bar{b}_{jj} &= \frac{w_j}{w} b_{jm} \\
\bar{b}_{kk} &= \frac{w_k}{w} b_{km} & \bar{b}_{jk} &= b_{jk} + \frac{w_k}{w} b_{jm} \\
\bar{c}_{ki} &= c_{ki} + \frac{w_i}{w} c_{km} & \bar{c}_{ji} &= c_{ji} + \frac{w_i}{w} c_{jm} \\
\bar{c}_{kj} &= c_{kj} + \frac{w_j}{w} c_{km} & \bar{c}_{jj} &= \frac{w_j}{w} c_{jm} \\
\bar{c}_{kk} &= \frac{w_k}{w} c_{km} & \bar{c}_{jk} &= c_{jk} + \frac{w_k}{w} c_{jm}
\end{aligned}
\tag{7.55}
$$

Note that, when the upstream weights are one, the equations simplify to general linear-finite elements.

If we now consider all the elements that have a common node i, and link the center of each element and the middle point of each side, we obtain a multi-angular subregion that is known as the exclusive subdomain of node i (see Figure 7.9) (Cooley, 1974; Neuman & Narasimhan, 1977). This subdomain is the mass balance region for the mass transport equation (Eq. 7.47).

Substituting Equations 7.53 and 7.54 into the left-hand side of the integrated transport equation and noting that the concentration as well as its partial derivatives have different representations at different parts in an element, the dispersion terms can be expressed as

$$
\int_S m\left[\left(D_{xx}\frac{\partial c}{\partial x} + D_{xy}\frac{\partial c}{\partial y}\right) dy - \left(D_{xy}\frac{\partial c}{\partial x} + D_{yy}\frac{\partial c}{\partial y}\right)\right] dx = \sum_{e_i} (\alpha^e_{ii}\, c_i + \alpha^e_{ij}\, c_j + \alpha^e_{ik}\, c_k),
\tag{7.56}
$$

where

$$
\alpha^e_{il} = -\frac{m_i}{4\Delta_e}[(D_{xx}\bar{b}_{kl} + D_{xy}\bar{c}_{kl})(b_i - b_j) + (D_{xy}\bar{b}_{kl} + D_{yy}\bar{c}_{kl})(c_i - c_j) + (D_{xx}\bar{b}_{jl} + D_{xy}\bar{c}_{jl})(b_i - b_k) + (D_{xy}\bar{b}_{jl} + D_{yy}\bar{c}_{jl})(c_i - c_k)],\ l = i, j, k
\tag{7.57}
$$

and

$$\begin{aligned} b_i &= y_j - y_k & c_i &= x_k - y_j \\ b_j &= y_k - y_i & c_j &= x_i - x_k \\ b_k &= y_i - y_j & c_k &= x_j - x_i \end{aligned} \tag{7.58}$$

Assuming $\nabla \cdot m\mathbf{v} = 0$ and separately evaluating the convection and time derivatives of Equation 7.47, the balance equations may be expressed as

$$\sum_{e_i} (A^e_{ii} c_i + A^e_{ij} c_j + A^e_{ik} c_k) + \sum_{e_i} \left(B^e_{ii} \frac{\partial c_i}{\partial t} + B^e_{ij} \frac{\partial c_j}{\partial t} + B^e_{ik} \frac{\partial c_k}{\partial t} \right) + R_i = 0 \tag{7.59}$$

where

$$A^e_{il} = -\alpha^e_{il} + \beta^e_{il} + \delta^e_{il}, \qquad l = i, j, k \tag{7.60a}$$

$$\beta^e_{il} = \frac{m_i}{4} \{v_x(\bar{b}_{jl} + \bar{b}_{kl}) + v_y(\bar{c}_{jl} + \bar{c}_{kl})\}, \qquad l = i, j, k \tag{7.60b}$$

$$\begin{aligned} B^e_{ii} &= \frac{m_i \Delta_e}{3} \left(\frac{1}{2} + \frac{w_i}{3w} \right) \\ B^e_{ij} &= \frac{m_i \Delta_e}{3} \left(\frac{1}{12} + \frac{w_j}{3w} \right) \\ B^e_{ik} &= \frac{m_i \Delta_e}{3} \left(\frac{1}{12} + \frac{w_k}{3w} \right) \end{aligned} \tag{7.60c}$$

and

$$\begin{aligned} \delta^e_{ii} &= \frac{\Delta_e}{3} \left(\frac{1}{2} + \frac{1}{3} \frac{w_i}{w} \right) \frac{\partial m_i}{\partial t} \\ \delta^e_{ij} &= \frac{\Delta_e}{3} \left(\frac{1}{12} + \frac{1}{3} \frac{w_j}{w} \right) \frac{\partial m_i}{\partial t} \\ \delta^e_{ik} &= \frac{\Delta_e}{3} \left(\frac{1}{12} + \frac{1}{3} \frac{w_k}{w} \right) \frac{\partial m_i}{\partial t} \end{aligned} \tag{7.60d}$$

The source/sink term of the equation is given as

$$R_i = \int\int_A M \, dx \, dy \tag{7.61}$$

Equation 7.59 can be expressed as

$$A_{ii}c_i + \sum_{j \neq i} A_{ij}c_j + B_{ii}\frac{\partial c_i}{\partial t} + \sum_{j \neq i} B_{ij}\frac{\partial c_j}{\partial t} + R_i = 0 \tag{7.62}$$

where

$$A_{ii} = \sum_{e_i} A^e_{ii}, \qquad B_{ii} = \sum_{e_i} B^e_{ii} \tag{7.63}$$

$$A_{ij} = \sum_{e_{ij}} A^e_{ij}, \qquad B_{ij} = \sum_{e_{ij}} B^e_{ij}$$

where $\sum_{j \neq i}$ represents the sum of all nodes that are neighboring node i and $\sum_{e_{ij}}$ is the sum of all elements that have common nodes i and j. Writing Equation 7.62 for each node when the concentration is unknown, and incorporating the boundary conditions, the response equations of the upstream weight multiple-cell balance method may be expressed as

$$\mathrm{A}\mathbf{c} + B\dot{\mathbf{c}} + \mathbf{r} = 0 \tag{7.64a}$$

Adopting the notation previously developed, the equations can be expressed as

$$P\dot{\mathbf{c}} + R\mathbf{c} + \mathbf{f} = \mathbf{0} \tag{7.64b}$$

where we have replaced A, B, and $\mathbf{r}$ with R, P, and $\mathbf{f}$.

When B^e_{ii}, B^e_{ij}, and B^e_{ik} are given by Equation 7.60c, the method is called upstream weight multiple cell balance (Sun & Yeh, 1983).

We note that if Equation 7.60c is replaced by

$$B^e_{ii} = \frac{1}{3} m_i \Delta_e$$

$$B^e_{ij} = 0 \tag{7.65}$$

$$B^e_{ik} = 0$$

Equation 7.64 reduces to that of the lumped mass Galerkin finite-element method. And, similarly, if Equation 7.60c is replaced by

$$B^e_{ii} = \frac{1}{6} m_i \Delta_e$$

$$B^e_{ij} = \frac{1}{12} m_i \Delta_e \tag{7.66}$$

$$B^e_{ik} = \frac{1}{12} m_i \Delta_e$$

then the response equations are the same as those obtained with the Galerkin method with linear basis functions.

The upstream weights of the model are expressed as a function of the local Peclet number. Defining $\bar{v}$ as the average velocity vector in element e and v_{ij}, v_{jk}, and v_{ki} as its projections on direction ij, jk, ki, then τ_{ij}, the local Peclet number between nodes i and j can be expressed as

$$\tau_{ij} = \frac{v_{ij}|ij|}{\alpha_l \bar{v}} \tag{7.67}$$

where α_l is the longitudinal dispersivity. The convention adopted in the model is that when $\tau_{ij} > 0$, node i is upstream of node j and vice versa. The Peclet numbers τ_{ik}, τ_{jk} have similar meanings. Further defining $\tau_i = \tau_{ij} + \tau_{ik}$, $\tau_j = \tau_{ji} + \tau_{jk}$, and $\tau_k = \tau_{ki} + \tau_{kj}$ as the upstream position of nodes i, j, k relating other two nodes in any elements, the upstream weights can be expressed as

$$w_i = 1 + \lambda \frac{v_{ij}|ij| - v_{ki}\,|ki|}{\alpha_l \bar{v}} \tag{7.68a}$$

$$w_j = 1 + \lambda \frac{v_{jk}\,|jk| - v_{ij}\,|ij|}{\alpha_l \bar{v}} \tag{7.68b}$$

$$w_k = 1 + \lambda \frac{v_{ki}\,|ik| - v_{jk}\,|jk|}{\alpha_l \bar{v}} \tag{7.68c}$$

where $|ij|$ is the distance between nodes i and j and λ is a positive, undetermined coefficient.

The importance of Equation 7.68 is again that the upstream weights are expressed functionally in terms of the velocity and local dispersion occurring in any element. The weights are automatically adjusted in the solution of the convection-dispersion equation. Numerical experiments have also demonstrated that the optimal value of λ is in the range 0.008-0.0012 for a variety of transport conditions (Sun & Yeh, 1983).

The multiple-cell balance method was applied to the one-dimensional convection-dispersion equation (Eq. 7.1). A total of 75 nodes and 96 two-dimensional triangular elements were used to discretize the spatial domain (see Figure 7.10); the simulation period extended from 0 to 50 days. The analytical solution for the numerical experiments is given by Equation 7.16. The results are summarized in Figures 7.11 through 7.15.

Figures 7.11 and 7.12 present the simulation results comparing the multiple-cell balance method (MCB), the general finite-element method (FEM), and the lumped mass finite-element method (LMFEM). All three numerical schemes produce satisfactory results. Specifically, the MCB gives the most accurate solutions

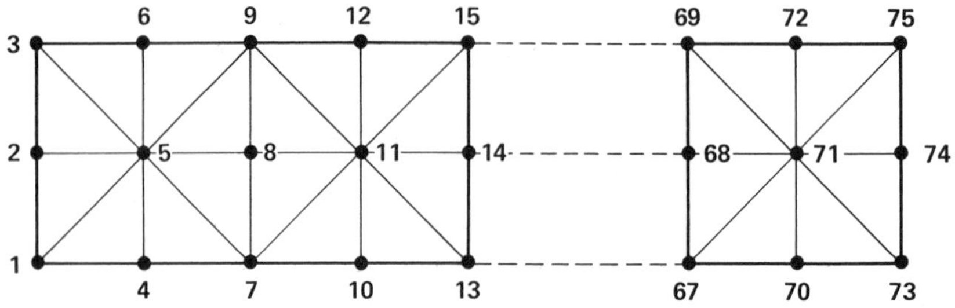

FIGURE 7.10 Elements and nodes of one-dimensional convection-diffusion problem (From N.Z. Sun and W.W-G. Yeh, "A Proposed Upstream Weight Numerical Method for Simulating Pollutant Transport in Groundwater." *Water Resources Research*, 19(6):1983, p. 1494. Copyright © 1983 by the American Geophysical Union)

for short times and the three methods generate almost identical solutions for the longer simulation period ($t = 40$ days).

Figure 7.13 presents the groundwater quality simulation results for flow with a large Peclet number (Pe = 100). The numerical solutions exhibit oscillations when the upstream weights are not introduced in the finite-element model. The LMFEM results are the worst, while the FEM solution has smaller overshoot than the MCB; but, the MCB simulates the front more accurately than the FEM. Oscillations in the numerical solution can be controlled or damped by including the upstream weights. The simulations shown in Figure 7.14 are for the case when

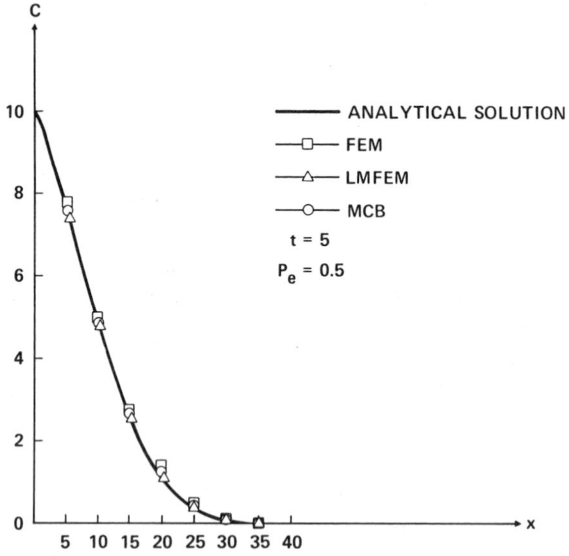

FIGURE 7.11 Comparison between analytical and numerical solutions (From N.Z. Sun and W.W-G. Yeh, "A Proposed Upstream Weight Numerical Method for Simulating Pollutant Transport in Groundwater." *Water Resources Research*, 19(6):1983, p. 1494. Copyright © 1983 by the American Geophysical Union)

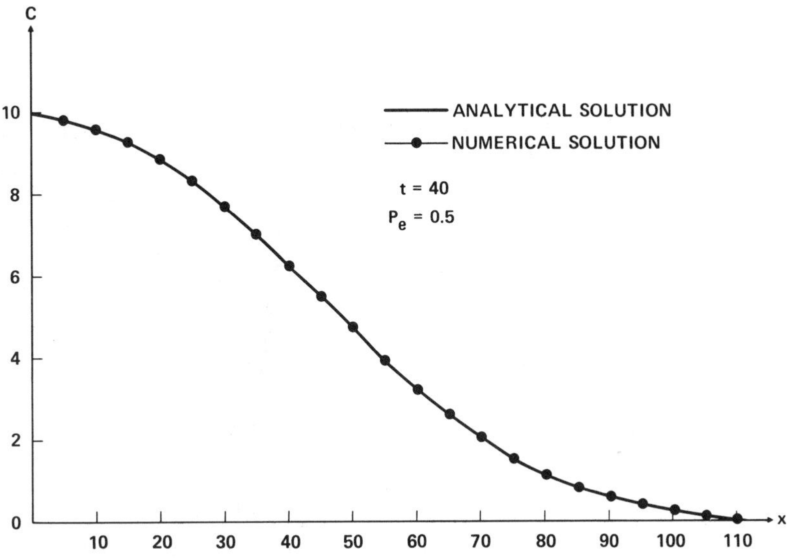

FIGURE 7.12 Comparison between analytical and numerical solutions (From N.Z. Sun and W.W-G. Yeh, "A Proposed Upstream Weight Numerical Method for Simulating Pollutant Transport in Groundwater." *Water Resources Research*, 19(6):1983, p. 1494. Copyright © 1983 by the American Geophysical Union)

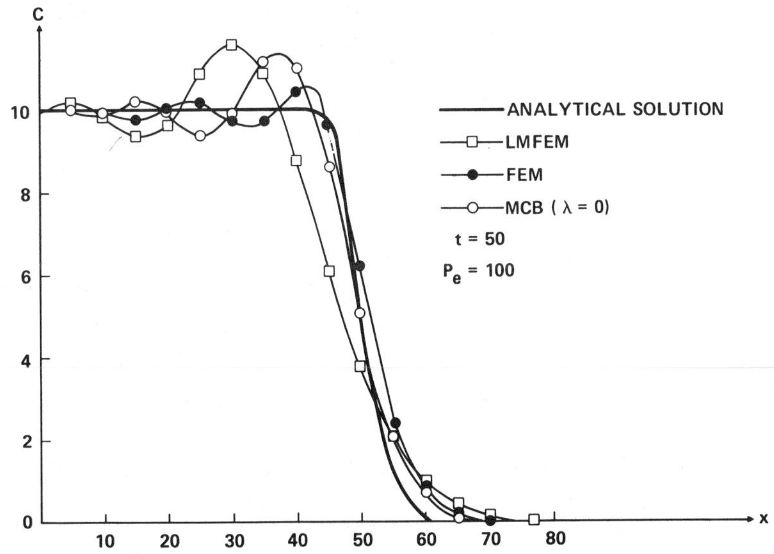

FIGURE 7.13 Comparison between analytical and numerical solutions (From N.Z. Sun and W.W-G. Yeh, "A Proposed Upstream Weight Numerical Method for Simulating Pollutant Transport in Groundwater." *Water Resources Research*, 19(6):1983, p. 1494. Copyright © 1983 by the American Geophysical Union)

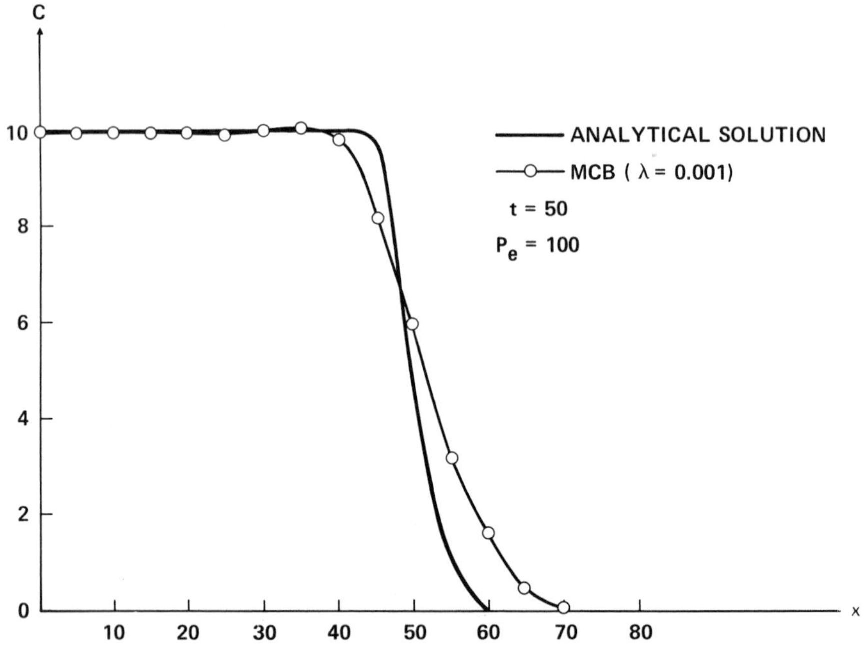

FIGURE 7.14 Comparison between analytical and MCB solutions with upstream weights (From N.Z. Sun and W.W-G. Yeh, "A Proposed Upstream Weight Numerical Method for Simulating Pollutant Transport in Groundwater." *Water Resources Research*, 19(6):1983, p. 1495. Copyright © 1983 by the American Geophysical Union)

λ is 0.001. The effects of varying the weighting factor on the simulation results is shown in Figure 7.15. We observe that the smaller upstream weights cannot control the oscillations of the solutions; in contrast, larger upstream weights introduce serious numerical dispersion in the concentration profiles. In all the cases that have been analyzed, λ tends to be approximately 0.001.

The following example problem illustrates the application of the multiple-cell balance method to a stream-aquifer groundwater contamination problem.

Example Problem 7.3 Simulation of Groundwater Quality in a Stream-Aquifer System (After Sun & Yeh, 1983)

Develop an upstream weight multiple-cell balance model to predict groundwater quality in the stream-aquifer system shown in Figure 7.16. The confined aquifer's west and east boundaries are impermeable while the northern and southern bound-

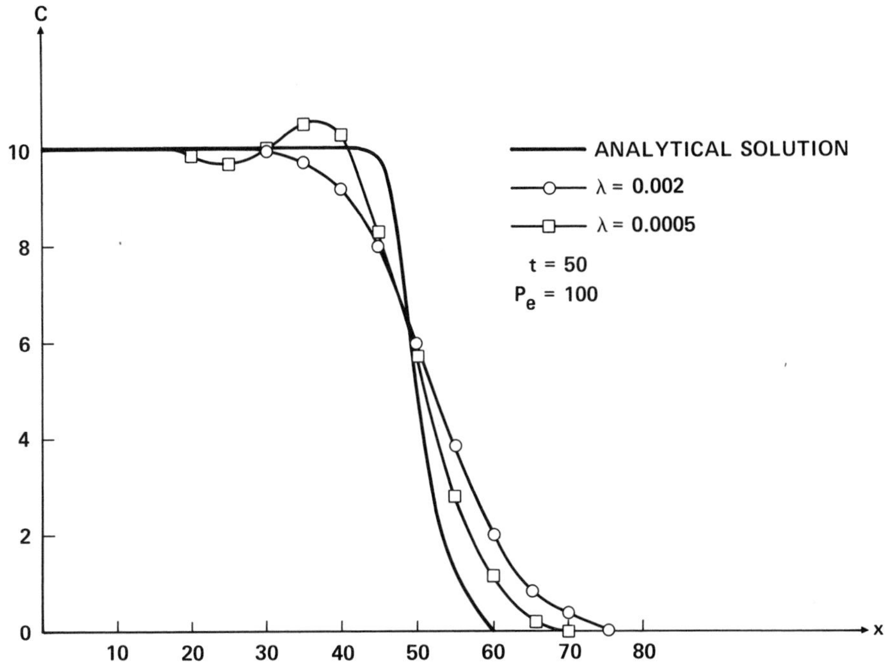

FIGURE 7.15 Upstream weighting MCB solutions with different constants of proportionality (From N.Z. Sun and W.W-G. Yeh, "A Proposed Upstream Weight Numerical Method for Simulating Pollutant Transport in Groundwater." *Water Resources Research*, 19(6):1983, p. 1495. Copyright © 1983 by the American Geophysical Union)

aries are streams that maintain a constant head level ($h = 100$m) in the aquifer. However, the northern stream is polluted; the concentration of the conservative waste constituent in the stream is 10 g/m^3. Assume that the pollutant does not affect the overall solution density of the aquifer, that the aquifer is homogeneous and isotropic, and that the flow field is approximately steady-state. The parameters of the problem are summarized in Table 7.2.

The transformation of the groundwater flow equation using the multiple-cell balance method is similar to the development of the convection-dispersion simulation model. We assume that the domain has been discretized into linear triangular elements. The head, $h(x, y, t)$, can then be represented by the linear equation

$$h(x, y, t) = N_i h_i + N_j h_j + N_k h_k \quad , \quad (x, y) \in \Delta\, ijk \tag{7.69}$$

where

$$N_l = \frac{1}{2\Delta_e}(a_l + b_l x + c_l y), \qquad l = i, j, k \tag{7.70a}$$

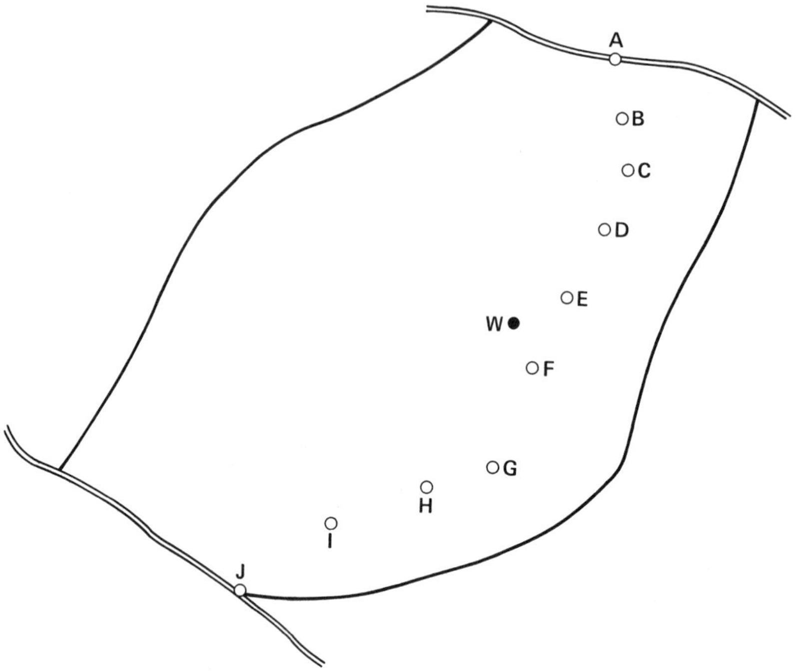

FIGURE 7.16 Aquifer system (From N.Z. Sun and W.W-G. Yeh, "A Proposed Upstream Weight Numerical Method for Simulating Pollutant Transport in Groundwater." *Water Resources Research*, 19(6):1983, p. 1496. Copyright © 1983 by the American Geophysical Union)

and

$$
\begin{aligned}
a_i &= x_j y_k - x_k y_j, \quad & b_i &= y_j - y_k, \quad & c_i &= y_k - x_j \\
a_j &= x_k y_i - x_i y_k, \quad & b_j &= y_k - y_i, \quad & c_j &= x_i - x_k \\
a_k &= x_i y_j - x_j y_i, \quad & b_k &= y_i - y_j; \quad & c_k &= x_j - x_i
\end{aligned}
\tag{7.70b}
$$

and Δ_e is the area of the triangular element.

TABLE 7.2 Aquifer Parameters

Hydraulic conductivity (m/day)	20
Storage coefficient	0.0001
Porosity	0.1
Thickness (m)	50
Longitudinal dispersivity (m)	50
Transverse dispersivity (m)	15
Initial concentration (g/m^3)	0
Pumping rate (m^3/day)	10,000

The partial derivatives of the head with respect to space and time can then be expressed as

$$\frac{\partial h}{\partial x} = \frac{1}{2\Delta_e}(b_i h_i + b_j h_j + b_k h_k) \tag{7.71}$$

$$\frac{\partial h}{\partial y} = \frac{1}{2\Delta_e}(c_i h_i + c_j h_j + c_k h_k) \tag{7.72}$$

$$\frac{\partial h}{\partial t} = N_i \frac{\partial h_e}{\partial t} + N_j \frac{\partial h_j}{\partial t} + N_k \frac{\partial h_k}{\partial t}, \qquad (x, y) \in \Delta ijk \tag{7.73}$$

Green's theorem is used to obtain the integrated form of the confined flow equation, or

$$\int_S m \left[\left(K_{xx}\frac{\partial h}{\partial x} + K_{xy}\frac{\partial h}{\partial y} \right) dy + \left(K_{xy}\frac{\partial h}{\partial x} + K_{yy}\frac{\partial h}{\partial y} \right) dx \right] = \iint_A \left(mS\frac{\partial h}{\partial t} + W \right) dxdy \tag{7.74}$$

Introducing Equations 7.71 through 7.73 into Equation 7.74, the discretized equations may be expressed as

$$\sum_{ei} (A^e_{ii} h_i + A^e_{ij} H_j + A^e_{ik} h_k) = \sum_{ei} \left(B^e_{ii}\frac{\partial h_i}{\partial t} + B^e_{ij}\frac{\partial h_j}{\partial t} + B^e_{ik}\frac{\partial h_k}{\partial t} \right) \tag{7.75}$$

where

$$A^e_{il} = -\frac{m_i}{4\Delta_e}[K_{xx} b_i b_l + K_{xy}(c_i b_l + b_i c_l) + K_{yy} c_i c_l], \tag{7.76a}$$

$l = i, j, k$, and m_i is the saturated thickness of the aquifer at node i. The B^e coefficients are defined as

$$B^e_{ii} = \frac{22}{108} S_i \Delta_e$$

$$B^e_{ij} = \frac{7}{108} S_i \Delta_e \tag{7.76b}$$

$$B^e_{ik} = \frac{7}{108} S_i \Delta_e$$

where S_i is the storage coefficient at node i.

Writing Equation 7.74 for each node and incorporating the boundary conditions, we obtain the hydraulic response equations

$$A\,\dot{\mathbf{h}} + B\,\mathbf{h} + \mathbf{g} = 0 \tag{7.77}$$

The mass transport response equations are given by Equation 7.64

$$P\,\dot{\mathbf{c}} + R\mathbf{c} + \mathbf{f} = 0 \tag{7.64}$$

Given the initial conditions for both problems, finite-difference methods are generally used for the solution of these coupled sets of ordinary differential equations. Simulation studies have demonstrated that a backward difference scheme provides the most accurate results for the mass transport and flow problems, although a Crank-Nicolson approximation of the time derivative minimizes the truncation error of the approximations (Pinder, 1973). Introducing the backward, implicit approximation, the response equation may be expressed as

$$A\left\{\frac{\mathbf{h}^k - \mathbf{h}^{k-1}}{\Delta t}\right\} + B\mathbf{h}^k + \mathbf{g} = 0$$

$$P\left\{\frac{\mathbf{c}^m - \mathbf{c}^{m-1}}{\Delta t_m}\right\} + R\mathbf{c}^m + \mathbf{f} = 0$$

where Δt and Δt_m are the lengths of the time steps for the flow and mass transport problems and k and m are running indices of the time level for the flow and mass transport problems.

The relation between the head values over successive time steps is formally given as

$$\mathbf{h}^k = \left\{\frac{A}{\Delta t} + B\right\}^{-1}\left\{\frac{A}{\Delta t}\mathbf{h}^{k-1} - \mathbf{g}\right\}, \quad \forall k$$

Similarly, for the transport problem, we have within any Δt

$$\mathbf{c}^m = \left\{\frac{P}{\Delta t_m} + R\right\}^{-1}\left\{\frac{P}{\Delta t_m}\mathbf{c}^{m-1} - \mathbf{f}\right\}, \quad \forall m \in \Delta t$$

Although the solutions have been expressed in terms of the matrix inverse, the inversion is rarely performed directly. Rather, Gaussian elimination methods (Thomas, ADI, or Hopscotch methods) or iterative schemes such as relaxation methods can be used for the efficient solution of the problem (see, for example, Lapidus & Pinder, 1982).

A common solution algorithm for the coupled flow and quality equations can be summarized by the following stages:

1. Solve the groundwater equations for the time step, Δt.
2. Evaluate, from the head distribution, the velocity components and dispersion coefficients. The R matrix in the mass transport equations can then be updated for the time interval, Δt_m.
3. Determine the concentration distribution by advancing the solution for the time step, Δt_m. Solve the quality equations for each time step within the flow-time period.
4. Advance the flow-time step and return to step 1 if the total simulation has not been completed.

As an example of the simulation methodology, let us return to the example aquifer system shown in Figure 7.16. The aquifer system has been divided into 97 linear triangular elements; there are a total of 62 nodal points; the finite-element

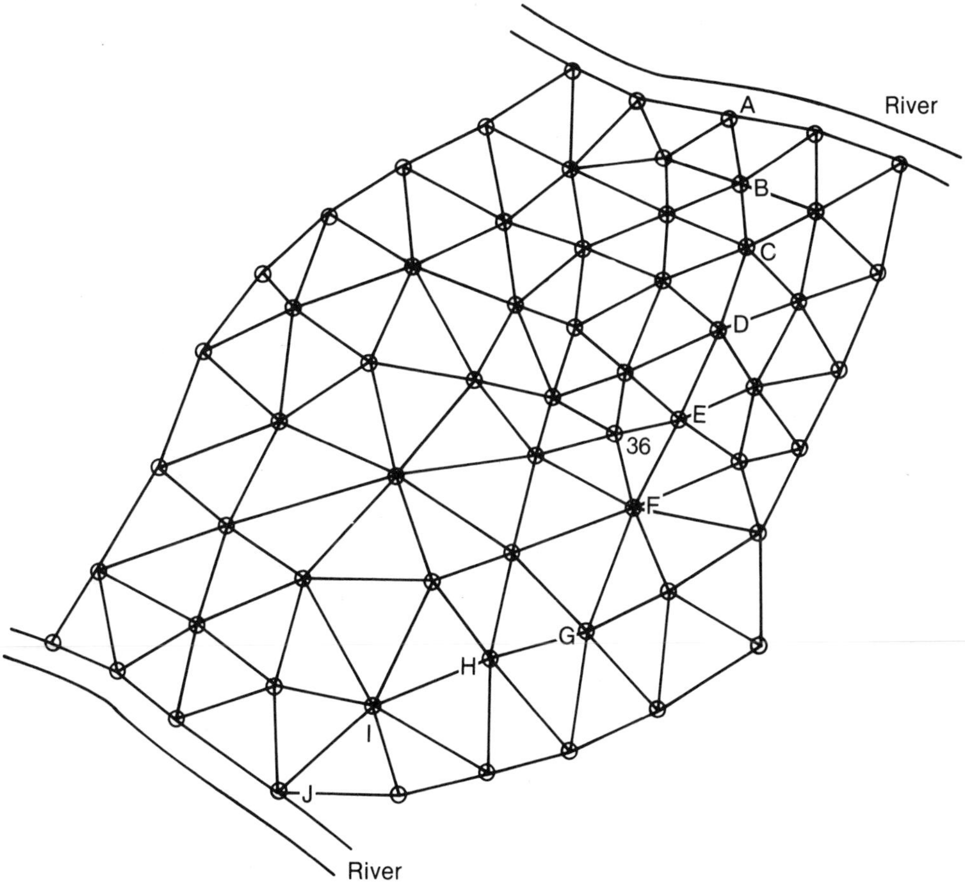

FIGURE 7.17 Finite-element grid

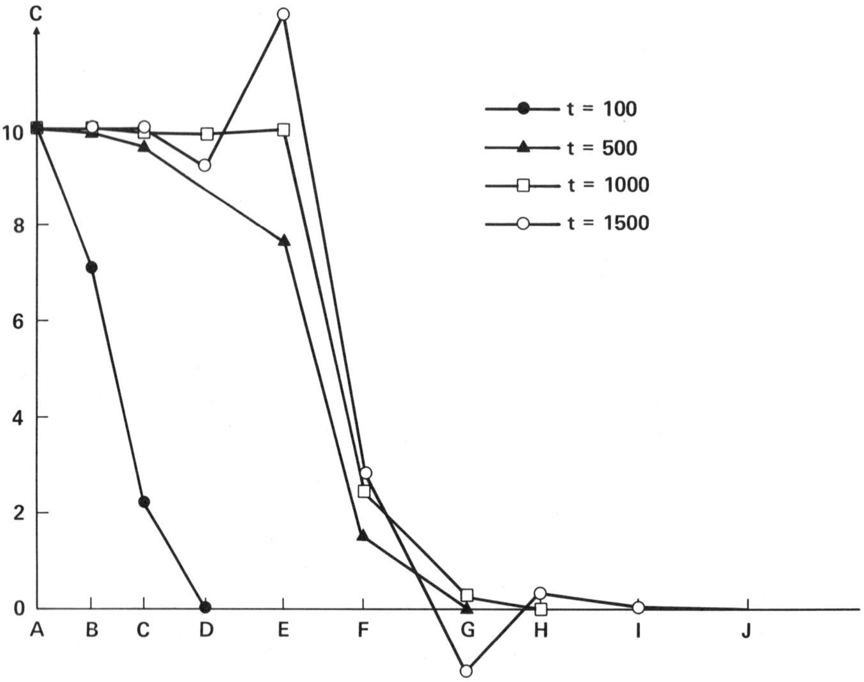

FIGURE 7.18 Concentration profiles obtained by MCB without upstream weights at observation wells A through J at different times (From N.Z. Sun and W.W-G. Yeh, "A Proposed Upstream Weight Numerical Method for Simulating Pollutant Transport in Groundwater." *Water Resources Research*, 19(6):1983, p. 1497. Copyright © 1983 by the American Geophysical Union)

grid is shown in Figure 7.17. Observation wells are located at points A through J. The pumping well, which extracts 10,000 m^3/day, is located at node 36.

Typical results of the simulation analysis are presented in Figures 7.18 and 7.19. Figure 7.18 details the concentration profiles at the observation wells for times ranging from 100 to 1500 days. Contour maps of the head and concentration distributions after 1500 days are shown in Figures 7.20 and 7.21.

The average Peclet number in the example does not exceed 5 and it would appear that overshoot and numerical dispersion are not important. If, however, $\lambda = 0$ in Equation 7.68, that is, we do not introduce any upstream weights, strong oscillations of the numerical solution are observed as the concentration front becomes very sharp. Figure 7.18 shows that there are no oscillations when the time is less than 1000 days, but the numerical solutions lose their significance because of strong oscillations when it becomes longer than 1500 days.

Introducing the upstream weights in the model (e.g., $\lambda = 0.001$) effectively controls the oscillations and numerical dispersion (see Figure 7.19). Approximately 32 machine unit seconds are required for the complete water quality simulation on an IBM 3033. Further examples of the simulation of the water quality in stream-

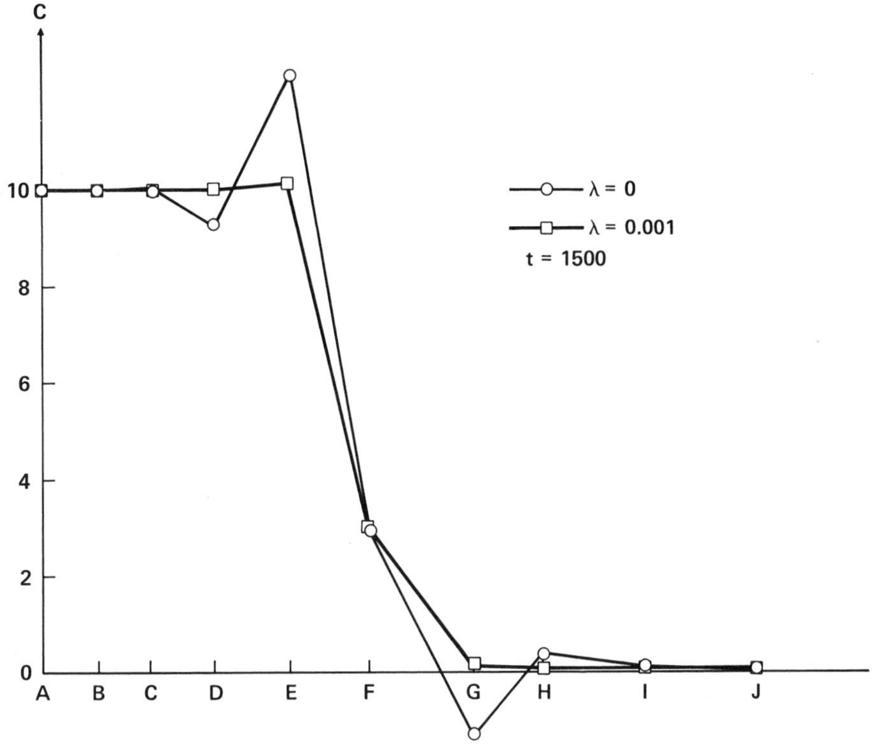

FIGURE 7.19 Concentration profiles obtained by MCB with and without upstream weights at observation wells A though J at t = 1500 days (From N.Z. Sun and W.W-G. Yeh, "A Proposed Upstream Weight Numerical Method for Simulating Pollutant Transport in Groundwater." *Water Resources Research,* 19(6):1983, p. 1497. Copyright © 1983 by the American Geophysical Union)

aquifer systems may be found in Cabrera and Mariño (1976) and Mariño (1978, 1981).

7.3.3 The Method of Characteristics

The method of characteristics is a numerical technique that minimizes numerical diffusion and stability problems concomitant with the solution of the convection-dispersion equation. The method is predicated on the characteristic equations of the mass transport equation. These equations can be expressed for any homogeneous, linear partial-differential equation

$$a_1\frac{\partial\phi}{\partial x} + a_2\frac{\partial\phi}{\partial y} + a_3\frac{\partial\phi}{\partial t} + a_4\frac{\partial\phi}{\partial z} = 0 \qquad (7.78)$$

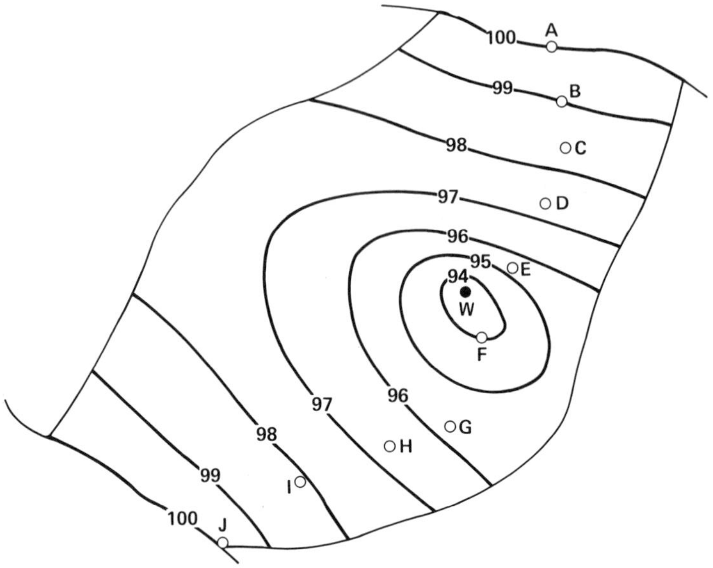

FIGURE 7.20 **Contour map of head distribution** (From N.Z. Sun and W.W-G. Yeh, "A Proposed Upstream Weight Numerical Method for Simulating Pollutant Transport in Groundwater." *Water Resources Research,* 19(6):1983, p. 1497. Copyright © 1983 by the American Geophysical Union)

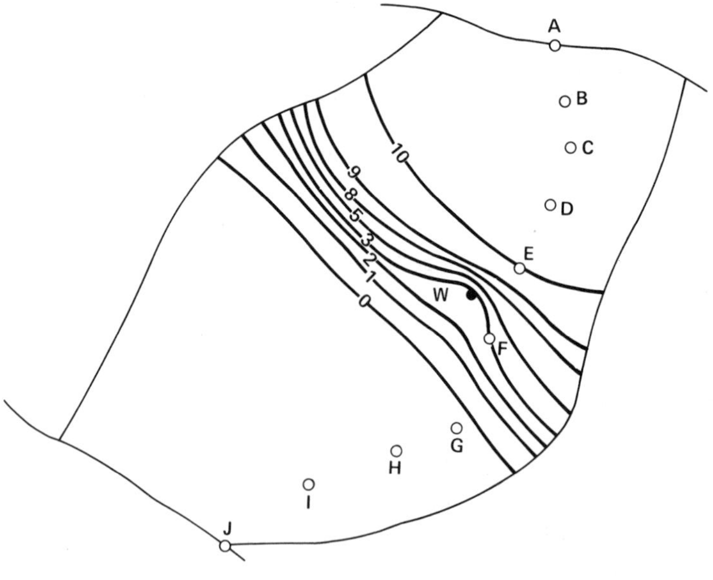

FIGURE 7.21 **Contour map of concentration distribution at t = 1500 days** (From N.Z. Sun and W.W-G. Yeh, "A Proposed Upstream Weight Numerical Method for Simulating Pollutant Transport in Groundwater." *Water Resources Research,* 19(6):1983, p. 1498. Copyright © 1983 by the American Geophysical Union)

with parameters (a_i), as the system of ordinary differential equations

$$\frac{dx}{ds} = a_1, \quad \frac{dy}{ds} = a_2, \quad \frac{dt}{ds} = a_3, \quad \frac{dz}{ds} = a_4 \tag{7.79}$$

The solutions of the equations $x = x(s)$, $y = y(s)$, $z = z(s)$, and $t = t(s)$ are called the *characteristic curves* of Equation 7.78, where s is an arbitrary curve parameter (for example, $s = t$). We also note that, along a characteristic curve, the solution is a constant, which can be demonstrated by differentiating the solution along the characteristic curves, or

$$\frac{d\phi}{ds} = \frac{\partial\phi}{\partial x}\frac{dx}{ds} + \frac{\partial\phi}{\partial y}\frac{dy}{ds} + \frac{\partial\phi}{\partial t}\frac{dt}{ds} + \frac{\partial\phi}{\partial z}\frac{dz}{ds} \tag{7.80a}$$

And, after substituting Equation 7.79, or

$$\frac{d\phi}{ds} = \frac{\partial\phi}{\partial x} a_1 + \frac{\partial\phi}{\partial y} a_2 + \frac{\partial\phi}{\partial t} a_3 + \frac{\partial\phi}{\partial z} a_4 = 0 \tag{7.80b}$$

To continue the analysis of the convection-dispersion equation, consider the one-dimensional mass transport equation (Eq. 7.3),

$$\frac{\partial c}{\partial t} = D^* \frac{\partial^2 c}{\partial x^2} - v_x \frac{\partial c}{\partial x} \tag{7.3}$$

The characteristic equations associated with Equation 7.3 can be expressed as

$$\frac{dx}{dt} = v_x \tag{7.81a}$$

$$\frac{dc}{dt} = D^* \frac{\partial^2 c}{\partial x^2} \tag{7.81b}$$

These equations can be solved by discretizing the domain into a stationary grid and introducing a set of moving points which are used to solve numerically the characteristic equations. For the one-dimensional problem, the stationary grid is the usual finite-difference grid, or $x_i = i\Delta x$, $i = 0, 1, \ldots$. Each moving point in turn corresponds to one characteristic curve, and values of x and c can be obtained as functions of t for each characteristic. If we define p as the index number identifying the point, and k as an index of the time step, then we require that $x_{p,k}$ and $c_{p,k}$ are to be calculated for each point at each time, t_k.

To begin the calculations, the set of moving points is uniformly distributed throughout the solution domain. Initial coordinates, $x_{p,0}$ and an initial concentration are then assigned to each point, p. We also assume that at time t_k the concentration at each stationary point is known. The difference equations of Equation 7.81 can then be used to obtain the new positions of the moving points according to the relation

$$x_{p,k+1} = x_{p,k} + \Delta t_{k+1/2}\, v_x(x_{p,k}) \tag{7.82}$$

After moving each point, the coordinates are examined to determine which finite-difference cell the point lies in. Each computational element is assigned a concentration, $c^*_{i,k}$, equal to the average of the concentrations, c_{pk}, of all the points that lie in the computational element after they are moved. The change in concentration due to dispersion can then be computed for each element as

$$\Delta c_{i,k+1/2} = \Delta t_{k+1/2} \left(\frac{D^*}{\Delta x^2}\right) \{c^*_{i+1,k} - 2c^*_{i,k} + c^*_{i-1,k}\} \tag{7.83}$$

Each moving point can then be assigned a new concentration, or

$$c_{p,k+1} = c_{p,k} + \Delta c_{i,k+1/2} \tag{7.84}$$

All points within an element at a given time undergo the same change in concentration due to dispersion. Finally, the concentrations at the stationary grid points can be estimated from the relation

$$c_{i,k+1} = c^*_{i,k} + \Delta c_{i,k+1/2} \tag{7.85}$$

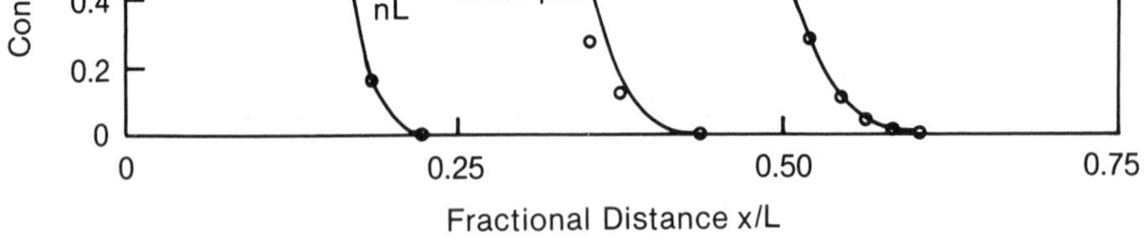

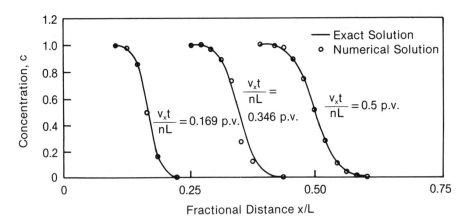

FIGURE 7.22 Numerical solutions of one-dimensional transport with dispersion (Adapted with permission from A. O. Gardner, Jr., D. W. Peaceman, and A. L. Pozzi, "Numerical Calculation of Multidimensional Miscible Displacement by the Method of Charactertistics." *Journal of the Society of Petroleum Engineers*, March 1964, p. 29. Copyright © 1964 by the Society of Petroleum Engineers of AIME)

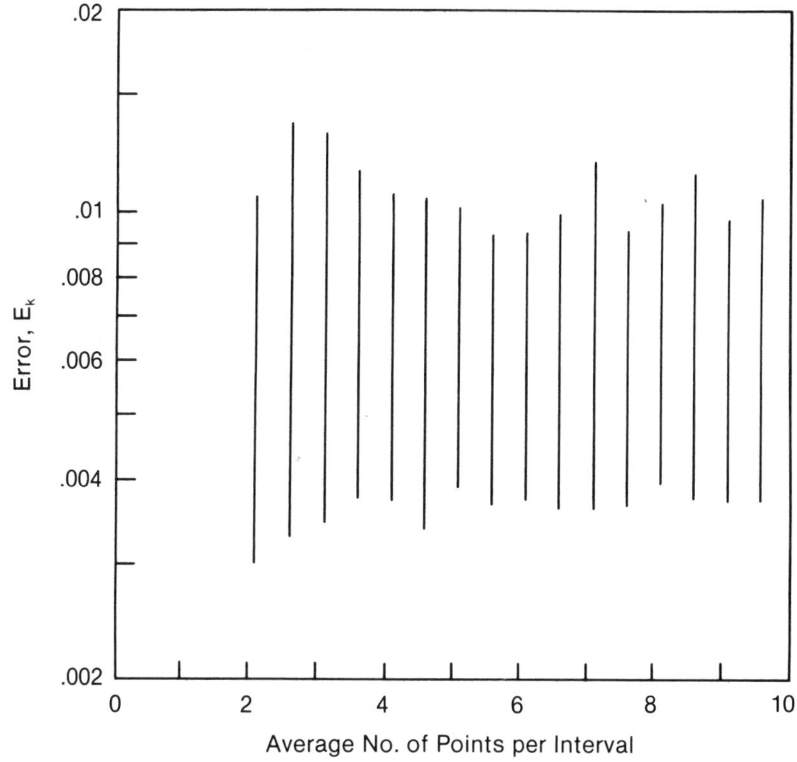

FIGURE 7.23 Effect of spacing of moving points on errors in numerical solution (Adapted with permission from A. O. Gardner, Jr., D. W. Peaceman, and A. L. Pozzi, "Numerical Calculation of Multidimensional Miscible Displacement by the Method of Characteristics." *Journal of the Society of Petroleum Engineers,* March 1964, p. 29. Copyright © 1964 by the Society of Petroleum Engineers of AIME)

The process then continues for the entire simulation period. Note that, in this formulation, an explicit solution method is used to estimate the dispersion occurring in the system.

Example simulations of the convection-dispersion equation using the method of characteristics are illustrated in Figure 7.22. In comparison with the analytical solution (Eq. 7.16), the average concentrations behave somewhat erratically because the points move from one subregion to another at random times. The concentrations do behave properly on the average, as the points scatter on both sides of the exact solution. While there are errors in the numerical solution, they do not amplify during the course of the simulation (Gardner, et al., 1964).

The effects of changing the number of moving points per interval are shown in Figure 7.23. For any time step, the error E_k is defined as the maximum deviation from the analytical solution $c_{i,\text{true}}$, $E_k = \max_i | c_i - c_{i,\text{true}} |$. The range of the error does not seem to be significantly affected. Also, the effects of differing dispersion values are illustrated in Figure 7.24. These results again demonstrate the accuracy of the numerical solutions for a wide range in dispersion coefficients.

The method of characteristics has been successfully applied to groundwater quality field problems (see, for example, Konikow & Bredehoeft, 1974; Konikow, 1977). However, the numerical method is difficult to program and, from a planning perspective, the response equations are not directly generated using the numerical model. For these reasons, we will rely on the Galerkin and multiple-cell balance methods for the development of optimization models of the groundwater quality system.

We conclude this section with an example of the simulation of density-dependent mass transport illustrating sea water intrusion in a coastal aquifer system.

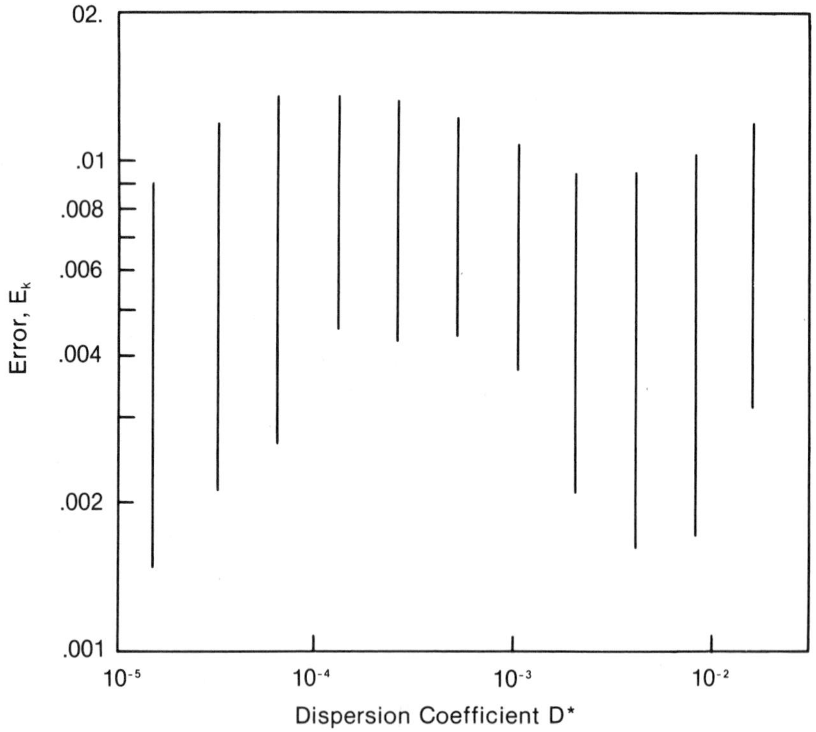

FIGURE 7.24 Simulation results using the method of characteristics (Adapted with permission from A. O. Gardner, Jr., D. W. Peaceman, and A. L. Pozzi, "Numerical Calculation of Multidimensional Miscible Displacement by the Method of Charactertistics." *Journal of the Society of Petroleum Engineers,* March 1964, p. 30. Copyright © 1964 by the Society of Petroleum Engineers of AIME)

Example Problem 7.4 Saltwater Intrusion

Using the method of characteristics, develop a numerical model to simulate the movement of the salt water front in the coastal aquifer system shown in Figure 7.8. Compare the results with the results obtained using the Galerkin finite-element model developed in Example Problem 7.2.

Using the method of characteristics, we first develop the numerical model for simulating salt water intrusion. Following Pinder and Cooper (1970), we assume that (1) the release of water from storage has a negligible effect on the movement of the salt water front, (2) the porosity and viscosity are constant, and (3) the dispersion coefficient is constant in space. With these assumptions, the governing flow and transport equations can be expressed as

$$\nabla \cdot \frac{k}{\mu} \nabla p - g\frac{\partial}{\partial z}\left(\frac{k\rho}{\mu}\right) = 0 \tag{7.86}$$

and,

$$\mathbf{D}\ \nabla^2 c - \frac{\mathbf{q}}{n} \cdot \nabla c = \frac{\partial c}{\partial t} \tag{7.87}$$

where c is the mass concentration of dissolved salt.

Again, the empirical relationship presented by Baxter and Wallace (1916) is used to relate the fluid density and the concentrations of fresh c_f and salt water.

$$\rho = c_f + c = \rho_0 + (1-E)c \tag{7.88}$$

where ρ_0 is the density of fresh water.

The solution of the pressure equation can be developed using the alternating direction implicit method (see Section 4.5.2). The method consists of solving the pressure equation alternately in the x and z directions. The finite-difference equation may be expressed as

$$\begin{aligned}
&\frac{k_{i,j+1/2}}{\mu\Delta x^2}(p^{m+1/2}_{i,j+1,k+1} - p^{m+1/2}_{i,j,k+1}) - \frac{k_{i,j\text{-}1/2}}{\mu\Delta x^2}(p^{m+1/2}_{i,j,k+1} - p^{m+1/2}_{i,j-1,k+1}) + \\
&\quad \frac{k_{i+1/2,j}}{\mu\Delta z^2}(p^{m+b}_{i+1,j,k+1} - p^{m+b}_{i,j,k+1}) - \frac{k_{i-1/2,j}}{\mu\Delta z^2}(p^{m+b}_{i-1,j,k+1} - p^{m+b}_{i-1,j,k+1}) - \\
&\quad \frac{g}{\mu\Delta z}(k_{i+1/2,j}\rho_{i+1/2,j,k} - k_{i-1/2,j}\rho_{i-1/2,j,k}) + aH(p^{m+1/2}_{i,j,k+1} - p^{m+b}_{i,j,k+1}) = 0
\end{aligned} \tag{7.89}$$

The parameters a and b define the direction of solution of the equations; $b = 0$, for example, implies that the equations are solved implicitly (in the x direction). When a is 1, the equations are solved implicitly in the x direction. If $a = -1$, the equations are solved implicitly in the z direction. H is a normalized iteration parameter.

Because the pressure equation is nonlinear, it is necessary to solve iteratively the finite-difference model. For each iteration, the finite-difference equation is solved once in the x direction and then once in the z direction. For each time step, iteration continues until the maximum pressure difference at each node is less than a predetermined error tolerance.

The characteristics equations of the mass transport model are given as

$$\frac{dx}{dt} = \frac{q_x}{n} \tag{7.90}$$

$$\frac{dz}{dt} = \frac{q_z}{n} \tag{7.91}$$

$$\frac{dc}{dt} = D_{xx}\frac{\partial^2 c}{\partial x^2} + D_{zz}\frac{\partial^2 c}{\partial z^2} \tag{7.92}$$

The solution of these ordinary differential equations is based on the movement of particles within the solution domain. In the two-dimensional case, a rectangular region is constructed around each nodal point (see Figure 7.25), x_i, z_j. Moving particles are placed within each rectangle if a particle at x,z lies within $R_{i,j}$

$$\begin{aligned} x_i - \frac{\Delta x}{2} &\leqslant x \leqslant x_i + \frac{\Delta x}{2} \\ z_j - \frac{\Delta z}{2} &\leqslant z \leqslant z_j + \frac{\Delta z}{2} \end{aligned} \tag{7.93}$$

The path of each moving particle corresponds to one characteristic curve.

As in the one-dimensional case, initially the particles are distributed uniformly through the rectangular $x-z$ region. Assuming that the concentration, $c_{i,j,k}$, and the pressure, $p_{i,j,k+1}$, are known at each stationary node, the average pore velocity, $\mathbf{v}$, can be computed from the finite-difference approximation to the pressure equation and the relation

$$\mathbf{v} = \mathbf{q}/n$$

The velocities at any point are assumed to be equal to the nearest velocity components calculated on the fixed-point grid. The updated coordinates of the particles can then be computed from the equations

$$x_{r,k+1} = x_{i,k} + \Delta t_{k+1/2}\, v_x(x_{r,k}, z_{r,k}) \tag{7.94}$$

$$z_{r,k+1} = z_{r,k} + \Delta t_{k+1/2}\, v_z\,(x_{r,k}, z_{r,k}) \tag{7.95}$$

The maximum distance traveled by a particle during a time step is limited to a preassigned value by computing the velocities from the stationary grid and then

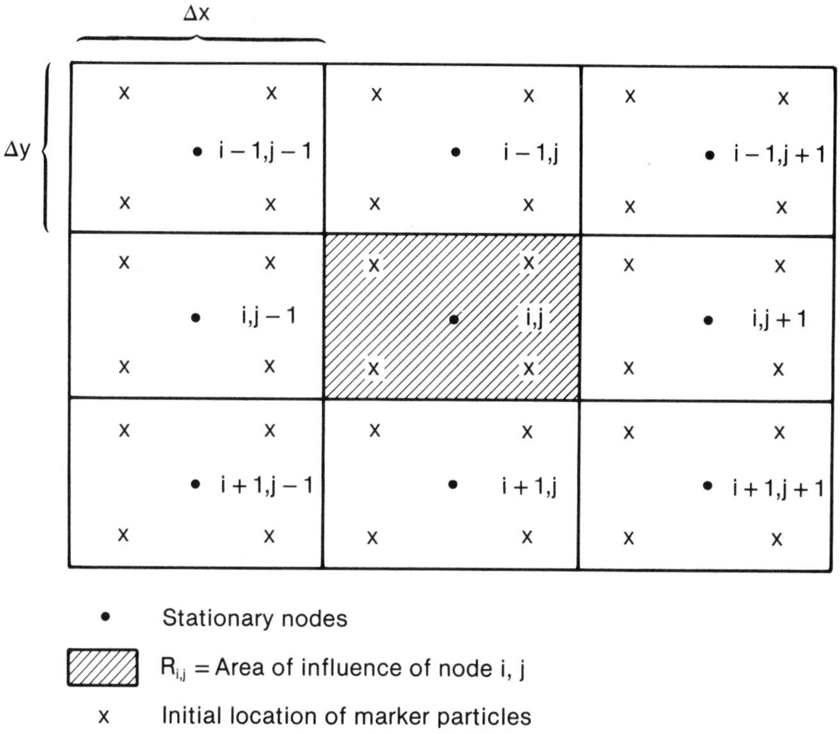

FIGURE 7.25 Nodal array for finite-difference network and initial location of marker particles (Adapted from G. F. Pinder and H. H. Cooper, Jr., "A Numerical Technique for Calculating the Transient Position of the Saltwater Front." *Water Resources Research*, 6(3), 1970, p. 880. Copyright © 1970 by the American Geophysical Union)

calculating the magnitude of the time step from the equation

$$\Delta t_{k+1/2} = \frac{\epsilon}{v_{\max,k+1}} \tag{7.96}$$

where ϵ is the maximum distance traveled during a time step.

Following the movement of each particle, the area of influence in which the particle is located can be determined. After all the particles have been moved, the concentration at each node is temporarily assigned a value, $c^*_{i,j,k}$, which is the mean concentration in the region $R_{i,j}$. The change in concentration as a result of dispersion is determined from the finite-difference approximation of Equation 7.87, or

$$\Delta c_{i,j,k+1/2} = \Delta t_{k+1/2}\left\{\frac{D_{xx}}{\Delta x^2}\left(c^*_{i,j+1,k} - 2c^*_{i,j,k} + c^*_{i,j-1,k}\right) + \frac{D_{zz}}{\Delta z^2}\left(c^*_{i+1,j,k} - 2c^*_{i,j,k} + c^*_{i-1,j,k}\right)\right\} \tag{7.97}$$

The updated particle concentration can then be expressed as

$$c_{r,k+1} = c_{r,k} + \Delta c_{i,j,k+1/2} \tag{7.98}$$

$$c_{i,j,k+1} = c^{*}_{i,j,k} + \Delta c_{i,j,k+1/2} \tag{7.99}$$

One time step of the simulation is completed by assigning concentrations to all stationary points according to the relation.

Figure 7.26 presents the results of the salt water intrusion simulation using the method of characteristics. As discussed by Pinder and Cooper (1970), two initial conditions on the salt distribution in the aquifer were assumed in the analysis. One scenario assumed that the aquifer was initially filled with fresh water, and the salt water moved landward toward the steady-state position. The second condition assumed that initially a sharp interface existed at its steady-state position, and that at $t = 0$, dispersion began that caused the front to begin retreating to a new steady-state position. Figure 7.26 shows the position of the 0.5 isochlor obtained from the numerical solution for each set of the initial conditions. The curve to the right of Henry's steady-state solution demonstrates that the salt water front advances rapidly during the early stages of the simulation and more slowly as the system approaches steady-state (Henry, 1962). The curves to the left of the 0.5 isochlor (corresponding

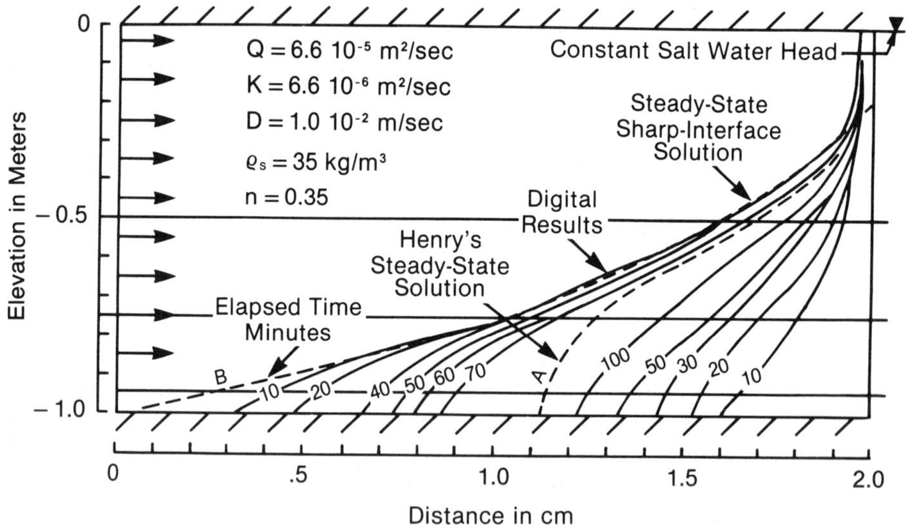

FIGURE 7.26 Transient two-dimensional flow with dispersion (Adapted from G. F. Pinder and H. H. Cooper, Jr. "A Numerical Technique for Calculating the Transient Position of the Saltwater Front." *Water Resources Research*, 6(3), 1970, p. 880. Copyright © 1970 by the American Geophysical Union)

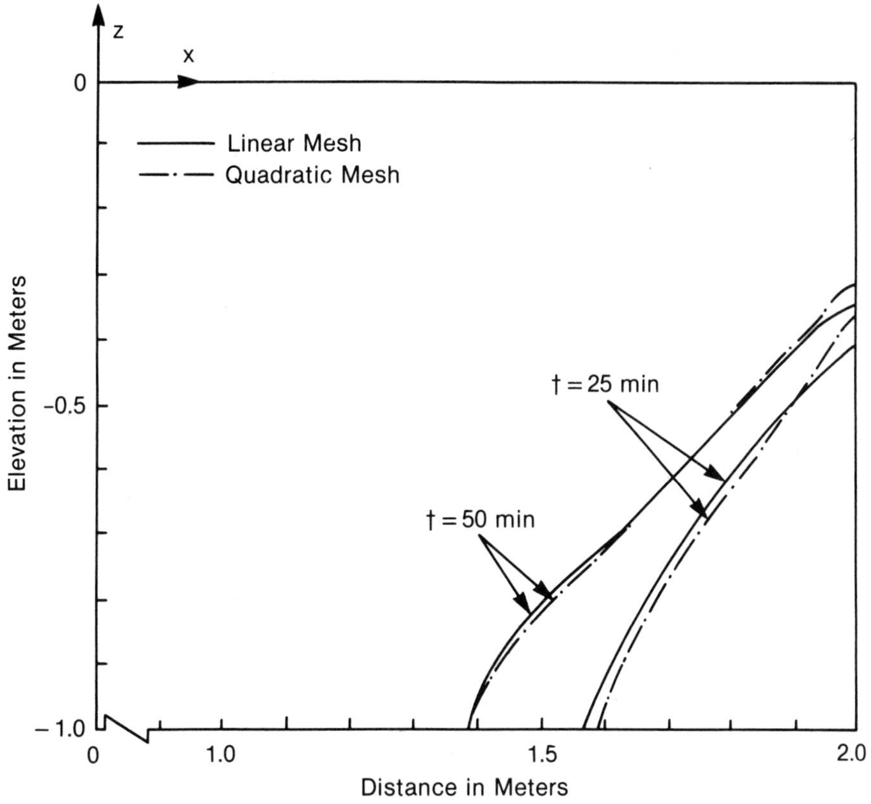

FIGURE 7.27 Position of 0.5 isochlor using linear and quadratic basis functions (Adapted from G. Segol, G. F. Pinder, and W. G. Gray, "A Galerkin-Finite Element Technique for Calculating the Transient Position of the Saltwater Front." *Water Resources Research*, 11(2), 343–347, 1975. Copyright © 1975 by the American Geophysical Union)

to the first set of initial conditions) depict the salt water wedge retreating as a result of dispersion. During the initial stages of the simulation, the isochlors are concave upwards. The isochlors do become convex near the top of the wedge, again as a result of hydrodynamic dispersion.

The Galerkin finite-element results, for the model presented in Example Problem 7.4, are also illustrated in Figures 7.27 and 7.28. Linear and quadratic basic functions were used for discretization of the aquifer system. The finite element grids are shown in Figure 7.29 (Segol, et al., 1975).

The solution of the nonlinear flow and transport equations, discussed in Example Problem 7.2, was obtained by the iterative solution of the system's equations. In the analysis, the solution density in the mass transport equation was assumed constant and the pressure variations in the continuity equation were also

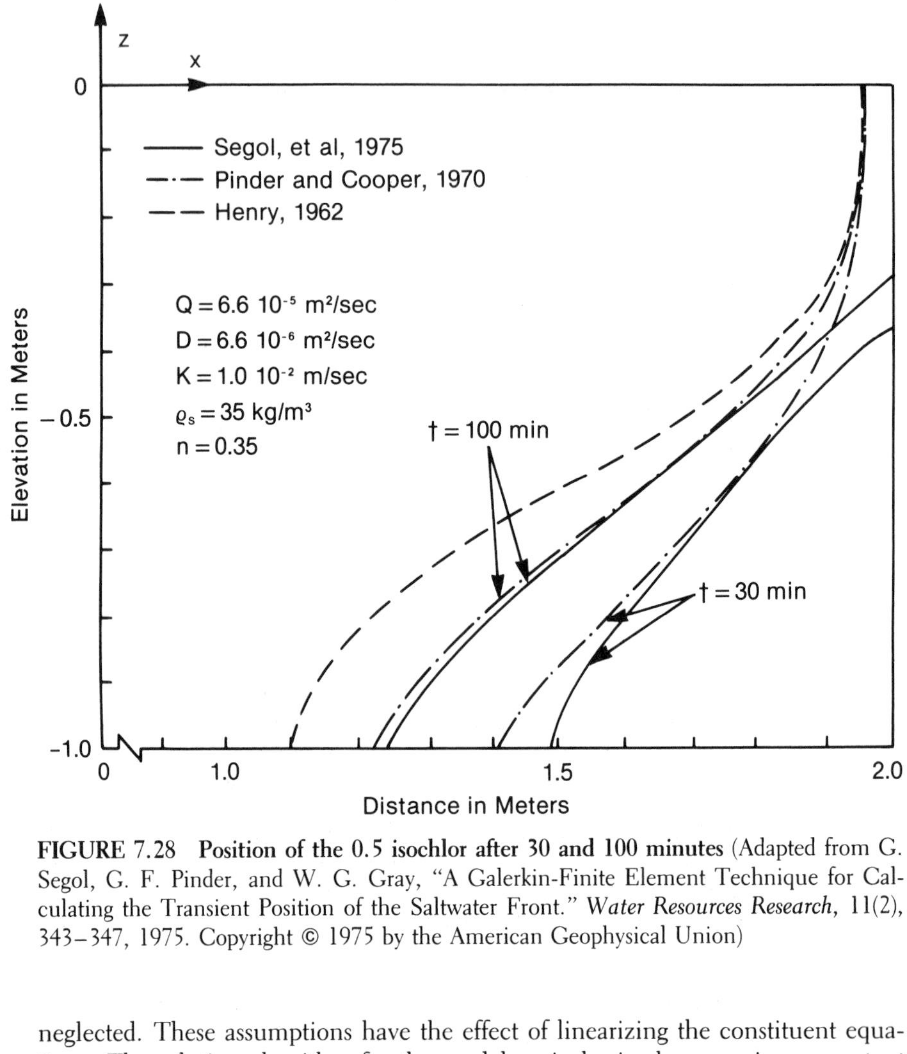

FIGURE 7.28 Position of the 0.5 isochlor after 30 and 100 minutes (Adapted from G. Segol, G. F. Pinder, and W. G. Gray, "A Galerkin-Finite Element Technique for Calculating the Transient Position of the Saltwater Front." *Water Resources Research*, 11(2), 343–347, 1975. Copyright © 1975 by the American Geophysical Union)

neglected. These assumptions have the effect of linearizing the constituent equations. The solution algorithm for the model again begins by assuming a constant initial density field. The Darcy velocities and the pressure distribution can then be determined from Equation 7.43. This information can then be used to evaluate the coefficient matrices of the mass transport equations. Although the velocity and concentration values have been determined, the values are only approximated because the velocity field is not compatible with the solute concentrations. The velocities must then be recalculated using the updated solution densities. The iterative process continues until there is negligible variation in the overall solution density for a given time step in the simulation period.

The results of the finite-element simulation shown in Figure 7.28 compare favorably with the results obtained with the method of characteristics and Henry's steady-state solution. The position of the 0.5 isochlor is nearly identical for both simulations. The accurate prediction of the location of the salt water front requires that the velocity field be re-evaluated for each concentration time step (Segol, et al., 1975). This demonstrates the importance of the density terms in the solution of flow equations.

The hydrodynamics of the salt water front are illustrated in Figure 7.30. As

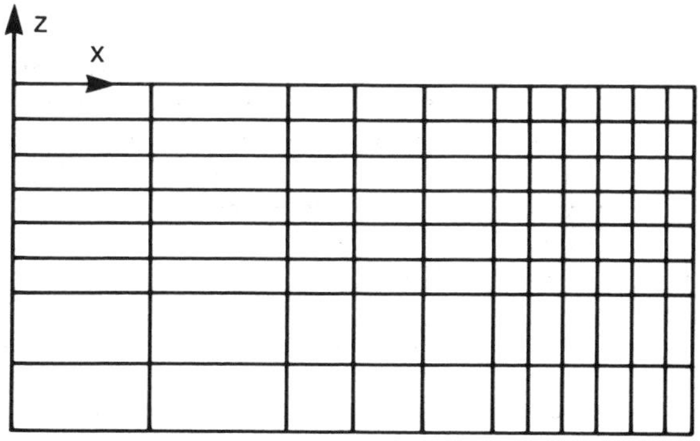

(a)

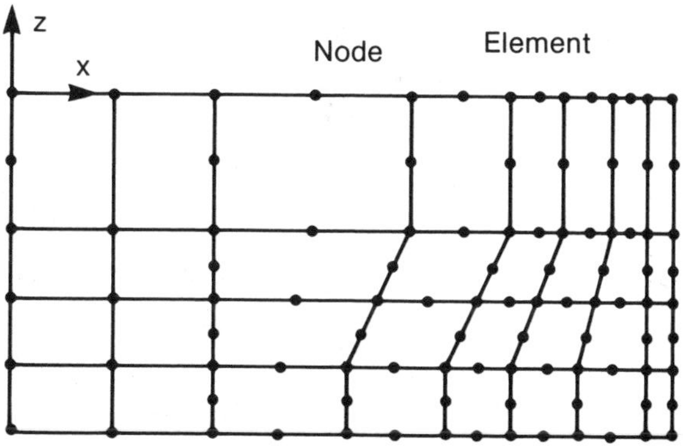

(b)

FIGURE 7.29 Finite-element discretizations of the rectangular aquifer of Henry's problem. (a) Linear mesh—108 nodal points and 88 elements. (b) Quadratic mesh—107 nodal points and 28 elements. (Adapted from G. Segol, G. F. Pinder, and W. G. Gray, "A Galerkin-Finite Element Technique for Calculating the Transient Position of the Saltwater Front." *Water Resources Research*, 11(2), 343–347, 1975. Copyright © 1975 by the American Geophysical Union)

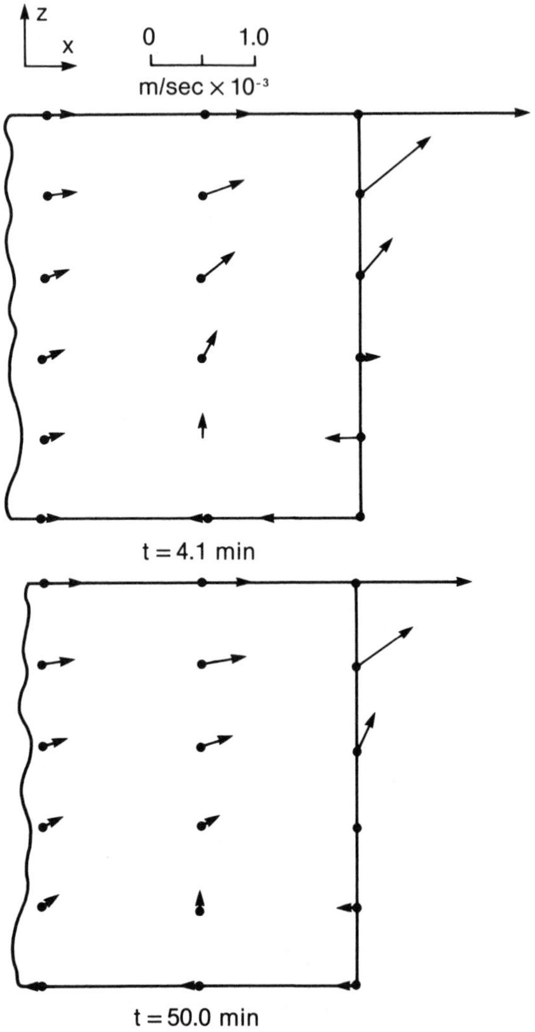

FIGURE 7.30 Velocity field computed using linear basis functions for elapsed time of 4.1 and 50.0 minutes (Adapted from G. Segol, G. F. Pinder, and W. G. Gray, "A Galerkin-Finite Element Technique for Calculating the Transient Position of the Saltwater Front." *Water Resources Research*, 11(2), 343–347, 1975. Copyright © 1975 by the American Geophysical Union)

the salt water wedge moves inland, the density-dependent velocity component decreases. Note also that q_z is not identically zero along the outflow boundary, $x = L$. This is a consequence of the averaging process of the finite-element method.

7.4 OPTIMAL GROUNDWATER QUALITY MANAGEMENT MODEL

The conjunctive management of groundwater supply and quality of regional aquifer systems is inherently a multi-objective planning problem, a problem characterized by conflicting objectives, constraints, and policies. The evaluation of the optimal policies and the analysis of the system trade-offs can be determined from optimization models of the groundwater resource system. These models are particularly useful in ascertaining the feasibility of wastewater injection for the control of sea water intrusion, in alleviating land subsidence problems, or in recharging over-developed groundwater basins with reclaimed wastewater. The central management issues characterizing the conjunctive use problem are to determine:

1. The optimal pumping schedules (well locations and pumping rates) to satisfy a given water demand.
2. The optimal injection schedules. This involves specifying the well locations and injection rates necessary to satisfy a wasteload demand.
3. The maximum waste input concentration that can be safely discharged (in the context of groundwater quality standards) to the basin. Equivalently, the model can determine the required surface wastewater treatment of the wasteload.
4. The hydrologic and environmental trade-offs associated with the waste disposal program. The trade-offs will define, for example, the variation in waste injection concentrations with changing groundwater quality standards or the relation between the input mass flux of the wasteload and the water demands.

We approach the analysis of these problems by developing an optimization model of the groundwater quality system. However, we assume from the onset that

1. All the physical and chemical processes are well defined; i.e., there is sufficient data to estimate the dispersion and kinetic reaction parameters.
2. The flow and mass transport models have been validated and calibrated for the basin; i.e., there is sufficient baseline data.
3. A groundwater monitoring network is in operation for the basin.
4. The solution density of the aquifer is constant. This has the effect of linearizing the hydraulic and water quality response equations of the problem.

The objective function of the planning model will reflect the use and allocation of the water supply and water resources of the groundwater system. As discussed in Chapter 6, a reasonable water supply objective is to maximize the net discounted benefits resulting from the allocation of the groundwater to municipal, industrial, or agricultural water uses. From a water quality or environmental perspective, minimization of the possible contamination of the aquifer or the costs of surface waste treatment and/or waste storage are also possible environmental and water quality objectives.

If we consider a dynamic model and assume (1) a single waste constituent and (2) that the planning horizon consists of T^* discrete time periods, the objective function of the optimization model can be expressed as the weighted sum of the individual objectives, or

$$\max z = \sum_{k=1}^{T*} \sum_{l} \lambda_l f_l (\mathbf{h}^k, \mathbf{c}_s^k, \mathbf{Q}_w^k, \mathbf{Q}_r^k) \tag{7.100}$$

f_l is the lth objective of the planning problem and we have assumed that a particular objective is a function of the state variables (the head and mass concentrations in planning period k) and the pumping ($\mathbf{Q}_w^k$) and injection ($\mathbf{Q}_r^k$) policies. The λ_l represent the preferences or weights associated with planning objective.

The policy variables—the heads, mass concentrations, and the pumping and injection rates—are constrained in each planning period by (1) the water supply requirements of the groundwater basin, or

$$\sum_{w \in \pi} Q_w^k \geqslant D^k, \quad \forall k \tag{7.101}$$

where D_k is the water demand in period k, (2) the wasteload disposal target

$$\sum_{r \in \psi} Q_r^k \geqslant WL^k, \quad \forall k \tag{7.102}$$

The index sets π and ψ define the feasible pumping and injection sites in the basin. WL_k is the magnitude of the wasteload in period k. As in the water supply optimization models, there are possible restrictions on the maximum pumping and injection rates and maximum or minimum head levels, or

$$Q_w^k \leqslant Q_w^*, \quad w \in \pi, \forall k \tag{7.103}$$

$$Q_r^k \leqslant Q_r^*, \quad r \in \psi, \forall k \tag{7.104}$$

$$h_i^k \geqslant h_i^*, \; i \in \Delta, \forall k \tag{7.105}$$

where Δ is an index set defining the control locations in the aquifer. The injection and head upper bounds are necessary to minimize possible clogging or fracturing of the disposal formation, and the excessive buildup of recharge mounds in the vicinity of the injection wells.

Groundwater quality constraints are also introduced at each pumping well site to prevent the degradation of the aquifer's water supply, or

$$c_{s,w}^{k} \leq \bar{c}_{s,w}^{*}, \quad w \in \pi, \ \forall k \tag{7.106}$$

Alternatively, we can also limit the injection concentrations, again to minimize possible clogging problems, or

$$c_{s,r}^{k} \leq \bar{c}_{s,r}^{*}, \quad r \in \psi, \ \forall k \tag{7.107}$$

Finally, the equations that link the management or control policies with the aquifer's state variables are the hydraulic and water quality response equations, or

$$A\,\dot{\mathbf{h}} + B\mathbf{h} + \mathbf{g}\left(Q_{w}^{k}, Q_{r}^{k}\right) = 0 \tag{7.108}$$

$$P\,\dot{\mathbf{c}} + R\mathbf{c} + \mathbf{f}\left(Q_{r}^{k}, c_{s,r}^{*}\right) = 0 \tag{7.109}$$

where $c_{s,r}^{*}$ is the input concentration of a constituent at injection well r. The planning model presented in Equations 7.100 through 7.109 allows the planner to determine the optimal waste injection concentrations and the pumping and injection rates that are compatible with the groundwater quality standards and water and waste load demands. The dual variables of the optimization model also define the environmental and hydrologic trade-offs associated with changing water demands and groundwater quality standards.

Computationally, the model is extremely difficult to solve for several reasons. First, the constraint set of the model is a nonconvex set. The pumping and injection decisions affect the velocity field and, consequently, the convective and dispersive mass transport occurring in the aquifer. In the mass transport equations, these terms are multiplied by the mass concentrations. Because the constraints and the response equations are equalities, the restrictions form a nonconvex feasible region (Mangasarian, 1969).

The size of the constraint set may also be inordinately large because of the temporal discretization of the response equations. Unfortunately, the discretization affects the optimality of the planning policies. The coarser the discretization of the problem, the more suboptimal are the planning policies. And, third, if the objectives are nonlinear or nonconvex, convergence properties of solution algorithms are difficult to prove. The ramification of these limitations is that local solutions to the planning problem can be expected using nonlinear programming algorithms. However, there also exists the possibility of obtaining only stationarity solutions to the optimization model. In this event, simulation may be a more viable solution approach.

In the following sections, we explore possible solution algorithms for the conjunctive groundwater supply and quality management problem.

7.5 SOLUTION ALGORITHMS FOR THE CONJUNCTIVE MANAGEMENT MODEL

The conjunctive management of groundwater supply and quality is a nonlinear, nonconvex optimization problem. In this section, we will examine how partitioning and dual methods may be used to obtain local solutions to the planning or operational problem. We first examine dual programming algorithms.

7.5.1 Dual Programming Algorithm

Dual methods are predicated on the dual programming problem rather than on primal methods such as MINOS, SUMT, or convex programming techniques. To simplify our discussion of these methods, we rewrite the conjunctive use problem, Problem P-1, as

$$\min z = f(\mathbf{x}, \mathbf{y})$$

subject to

$$\mathbf{g}(\mathbf{x}, \mathbf{y}) \leq 0, \; x \in X$$

where the $\mathbf{x}$ vector contains the vector of head values and the pumping and injection decisions, $(\mathbf{h}^k, Q^k, \forall_k)$. The components of the $\mathbf{y}$ vector are the mass concentrations and the waste input concentrations. The set X contains the hydraulic response equations, the pumping and injection upper bounds, and the hydraulic head control constraints. The constraint vector $\mathbf{g}$ contains the sets of concentration response equations and the groundwater quality standards.

The dual of the primal problem can be expressed as (Lasdon, 1970),

$$\max_{u \in D} h(\mathbf{u})$$

where $\mathbf{u}$ is the vector of Lagrangian multipliers and the dual function $h(\mathbf{u})$ is defined as

$$h(\mathbf{u}) = \min L(\mathbf{x}, \mathbf{y}, \mathbf{u}) \; x \in X$$

The domain of D is

$$D(\mathbf{u}) = \{\mathbf{u} | \mathbf{u} \geq 0, \; \min L \text{ exists}\}$$

The maximin dual problem is then given as (Rockafellar, 1967)

$$\max_{\mathbf{u}\in D} \min_{\mathbf{x}\in X} \ L(\mathbf{x}, \mathbf{y}, \mathbf{u})$$

In this form, the dual programming problem is actually two optimization problems. The first subproblem (Problem P-2) optimizes the **x** and **y** policy variables for a given set of Lagrangian multipliers, or

$$\min_{\mathbf{x}\in X} L = f(\mathbf{x}, \mathbf{y}, \mathbf{u}^*) + \mathbf{u}^*\mathbf{g}$$

Given the optimal solutions of the subproblem $\mathbf{x}^*(\mathbf{u}), \mathbf{y}^*(\mathbf{u})$, the master or outer maximization (Problem P-3) is then

$$\max_{\mathbf{u}\in D} h(\mathbf{u})$$

Dual algorithms solve these two problems iteratively, until at least a local solution is identified.

The algorithm can be summarized as follows:

1. Select initial values of the Lagrangian multipliers, $\mathbf{u}^* \geqslant 0$.
2. Solve the subproblem, P-2, with $\mathbf{u} = \mathbf{u}^*$; identify the optimal $\mathbf{x}^*$, $\mathbf{y}^*$. Because the model has a nonlinear objective and a convex constraint set, penalty function, convex simplex methods, or Rosen's gradient projection method can be used for solution of the subproblem.
3. Form the dual function and evaluate the gradient. Assuming that X is closed and bounded and that f and g are continuous and differentiable, the gradient of the dual function is (Danskin, 1966)

$$\nabla h(\mathbf{u}) = \mathbf{g}\ (\mathbf{x}, \mathbf{y}, \mathbf{u})$$

4. Define the gradient direction as

$$d_i^n = \begin{cases} \dfrac{\partial h}{\partial u_i} & \text{if} \quad u_i^n > 0 \\[2ex] \max \left\{0, \dfrac{\partial h}{\partial u_i}\right\} & \text{if} \quad u_i^n = 0 \end{cases}$$

5. Using gradient-based methods, update the Lagrangian multipliers, or

$$\mathbf{u}^{n+1} = \mathbf{u}^n + \tau\mathbf{d}^n$$

where τ, the step size, is selected such that

$$h(\mathbf{u}^{n+1}) \geqslant h(\mathbf{u}^n)$$

6. Check for convergence of the algorithm by examining the difference in the multipliers over successive iterations or when the step size is approximately zero, or

$$|\mathbf{u}^{n+1} - \mathbf{u}^n| < \delta, \text{ or}$$

$$\tau \simeq 0$$

If the convergence criteria are not satisfied, return to step 2 with the updated Lagrangian multipliers.

Dual algorithms also have several important advantages over conventional primal optimization methods. The iterative algorithm will produce a sequence of multipliers that will converge to the solution of the dual algorithm. The algorithm also generates a sequence of lower bounds ($\mathbf{u}^n$) and a set of points defining the optimum solution of the Lagrangian subproblems. As shown by Everett (1963), each point, $\mathbf{x}(\mathbf{u}), \mathbf{y}(\mathbf{u})$, solves the modified primal problem,

$$\min \ z = f(\mathbf{x}, \mathbf{y})$$

$$g_j(\mathbf{x}, \mathbf{y}) \leqslant z_j^n, \quad \forall j, \ \mathbf{x} \epsilon X$$

where

$$z_j^n = g_j(\mathbf{x}^n, \mathbf{y}^n), \quad u_j^n > 0$$

$$z_j^n \geqslant g_j(\mathbf{x}^n, \mathbf{y}^n), \quad u_j^n = 0$$

Free sensitivity information is obtained from the dual algorithm regarding perturbations of the right-hand sides of the optimization model.

7.5.2 Partitioning Methods

A variation of the dual algorithms are partitioning methods for nonlinear programs (Geoffrion, 1972). These techniques partition the decision variables of the model into two subsets. First, the algorithm changes the values of the policy variables in one subset; then the other variables are varied in such a way so as to increase or decrease the objective function of the problem. In this approach, the conjunctive use problem can be solved by minimizing, for fixed $\mathbf{x}$ values over $\mathbf{y}$, the concentration decision variables and then minimizing the result over **x**. The programming problem can then be expressed as

$$\min_{x \in S} \min_{\mathbf{y}} \{f(\mathbf{x}, \mathbf{y})\}$$

where S is the set of **x** values for each of the inner maximizations is feasible, or

$$S = \{\mathbf{x} \,|\mathbf{g}(\mathbf{x}, \mathbf{y}) \leq 0 \text{ and } \mathbf{x} \in X\}$$

The solution of the inner subproblem,

$$v(\mathbf{x}) = \min_{\mathbf{y}} f(\mathbf{x}, \mathbf{y}), \quad \mathbf{g}(\mathbf{x}, \mathbf{y}) \leq 0$$

determines the optimal dual multipliers and mass concentrations, $\mathbf{u}^*$, $\mathbf{y}^*$. The dual of the inner problem (Problem P-4) for fixed **x** values can be expressed as (Wolfe, 1961)

$$v(\mathbf{x}) = \max L(\mathbf{x}, \mathbf{y}, \mathbf{u}), \quad \nabla_{\mathbf{y}} L = 0, \quad \mathbf{u} \geq 0$$

The equivalency of this problem with the dual problem defined in Problem P-2 can be shown if f and g are convex and differentiable (Mangasarian, 1969). The master problem can then be expressed as

$$\min z = v(\mathbf{x}), \quad \mathbf{x} \in S$$

Gradient-based algorithms can be used for solution of the problem. Provided that S is compact, the gradient of v is again given as (Danskin, 1966)

$$\frac{\partial v}{\partial \mathbf{x}} = \nabla_{\mathbf{x}} L$$

The partitioning algorithm can then be summarized as

1. Select an initial set of feasible **x** vectors.
2. Solve the Lagrangian problem and identify the optimal $\mathbf{y}^*$ and Lagrangian multipliers.
3. Evaluate the gradient of $v(\mathbf{x})$, or

$$\frac{\partial v}{\partial \mathbf{x}} = \frac{\partial f}{\partial \mathbf{x}} + \mathbf{u} \frac{\partial g}{\partial \mathbf{x}}$$

4. Using gradient methods, update $\mathbf{x}^n$ using the relation

$$\mathbf{x}^{n+1} = \mathbf{x}^n + \tau \quad \mathbf{d}^n$$

where again τ is the step size and $\mathbf{d}^n$ is the gradient direction.

5. Check for convergence of the algorithm by examining the differences in the x vector over successive iterations,

$$|\mathbf{x}^{n+1} - \mathbf{x}^n| \leq \delta$$

If the condition is not satisfied, return to step 2.

Although these algorithms are relatively straightforward to apply to many problems (see, for example, Rossman & Vanecek, 1978), the overall optimization model is still a nonconvex programming problem. The algorithm cannot guarantee that a global minimum or maximum will be found. The existence of alternative local optima can be determined by varying the initial starting vectors for each problem. Moreover, each solution produced by the algorithm can also be perturbed to verify that the solution is a local minimum or maximum and not a stationary point of the Lagrangian.

The following example problems demonstrate how a particular variant of these algorithms—a decomposition procedure—can be used to generate useful solutions to the conjunctive groundwater supply and quality problem.

Example Problem 7.5 Steady-State Conjunctive Use Planning Model

The example aquifer shown in Figure 7.31 is conjunctively used as a source of agricultural water supply and as a disposal medium for a conservative waste constituent. The aquifer is bounded by two streams; the water levels in the streams, h_L and h_r, are assumed to be unaffected by pumping or recharge. Well log data suggests that the aquifer consists of three different hydraulic conductivity zones. Within each zone, the aquifer is homogeneous and isotropic. Injection is assumed to occur only in zone 3 of the aquifer system. Assuming pumping can occur at any internal nodal point in the groundwater basin, develop a steady-state planning model for the conjunctive use problem. The parameters of the aquifer system are summarized in Table 7.3.

The steady-state planning model has the same general form as the model presented in Equations 7.100 through 7.109. However, in lieu of cost and benefit data for the problem, we will define the objectives of the problem in terms of the state variables of the groundwater system. Specifically, we assume the objectives are to

1. Maximize the head levels in the aquifer, a linear surrogate for minimizing the groundwater extraction costs.

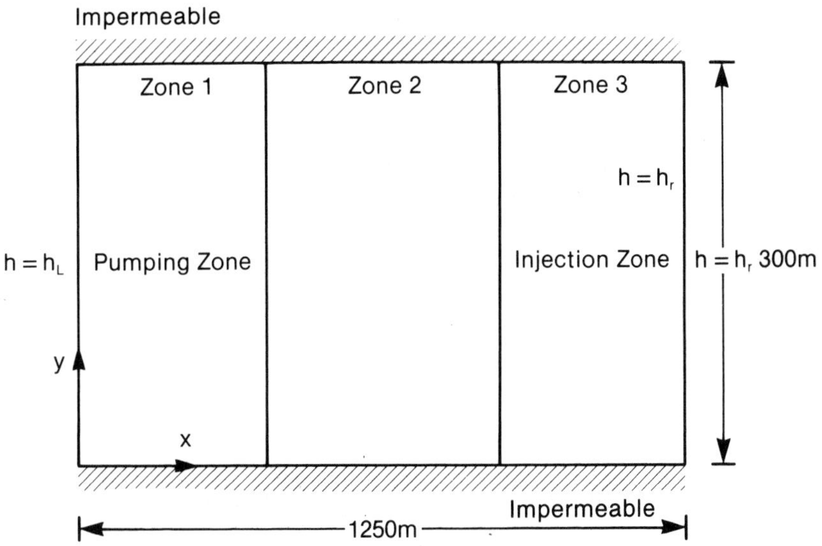

FIGURE 7.31 Example aquifer

2. Maximize the injected waste input concentration. This objective minimizes the cost of waste treatment.

The objective function of the planning model can then be expressed as

$$\max z = \lambda_1 \sum_{w \in \pi} h_w + \lambda_2 c_s^* \tag{7.110}$$

where λ_1 and λ_2 are the relative weights for the objectives and c_s^* is the input mass concentration.

The constraints of the model are again the water supply requirement,

$$\sum_{w \in \pi} Q_w \geq D \tag{7.111}$$

TABLE 7.3

	Zone		
Parameter	*1*	2	3
K (m/day)	200	100	20
Dispersivity (m)			
α_L	10	10	10
α_T	5	5	5
Δx, Δy (m)	50	50	50

the wasteload disposal target,

$$\sum_{r\in\psi} Q_r \geq WL \tag{7.112}$$

and upper bounds on injection and pumping

$$Q_r \leq Q_r^*, \quad r\in\psi \tag{7.113}$$

$$Q_w \leq Q_w^*, \quad w\in\pi \tag{7.114}$$

The groundwater quality constraints for each pumping well are

$$c_{s,w} \leq \bar{c}_{s,w}^*, \quad w\in\pi \tag{7.115}$$

and, finally the steady-state response equations,

$$B\mathbf{h} + \mathbf{g}(\mathbf{Q}_w, \mathbf{Q}_r) = 0 \tag{7.116}$$

and

$$R(\mathbf{Q}_w, \mathbf{Q}_r)\,\mathbf{c} + \mathbf{f}(c_s^*, \mathbf{Q}_w, \mathbf{Q}_r) = 0 \tag{7.117}$$

The planning problem is a nonconvex programming problem. We can, however, decompose the problem into two interdependent linear programming problems provided additional assumptions are made regarding the relative weighting of the water supply and water quality objectives. If, for example, the groundwater basin is used primarily as a source of water supply; that is if the assimilative waste capacity is of secondary importance, then we can reasonably assume that $\lambda_1 >> \lambda_2$. The planning problem can then be decomposed into two linear subproblems. In the first problem, the optimal pumping and injection policies are found from solution of the model,

$$\max z = \sum_{w\in\pi} h_w \tag{7.118}$$

subject to

$$\sum_{w\in\pi} Q_w \geq D \tag{7.119}$$

$$\sum_{r\in\psi} Q_r \geq WL \tag{7.120}$$

$$Q_w \leq Q_w^*, \; w\in\pi \tag{7.121}$$

$$Q_r \leq Q_r^*, \ r\epsilon\psi \tag{7.122}$$

$$\mathbf{B}\mathbf{h} + \mathbf{g}\,(\mathbf{Q}_w, \mathbf{Q}_r) = 0 \tag{7.123}$$

The optimal pumping and injection policies, $\mathbf{Q}_w^*$ and $\mathbf{Q}_r^*$, are then parameters in the quality optimization problems. This model can be expressed as

$$\max z = c_s^* \tag{7.124}$$

subject to

$$c_{s,w} \leq \bar{c}_{s,w}^*, \ w\epsilon\pi \tag{7.125}$$

$$R(\mathbf{Q}_w^*, \mathbf{Q}_r^*)\, c_s + \mathbf{f}\,(c_s^*, \mathbf{Q}_w^*, \mathbf{Q}_r^*) = 0 \tag{7.126}$$

The response equations in both problems are now linear function of the decision variables; both subproblems are linear programming models.

The results of applying this decomposition procedure to the aquifer system shown in Figure 7.31 are presented in Figures 7.32 through 7.34. The linear programming data are summarized in Table 7.4.

Initially, the water supply subproblem was solved using the CDC APEX-III large-scale optimization package. The optimal pumping and injection locations are summarized in Table 7.5. Next, parametric programming was used to determine the impact of changing wasteloads and water requirements on the water supply management objectives. As shown in Figure 7.32, the head objective is a concave function of the water demand for various waste load requirements. The trade-off or slope of the curve increases as the demand increases. The concentration of the waste constitutent in the injected water is a linear function of the prevailing groundwater quality standard (see Figure 7.33). The slope of the curve, 0.2, represents the capacity of the aquifer to assimilate or contain the wasteload.

The relation between the mass input (gm/day) and the water demand is shown in Figure 7.34. The mass discharged to the aquifer is a convex function of the water demand. As the water demand increases, the allowable mass input decreases since the increased pumping also increases the velocity, circulation, and transport occurring in the basin. As a result, there is a more dramatic increase in the concentration of the waste contaminant at the water supply wells in the aquifer system.

As a check on the consistency of the optimization results, the optimal flow and quality decisions were simulated using a generalized flow and groundwater quality simulation model (Haji-Djafari & Wiggert, 1976). The results of the simulations and the optimization results are also shown in Table 7.5. The results are quite similar, indicating that at least the errors in the optimization and simulation modeling are comparable.

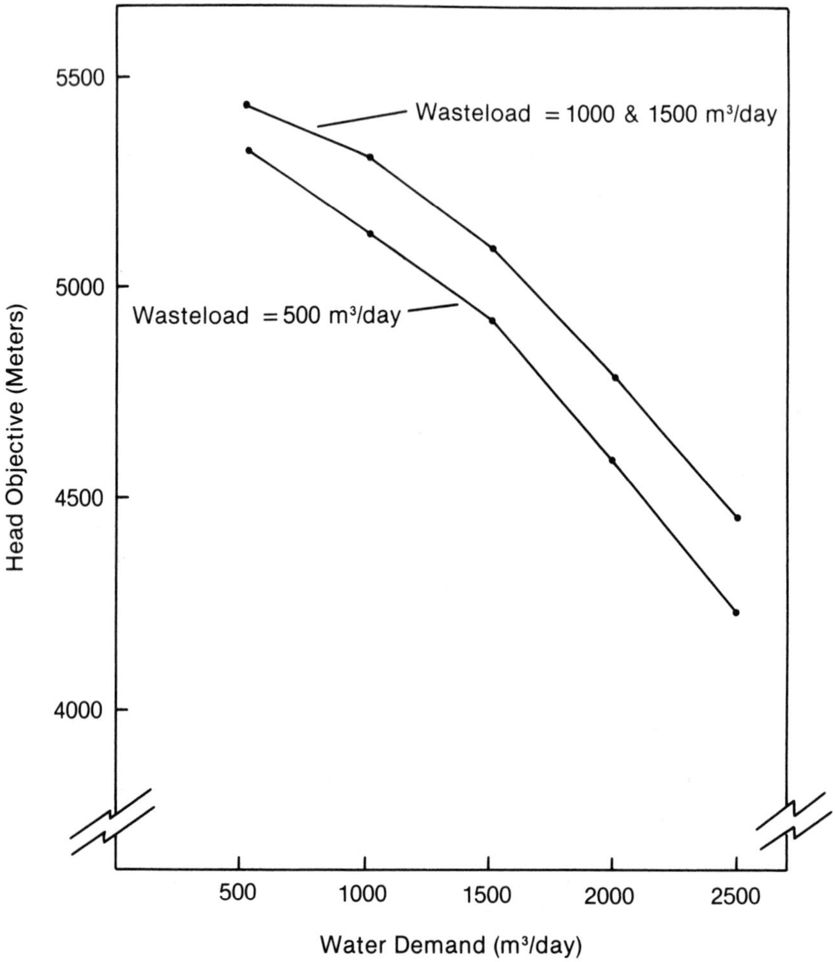

FIGURE 7.32 Head of objective versus pumping rate

TABLE 7.4 Linear Programming Problems

	Number of Elements	*Number of Nodes*	*LP Subproblem Constraints*	*Decision Variables*	*CPU Time (Sec)*
Hydraulic	150	182	394	280	40
Water quality	150	182	176	169	100

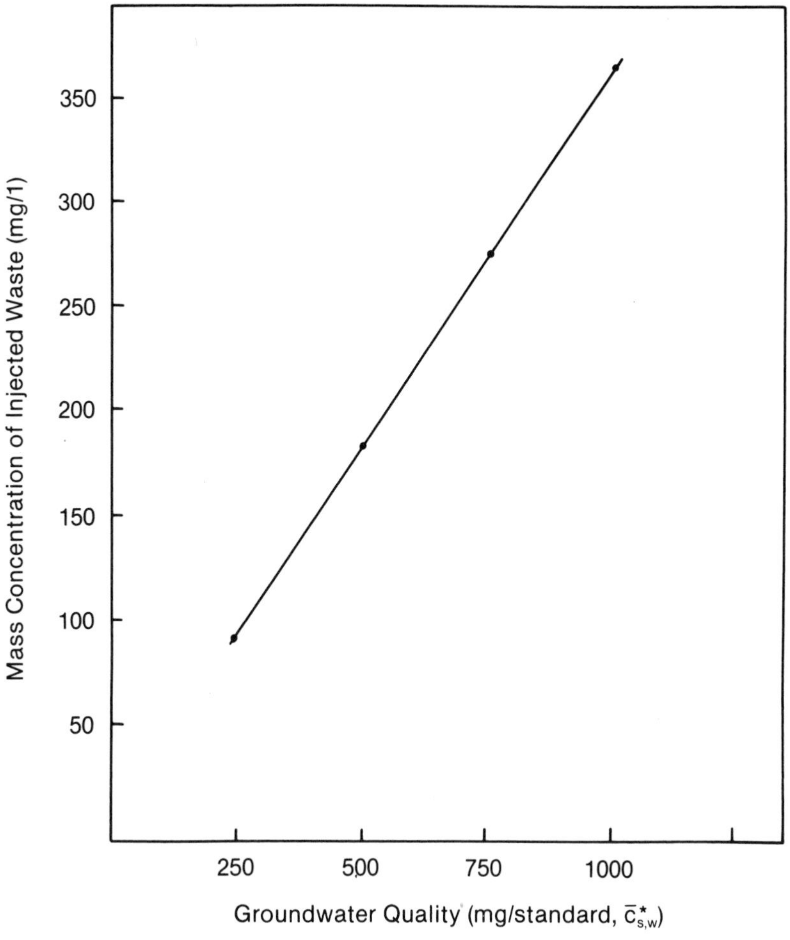

FIGURE 7.33 Mass input concentration versus upper bound on concentration

Example Problem 7.6 Groundwater Quality Conjunctive Waste Treatment Model

The unconfined aquifer system shown in Figure 7.35 is used conjunctively as a source of agricultural supply and for the disposal of municipal or agricultural waste. The assimilative waste capacity of the aquifer system and the wastewater treatment plant are considered as the major components of the regional waste treatment system. Imported water is also available for dilution of the waste flows. The quality

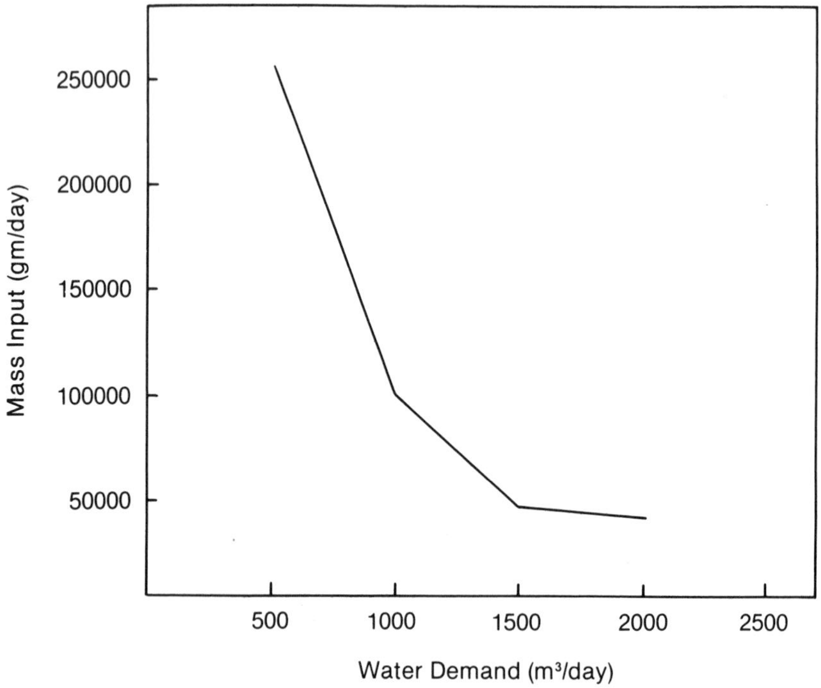

FIGURE 7.34 Flux versus pumping rate—injection rate at 1500 cubic meters per day; upper bound on concentration at 250 mg per liter

of the dilution water is summarized in Table 7.6. The design of the waste treatment facility will involve determining the design capacity of the plant, Q, and the optimal unit processes. The feasible unit processes of the waste treatment plant are given in Table 7.7. Assuming that the Dupuit-Forcheimer assumptions are valid and that steady-state flow and quality conditions exist in the aquifer system, develop a planning model to determine the optimal design of the waste treatment plant and the waste injection concentrations that are compatible with the groundwater quality standards of the basin. The characteristics of the wasteload that is to be treated and injected into the aquifer system are summarized in Table 7.8. The kinetic reaction parameter data for BOD, coliforms, ammonia, and nitrite nitrogen are given in Table 7.9 (McKee & McMichael, 1966). The hydraulic parameters of the aquifer and the assumed pumping and injection rates are summarized in Table 7.10.

The planning model is again predicated on the flow and mass transport equations of the groundwater system. The flow model is described by the Boussinesq

TABLE 7.5 Optimal Policies and Groundwater Conditions for a Water Supply Target of 1500 m³/day and a Waste Load of 500 m³/day

Node Number	Optimal Pumping/Injection Rate (m^3/day)	Head (m)		Concentration (mg/l)	
		Simulation Model	APEX-IV	Simulation Model	APEX-IV
8	−200.	84.8	84.9	16.79	16.79
9	−200.	88.9	88.3	25.55	25.55
10	−200.	88.5	88.5	34.20	34.20
11	−200.	88.5	88.5	41.17	41.18
12	−200.	88.5	88.5	33.36	33.36
13	−200.	88.2	88.2	23.84	23.84
14	−200.	84.8	84.8	15.58	15.58
15		87.3	87.4	141.59	141.60
16		87.4	87.5	107.47	107.48
17		87.6	87.7	133.72	131.73
18	−100.	86.0	86.0	250.00	250.00
19		87.6	87.6	126.84	126.85
120	100.	95.1	95.1	3009.74	3010.00
121	100.	93.2	93.2	2376.66	2376.88
122	100.	92.4	92.5	2262.11	2262.32
123	100.	91.7	91.7	2077.89	2078.08
124	100.	90.6	90.7	2119.44	2119.64
125		88.0	88.1	1595.45	1595.60
126		87.5	87.6	1658.58	1658.73

equation (Eq. 2.82), or

$$\frac{\partial}{\partial x}\left(K_{xx}\frac{\partial h^2}{\partial x}\right)+\frac{\partial}{\partial y}\left(K_{yy}\frac{\partial h^2}{\partial y}\right)\pm\sum_{w\in\Omega}2\,Q_w\,\delta(x-x_w,\,y-y_w)=0 \quad (7.127)$$

where Ω defines the location of all pumping and injection wells. Neglecting dispersional effects, the mass transport equation for the constituents can be expressed as

$$-\frac{\partial}{\partial x}(c_s q_x m)-\frac{\partial}{\partial y}(c_s q_y m)-k^s mnc_s-k_a^s m(1-n)R^s c_s + \sum_{r\in\psi}\frac{Q_r c^*_{s,r}}{m}\delta(x-x_r,\,y-y_r)=0 \quad (7.128)$$

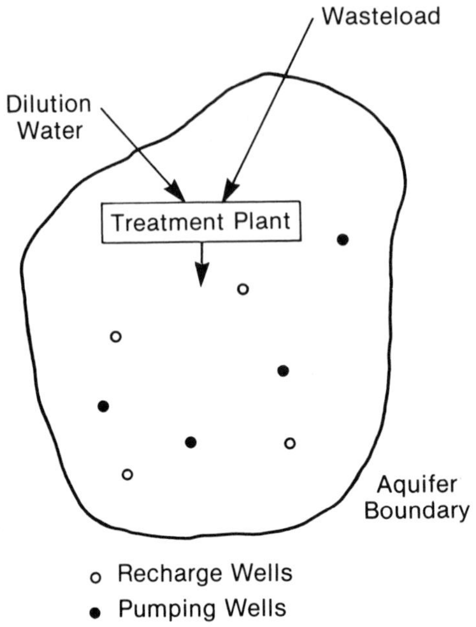

FIGURE 7.35 Example aquifer

where $c^*_{s,r}$ is the injected mass concentration of constitutent s, m is the thickness of the aquifer, n is the porosity, k^s_a is the kinetic reaction parameter in the adsorbed phase, k^s is the reaction parameter in the solution phase, and R^s is the distribution coefficient for constitutent s.

TABLE 7.6

Constituents	*Quality*
BOD	200 mg/l
Suspended solids	100 mg/l
Coliforms	1×10^4 MPN/100 ml
NO_3	5 mg/l
DDT	trace
Boron	2 mg/l
Chloride	50 mg/l
TDS	100 mg/l
Arsenic	trace
NH_4	5 mg/l

TABLE 7.7 Feasible Unit Processes

Processes	*Notations*	*Unit Process Set*
Primary	P	1
Primary and coagulation and sedimentation	P, CS	2
Primary, coagulation and sedimentation, and mixed media filtration	P, CS, MMF	3
Primary, coagulation and sedimentation, mixed media filtration, and activated carbon	P, CS, MMF, AC	4
Primary, coagulation and sedimentation, mixed media filtration, activated carbon, and nitrogen stripping	P, CS, MMF, AC, NS	5
Activated sludge	AS	6
Activated sludge and mixed media filtration	AS, MMF	7
Activated sludge and coagulation and sedimentation	AS, CS	8
Activated sludge, coagulation and sedimentation, and mixed media filtration	AS, CS, MMF	9
Activated sludge, coagulation and sedimentation, mixed media filtration, and activated carbon	AS, CS, MMF, AC	10
Activated sludge, coagulation and sedimentation, mixed media filtration, activated carbon, and nitrogen stripping	AS, CS, MMF, AC, NS	11
Trickling filter	TF	12
Trickling filter and mixed media filtration	TF, MMF	13
Trickling filter and coagulation and sedimentation	TF, CS	14
Trickling filter, coagulation and sedimentation, mixed media filtration, and activated carbon	TF, CS, MMF, AC	15
Trickling filter, coagulation and sedimentation, mixed media filtration, and activated carbon	TF, CS, MMF, AC	16
Trickling filter, coagulation and sedimentation, mixed media filtration, activated carbon, and nitrogen stripping	TF, CS, MMF, AC, NS	17

Typical boundary conditions for the problem are

$$h = h^*, \quad \frac{\partial h}{\partial n} = 0 \tag{7.129}$$

$$c_s = \bar{c}_s, \quad \frac{\partial c_s}{\partial n} = 0 \tag{7.130}$$

where n is the unit vector normal to the boundary and $\bar{c}_s$ and h^* are the known concentration and head values.

TABLE 7.8

Constituents	*Concentration*
BOD	300 mg/l
Suspended solids	100 mg/l
Coliforms	5×10^7 MPN/100 ml
NO_3	15 mg/l
DDT	0.001 mg/l
Boron	5 mg/l
Chloride	200 mg/l
TDS	1000 mg/l
Arsenic	0.001 mg/l
NH_4	40 mg/l

In the example problem, the equation will be transformed using finite-difference methods. Defining i and j as indices of the x, y spatial variables, the hydraulic finite-difference equation may be expressed as

$$A_{i+1,j}h^2_{i+1,j} + B_{i,j}h^2_{i,j} + C_{i-1,j}h^2_{i-1,j} + D_{i,j+1}h^2_{i,j+1} + E_{i,j-1}h^2_{i,j-1} \pm S^*_{i,j} = 0 \tag{7.131}$$

where

$$A_{i+1,j} = \frac{K^{i,j}_{xx} + K^{i+1,j}_{xx}}{2\Delta x^2}$$

$$B_{i,j} = -\frac{K^{i+1,j}_{xx} - 2K^{i,j}_{yy} + K^{i-1,j}_{yy}}{2\Delta x^2} - \frac{K^{i,j+1}_{yy} - 2K^{i,j}_{yy} + K^{i,j-1}_{yy}}{2\Delta y^2}$$

$$C_{i-1,j} = \frac{K^{i,j}_{xx} + K^{i-1,j}_{xx}}{2\Delta x^2}$$

$$D_{i,j+1} = \frac{K^{i,j+1}_{yy} + K^{i,j}_{yy}}{2\Delta y^2}$$

$$E_{i,j-1} = \frac{K^{i,j}_{yy} + K^{i,j-1}_{yy}}{2\Delta y^2}$$

TABLE 7.9

Constituents	$\bar{k}$, *1/s*
BOD	6×10^{-6}
Coliforms	9×10^{-6}
NH_4	1.66×10^{-5}
NO_2	1.92×10^{-5}

TABLE 7.10 Summary of Aquifer Parameters

Parameters	*Value*
Hydraulic conductivity (ft/s)	
K_x	1.54×10^{-3}
K_y	7.0×10^{-4}
Pumping wells (ft^3/s)	
Q_1	0.134
Q_2	0.112
Injection wells (ft^3/s)	
Q_1	0.223
Q_2	0.223
Δx (ft)	550
Δy (ft)	500

and $S^*_{i,j}$ is the pumping or injection occurring at node i, j. The finite-diference approximation of the mass transport equation can be expressed as

$$A^1_{i+1,j}c_{s,i+1,j} + B^1_{i,j}c_{s,i,j} + C^1_{i-1,j}c_{s,i-1,j} + D^1_{i,j+1}c_{s,i,j+1} + E^1_{i,j-1}c_{s,i,j-1} + \overline{S}^{*s}_{i,j} = 0 \quad (7.132)$$

where

$$A^1_{i+1,j} = -\frac{q_{x,i+1,j}m}{2\Delta x}$$

$$B^1_{i,j} = -k^s mn - k^s_a m(1 - n)R^s$$

$$C^1_{i-1,j} = \frac{q_{x,i-1,j}m}{2\Delta x}$$

$$D^1_{i,j+1} = -\frac{q_{y,i,j+1}m}{2\Delta y}$$

$$E^1_{i,j-1} = \frac{q_{y,i,j-1}m}{2\Delta y}$$

where $\overline{S}^{*s}_{i,j}$ is the recharge concentration discharged to the groundwater system. The truncation error of the approximation is $0(\Delta x^2 + \Delta y^2)$.

Again, writing the equation for all internal nodes in the system and incorporating the known boundary conditions, the hydraulic and mass transport equations may be expressed for each constituent as

$$\mathbf{Bh} = -\mathbf{g} \quad (7.133)$$

$$\mathbf{A}^s\mathbf{c}^s = \mathbf{l}^s, \quad s \epsilon \chi \quad (7.134)$$

where $\mathbf{l}^s$ again contains the input mass concentration to the groundwater system.

The input concentration of a constitutent to the aquifer system is a function of the removal efficiency of the wastewater treatment plant for a particular constituent, ϵ^s, and the concentration of the material in the wasteload, $\bar{l}^s$. Defining v_l as the wasteload volume, d as the volume of dilution water, and $\bar{l}_d^s$ as the concentration of the constituent in the dilution water, the input concentration can be expressed as

$$c_s^* = [(1 - \epsilon^s)(v_l \bar{l}^s + d\bar{l}_d^s)]/(v_l + d) \qquad (7.135)$$

The cost objective of the planning model consists of the cost of wastewater treatment and the cost of importing the dilution water supply. For each set of units processed (shown in Table 7.7), the annual cost can be expressed as (California Department of Water Resources, 1973)

$$C_l = a_l Q + b_l Q \log Q, \quad l \epsilon \tau \qquad (7.136)$$

where Q is the treatment plant capacity (MGD) and τ is the set of all unit processes.

The cost for transmission of the dilution water can be expressed as (Linaweaver & Clark, 1964)

$$C_d = \delta d^{0.598} \qquad (7.137)$$

where δ is a cost parameter reflecting the capacity of the pipeline, the distance and the method of water transmission.

Introducing the (0, 1) integer variables, x_l, which defines the feasible unit process combinations of the wastewater treatment plant, the objective of the planning and design model can be expressed as

$$\min z = \sum_{l \epsilon \tau} x_l C_l(Q) + C_d(d) \qquad (7.138)$$

The constraints of the model include

1. The capacity limitations of the wastewater treatment plant

$$v_l + d \leqslant Q \qquad (7.139)$$

2. Groundwater quality constraints at each recharge and injection well site

$$c_{s,w} \leqslant \bar{c}_{s,w}^*, \qquad w \epsilon \pi, \quad s \epsilon \chi \qquad (7.140a)$$

$$c_{s,r} \leqslant \bar{c}_{s,r}^*, \qquad r \epsilon \psi, \quad s \epsilon \chi \qquad (7.140b)$$

3. The water quality response equations,

$$\mathbf{A}^s \mathbf{c}_s = \mathbf{l}^s, \quad s \epsilon \chi \qquad (7.141)$$

4. The restriction that, at most, one unit process combination can be selected

$$\sum_{l \in T} x_l = 1 \tag{7.142}$$

We have assumed, throughout the analysis, that the optimal well injection and pumping sites and rates have been determined in a prior simulation analysis. The planning model is a mixed-integer programming problem.

The planning model can be solved readily by decomposing the overall problem into individual subproblems involving decisions on the amount of dilution water for each unit process combination. For a fixed-unit process, l, the objective is a concave function of the decision variables Q and d. The constraint equations are then linear functions of the decision variables. The solution of the problem occurs on the boundary of the feasible region, and locally optimal solutions can be identified using linear programming (Tui, 1964). The solution of each subproblem can then be compared to determine the overall minimum cost solution.

The results of the optimization are summarized in Table 7.11 (Willis & Dracup, 1973). The optimal unit process is a combination of primary and secondary treatment (trickling filter). The restriction on suspended solids and BOD requires some form of secondary or advanced wastewater treatment. The secondary waste

TABLE 7.11 Results of the Optimization

Unit Process Set	*Dilution Water, mgd*	*Treatment Plant Capacity, mgd*	*Daily Cost,* dollars*
1	14	15	1182
2	6	7	1030
3	6	7	1445
4	6	7	2382
5	1.25	2.25	1041
6	6	7	1154
7	6	7	1568
8	6	7	1504
9	6	7	1918
10	6	7	2856
11	1.25	2.25	1231
12	6	7	954
13	6	7	1369
14	6	7	1304
15	6	7	1719
16	6	7	3248
17	1.25	2.25	1156

* Parameter ρ is 6%, and n is 50 years.

treatment removes approximately 80 percent of the organic matter, up to 98 percent of the coliforms, and approximately 90 percent of the suspended solids. The balance of the BOD and coliforms entering the groundwater system is attenuated before the pumping wells are reached. Forms of nitrogen are oxidized to nitrate, a conservative, noninteracting constituent.

The optimal treatment plant capacity is 7 MGD, and the annual cost is $340,000. The volume of dilution water is approximately six times the original waste volume. The marginal cost of importing the dilution water for reducing the concentration of conservation constituents is less than the additional cost of incorporating other forms of advanced wastewater treatment in the treatment facility. In the absence of an adequate supply of dilution water, other more costly unit processes would need to be considered (reverse osmosis, electrodialysis, or ion exchange).

The results of the planning model demonstrate the feasibility of using secondary wastewater treatment in conjunction with assimilative capacity of the aquifer system for wastewater degradation and disposal.

Example Problem 7.7 Transient Groundwater Quality Management

The confined aquifer system shown in Figure 7.36 provides the principal water supply for a large agricultural area. However, the basin is considered a possible waste disposal site. The projected wasteloads and water demands for a 400 day operational period are shown in Table 7.12.

Preliminary data indicate that the groundwater system consists of two hydraulic

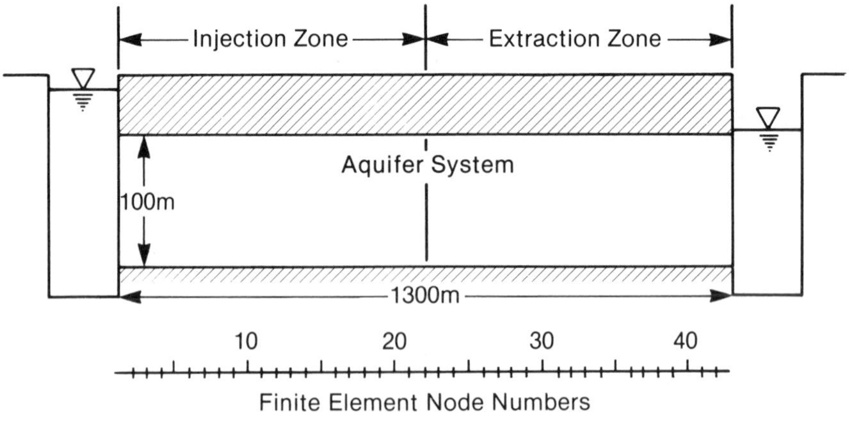

FIGURE 7.36 Optimal mass flux versus groundwater quality standard

TABLE 7.12

	Extraction Rates (10^3 m^3/day)				
	Schedule				
Planning Period	1	2	3	4	5
1	5.0	10.0	15.0	20.0	25.0
2	7.5	12.5	17.5	22.5	27.5
3	10.0	15.0	20.0	25.0	30.0
4	7.5	12.5	17.5	22.5	27.5
Total	30.0	50.0	70.0	90.0	110.0

	Injection Rates (10^3 m^3/day)				
	Schedule				
Planning Period	1	2	3	4	5
1	5.0	10.0	15.0	20.0	25.0
2	7.5	12.5	17.5	22.5	27.5
3	10.0	15.0	20.0	25.0	30.0
4	7.5	12.5	17.5	3.5	27.5
Total	30.0	50.0	70.0	90.0	110.0

conductivity zones; within each zone the aquifer is homogeneous and isotropic. The hydraulic conductivity for the pumping and injection zones, which have been determined using well log and geophysical data, are 850m/day and 8500m/day, respectively. The storage coefficient of the aquifer is 0.01. The porosity is 0.40. Dirichlet boundaries are maintained on the left and right boundaries of the basin (140m and 100m, respectively). The thickness of the aquifer is 100m. The aquifer's length is 1300m and a finite-difference model has been used to approximate the numerical solution of the flow and mass transport equations. The system consists of 40 nodal points; the discretized spatial interval is 30m. The initial concentration of the conservative waste constituent in the aquifer is zero.

Develop an optimization model to determine (1) the optimal pumping and injection policies (the well locations and the pumping and injection rates), (2) the maximum injection waste concentration for a conservative constituent, and (3) the trade-offs associated with the injection concentrations and the prevailing groundwater quality standards. Assume that pumping or injection can occur at any interval nodal point in the groundwater system.

The groundwater quality management model is described by Equations 7.100 through 7.109. We assume again that the planning objectives can be represented by the state variables of the aquifer system. The three planning objectives considered in the conjunctive use model are to

1. Maximize, over time, the head levels at all pumping well sites.
2. Minimize, over the planning horizon, the injection heads. This minimizes the injection pressures and possible fracturing or clogging problems.
3. Maximize the waste injection concentrations.

The objective function of the planning model can then be expressed as

$$\max z = \sum_k \left\{ \lambda_1 \sum_{w \in \pi} h_w^k - \lambda_2 \sum_{r \in \psi} h_r^k \right\} + \lambda_3 c_s^* \tag{7.143}$$

where λ_i is the weights or preferences associated with each objective, k is the index of the planning period, and c_s^* is the input mass concentration of constituents.

The hydraulics of the one-dimensional aquifer system are described by Equations 2.55 and 3.16,

$$T\frac{\partial h}{\partial x} + \sum_{w \in \Omega} Q_w \, \delta(x - x_w) = 0 \tag{7.144}$$

and

$$\frac{\partial}{\partial x}\left(D_{xx} m \frac{\partial c_s}{\partial x}\right) - \frac{\partial}{\partial x}(q_x \, c_s m) + \sum_{r \in \psi} \frac{Q_r c_s^*}{m} \, \delta(x - x_r) = \frac{\partial}{\partial t}(n c_s m) \tag{7.145}$$

where h is the hydraulic head, c_s is the mass concentration, m is the aquifer's thickness, and $c_{s,r}^*$ is the injection concentration.

Following the Galerkin transformation of the equations, the response equations may be expressed as

$$\mathbf{h}(t) = A_1(t)\mathbf{h}_0 + A_2(t)\mathbf{g}, \quad 0 \leq t \leq T \tag{7.146}$$

$$\mathbf{c}_s(t) = B_1(t)\mathbf{c}_{s,0} + B_2(t)\mathbf{f}, \quad 0 \leq t \leq T \tag{7.147}$$

where A_1, A_2, B_1, B_2 are time-dependent matrices and $\mathbf{h}_0$ and $\mathbf{c}_0$ are the initial conditions. The boundary conditions and decision variables are contained in the **f** and **g** vectors.

In the optimization model, the response equations are generated for each time period or step within the operational horizon. The time interval for recursively generating the equations is based, in part, on the overall accuracy and continuity of the hydraulic problem. Assuming the planning horizon consists of m operational periods, each of length $\bar{t}$, and t_k as the time defining the end of period k, the hydraulic response equations can be expressed as

$$\mathbf{h}(t_1) = A_1(\bar{t})\mathbf{h}_0 + A_2(\bar{t})\mathbf{g}_1 \quad (7.148a)$$

$$\mathbf{h}(t_2) = A_1(\bar{t})\mathbf{h}(t_1) + A_2(\bar{t})\mathbf{g}_2 \quad (7.148b)$$

$$\vdots$$

$$\mathbf{h}(t_m) = A_1(\bar{t})\mathbf{h}(t_{m-1}) + A_2(\bar{t})\mathbf{g}_m \quad (7.148c)$$

Similarly, the mass transport equations can be represented as

$$\mathbf{c}_s(t_1) = B_1^1(\bar{t})\mathbf{c}_{s,0} + B_2^1(\bar{t})\mathbf{f}_1 \quad (7.149a)$$

$$\mathbf{c}_s(t_2) = B_1^2(\bar{t})\mathbf{c}_s(t_1) + B_2^2(\bar{t})\mathbf{f}_2 \quad (7.149b)$$

$$\vdots$$

$$\mathbf{c}_s(t_m) = B_1^m(\bar{t})\mathbf{c}_s(t_{m-1}) + B_2^m(\bar{t})\mathbf{f}_m \quad (7.149c)$$

In constrast to the water supply problem, the B_1 and B_2 matrices are different for each operational period because of the temporal variation in the convective and dispersive mass transport occurring in the aquifer system. This is caused by the variations in the velocity field resulting from pumping and waste injection.

The optimization model is a highly nonconvex programming problem. However, we simplify the problem by assuming that the dominant use of the aquifer is as a source of water supply. Because the waste disposal is of secondary importance, the weights or preferences have the property that

$$\lambda_1, \lambda_2 >> \lambda_3,$$

This decomposes the model into flow and quality linear optimization problems.

The basis of the optimization model is the numerical model of the aquifer's response equations. In this example, the equations were generated using the Galerkin finite-element method (see Section 4.3). The aquifer was divided into 42 linear simplex elements for both the hydraulic and quality problems. The model characteristics are shown in Table 7.13.

The optimal pumping and injection schedules and waste injection concentrations are summarized in Table 7.14 for four different extraction targets. Table 7.14 details the optimal well sites and the variation in the waste input concentration over the entire planning horizon. The groundwater quality standard is 500 mg/l ($\bar{c}_{s,w}^*$).

TABLE 7.13

	Number of Elements	Number of Nodes	LP Submodel Constraints*	Decision Variables	Execution Time (CPU sec)**
Flow Submodel	42	43	178	188	14.0
Quality Submodel	42	43	167	168	12.0

* Not including nonnegativity constraints, upper, lower bounds.

**CYBER 170/720

The effect of the groundwater quality standard on the maximum injection flux is shown in Figure 7.37. The results, which were determined using parametric programming, indicate that the change in the mass input per unit change in the standard is approximately 0.5 [(gm/day)/(mg/l)]. The trade-off is again a measure of the assimilative capacity of the basin.

Parametric programming was also used to assess the interaction between changing water demands and the mass discharged to the aquifer. A typical result is shown in Figure 7.38 for injection schedule 1 and a groundwater quality standard of 500 mg/1. The total mass input is a piecewise linear function of the total water

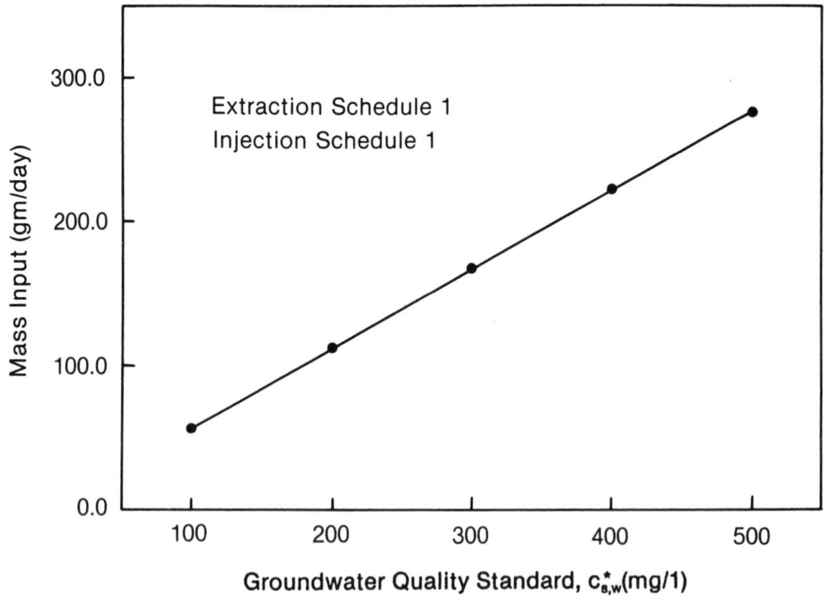

FIGURE 7.37 Total mass flux versus total extraction rate for entire planning horizon

TABLE 7.14 Optimal Operation Policies* ($\bar{c}_w^* = 500$ mg/l)

Extraction Schedule	*Node*	*Planning Period* 1	2	3	4
1	5				
	11				
	17	5000.0	7500.0	10000.0	7500.0
	27	−5000.0	−7500.0	−10000.0	−7500.0
	33				
	39				
	c_s^*	15.5	14.8	8.2	10.9
2	5				
	11				
	17	5000.0	7500.0	10000.0	7500.0
	27	−10000.0	−10000.0	−10000.0	−10000.0
	33				
	39				
	c^*	10.6	13.9	11.6	15.5
3	5				
	11				
	17	5000.0	7500.0	10000.0	7500.0
	27	−10000.0	−10000.0	−10000.0	−10000.0
	33				
	39		−2500.0	−5000.0	−2500.0
	c_s^*	15.1	18.5	10.8	14.3
4	5				
	11				
	17	5000.0	7500.0	10000.0	7500.0
	27	−10000.0	−10000.0	−10000.0	−10000.0
	33				
	39	−10000.0	−7500.0	−10000.0	−7500.0
	c_s^*	17.2	24.0	14.7	19.6
5	5				
	11				
	17	5000.0	7500.0	10000.0	7500.0
	27	−10000.0	−10000.0	−10000.0	−10000.0
	33	−5000.0	−7500.0	−10000.0	−7500.0
	39	−10000.0	−10000.0	−10000.0	−10000.0
	c_s^*	26.2	34.4	16.9	22.6

*Injection and extraction rates (m^3/day)
Injection concentration (mg/l)

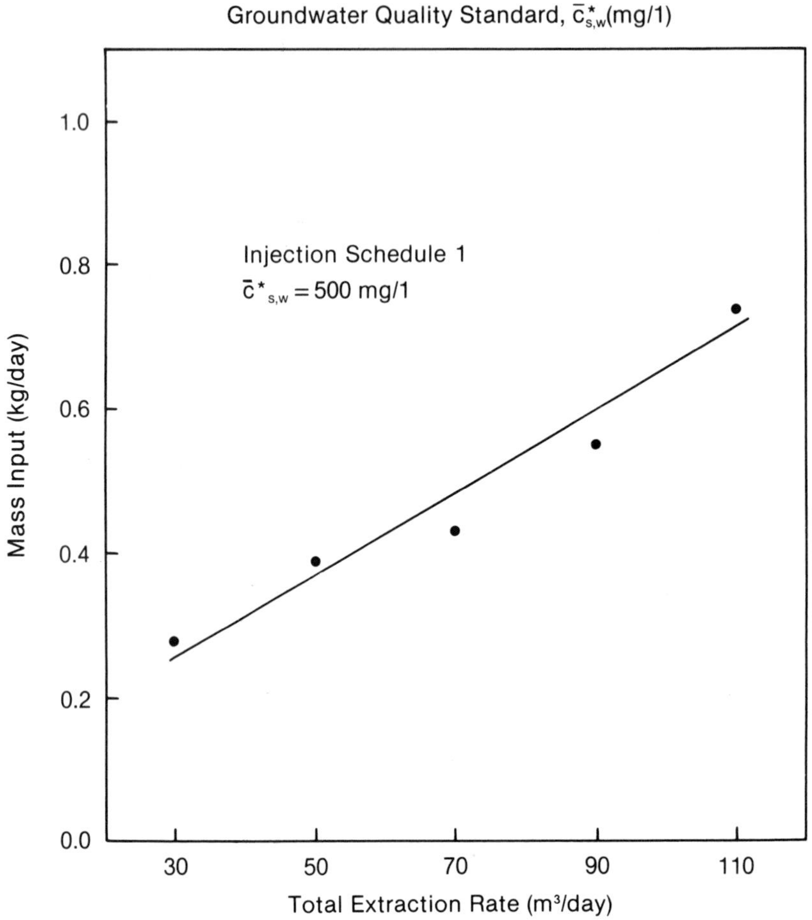

FIGURE 7.38 Hydraulic head and concentration profiles—extraction schedule 1, injection schedule 1

demand. The increase in waste discharged to the system with water demand can be attributed, in part, to the increase in magnitude of convective and dispersive mass transport, resulting from increased extraction rates. This is illustrated in Figures 7.39 through 7.42, which show the head and concentration profiles over the planning horizon.

Finally, the variation in the injection concentrations over the entire planning horizon is shown in Figure 7.43 for a range of water demands.

Before concluding this section, we examine how the general problem of well design and groundwater quality control can be solved using a combination of simulation and optimization modeling, a methodology presented by Gorelick, et al. (1984).

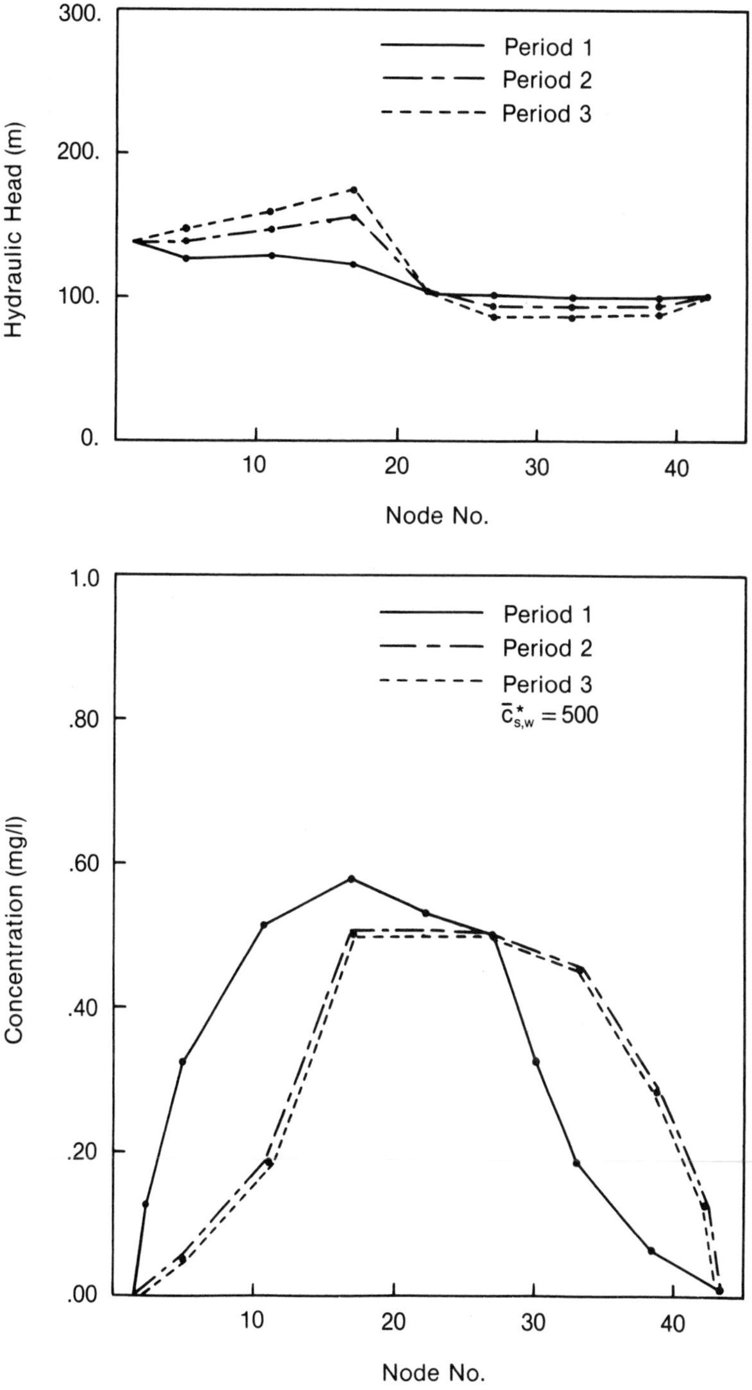

FIGURE 7.39 Hydraulic head and concentration profiles—extraction schedule 2, injection schedule 1

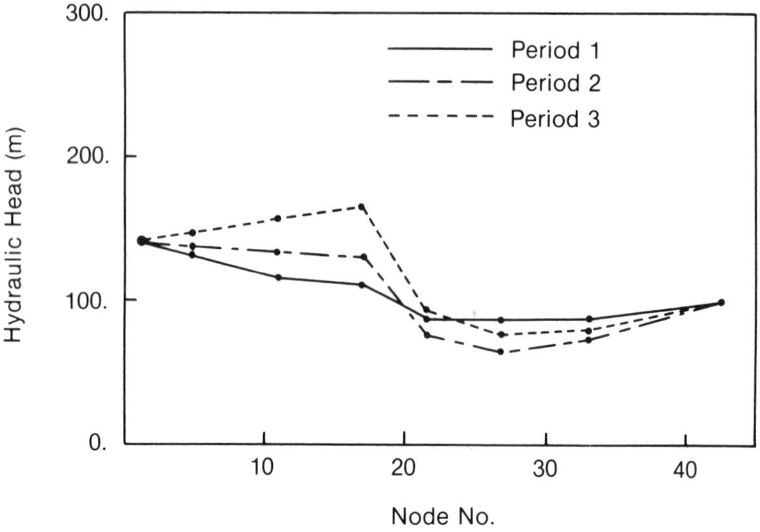

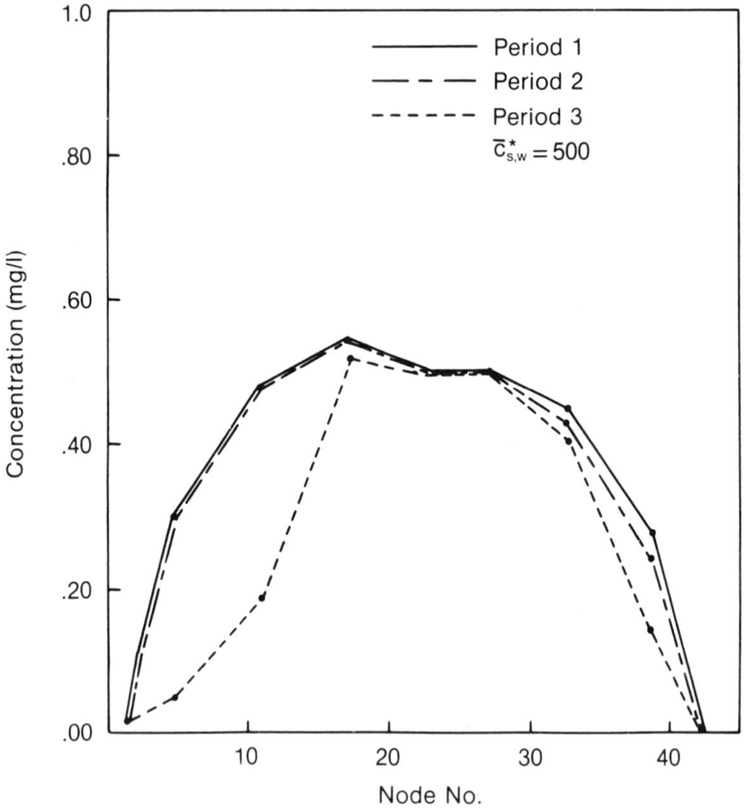

FIGURE 7.40 Hydraulic head and concentration profiles—extraction schedule 3, injection schedule 1

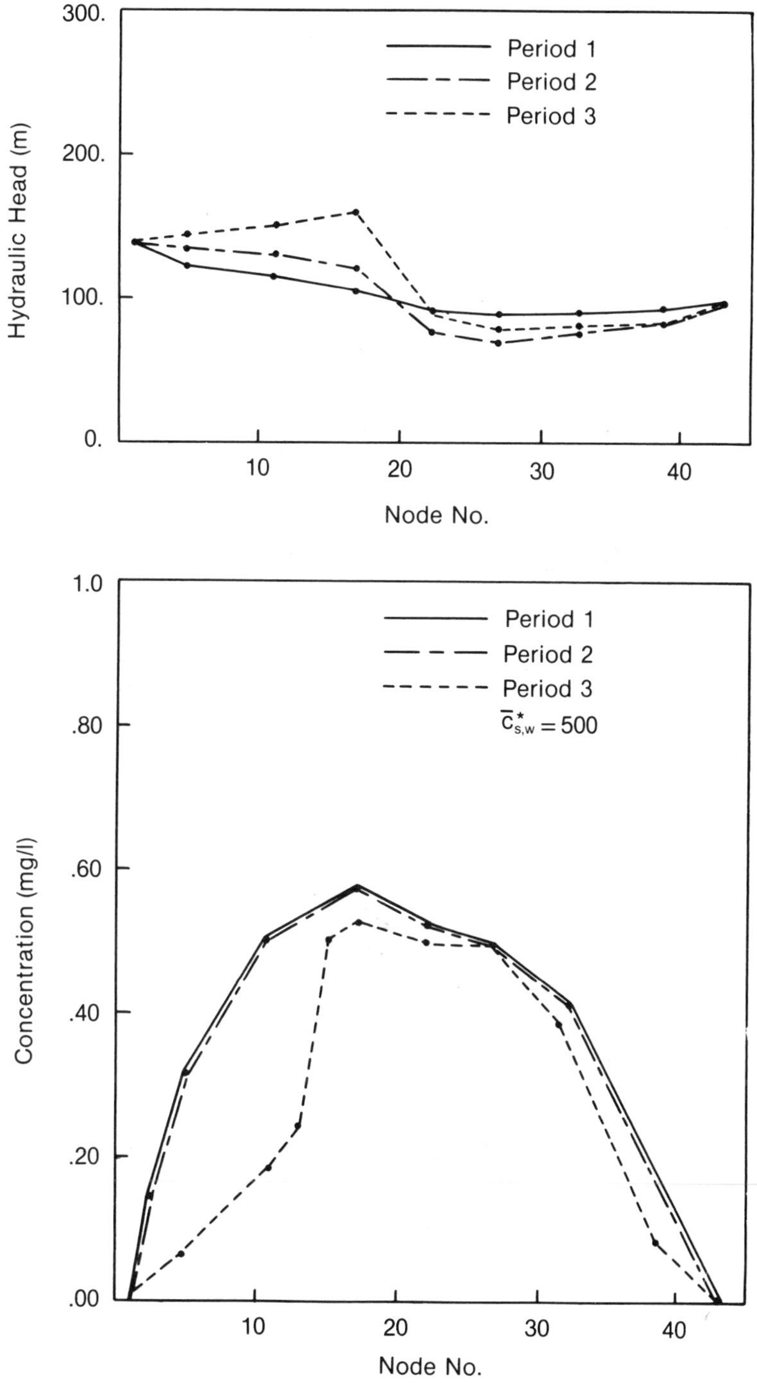

FIGURE 7.41 Hydraulic head and concentration profiles—extraction schedule 4, injection schedule 1

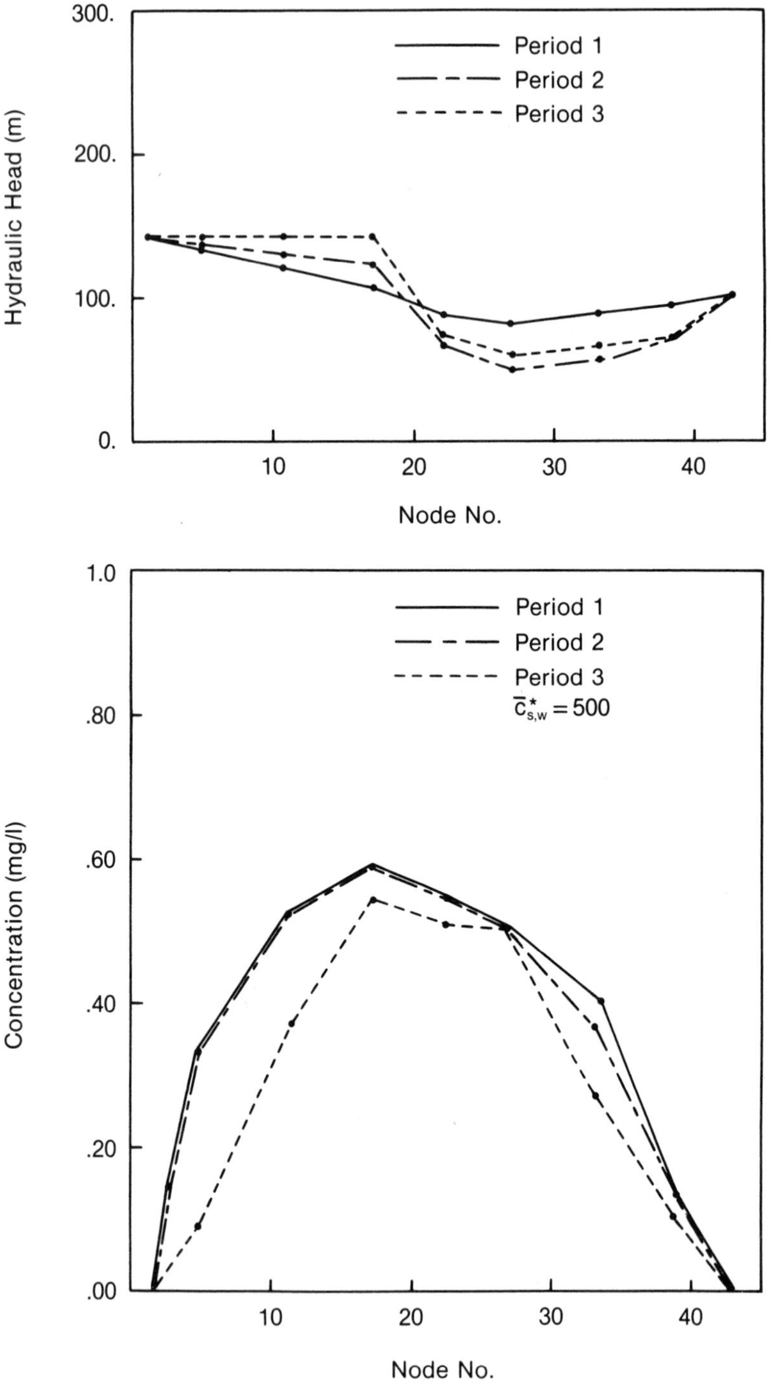

FIGURE 7.42 Mass flux versus planning period

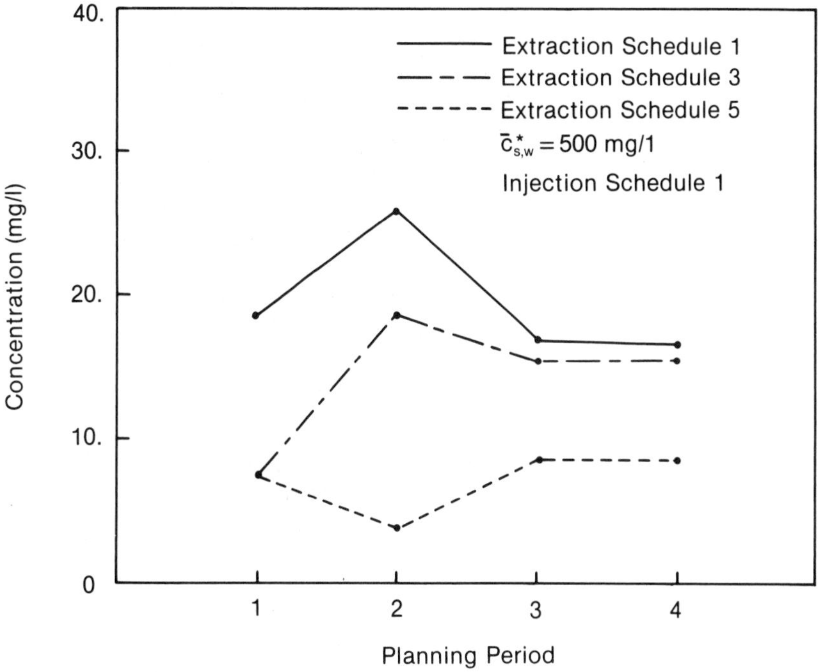

FIGURE 7.43 Example aquifer (Adapted from S. Gorelick, C. I. Voss, P. E. Gill, W. Murray, M. A. Saunders, and M. H. Wright, "Aquifer Reclamation Design: The Use of Contaminant Transport Simulation Combined with Nonlinear Programming." *Water Resources Research*, 20(4), 1984, p. 423. Copyright © 1984 by the American Geophysical Union)

Example Problem 7.8 Simulation/Optimization Methodology

Develop a management model to identify the optimal well locations and pumping and recharge rates necessary to capture or disperse a migrating contaminant plume. Water quality standards are to be maintained at the water supply wells. The aquifer system is shown in Figure 7.44; the aquifer parameters are summarized in Table 7.15. Assume that the contaminant exhibits first-order decay in the aquifer system.

The management model for the example problem is described by Equations 7.100 through 7.109. As presented by Gorelick et al. (1984), the optimization model can be used to analyze different design strategies for aquifer reclamation. Contaminant removal can be achieved, in part, through pumping and withdrawal of the contaminated groundwater followed by treatment and, possibly, re-use. The decision variables associated with this alternative are the locations and extraction rates of all wells in the aquifer system; injection wells are not considered in the

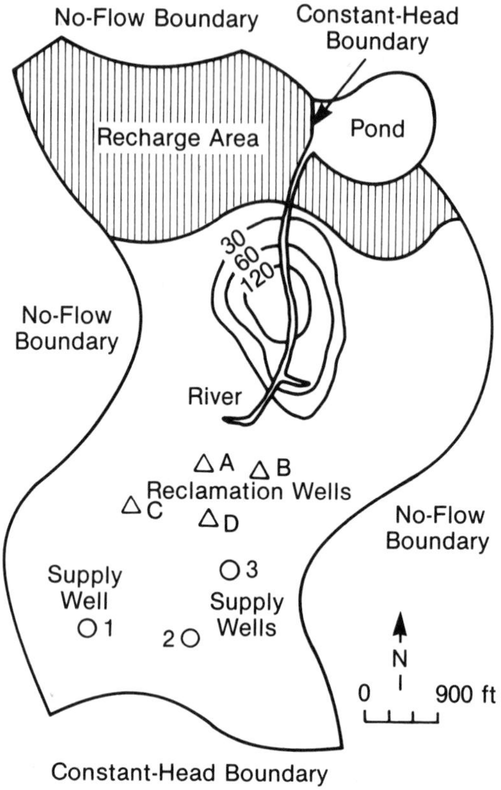

FIGURE 7.44 Optimization results (Adapted from S. Gorelick, C. I. Voss, P. E. Gill, W. Murray, M. A. Saunders, and M. H. Wright, "Aquifer Reclamation Design: The Use of Contaminant Transport Simulation Combined with Nonlinear Programming." *Water Resources Research*, 20(4), 1984, p. 423. Copyright © 1984 by the American Geophysical Union)

analysis. The objective of the planning model can then be expressed as

$$\min z = \sum_{w \in \pi} Q_w$$

where π is an index set defining the feasible well locations and the extraction rates are assumed time-invariant.

The second alternative considered in the example problem is in-ground dilution of the contaminant. The decision variables of the planning model are the well locations and the injection rates that are used to control the movement and the distribution of the contaminant plume. The objective of the planning model

TABLE 7.15 Summary of Data for Transient Model

Range of Transmissivities	0.00625 – 0.01250 ft^2/s
Range of Porosities	0.1 – 0.2
Longitudinal Dispersivity	66 ft
Transverse Dispersivity	13 ft
Inflow Through Recharge Area	1.6 cfs
Inflow Through River	0.13 cfs
Discharge Rate at Each Supply Well	0.2 cfs
Hydraulic Head at Pond	310 ft
Hydraulic Head at Southern Boundary	0 ft
Number of Finite Elements	278
Number of Finite-Element Nodes	315

From J.M. Gorelick, O.I. Voss, P.E. Gill, W. Saunders, and M.H. Wright, "Aquifer Reclamation Design: The Use of Contaminant Transport Simulation Combined with Nonlinear Programming." *Water Resources Research,* 20(4):415-427, 1984. Copyright 1984 by the American Geophysical Union. Reprinted with permission.

can then be expressed as

$$\min z = \sum_{j \in \psi} Q_j$$

where the index set, ψ, defines the feasible injection well locations in the aquifer system.

The hydraulic and water quality response equations of the aquifer system can again be developed using finite-element or multiple-cell balance methods. These constraints are described by Equations 7.108 and 7.109.

The solution of the management model is complicated by (1) the nonconvexity of the optimization problem, (2) the large size of the constraint set (the response equations), and (3) the unknown arrival times of the contaminant plume peak at the water supply wells. Gorelick, et al. (1984) present a planning methodology that resolves these problems by using a combination of simulation and optimization modeling. Rather than incorporating the simulation equations directly into the optimization model, the methodology uses the simulation model as a subroutine that is called by the optimization algorithm. In this example, MINOS, a projected Lagrangian algorithm discussed in Chapter 5, can be used to solve the groundwater optimization model. The advantages of the approach are that (1) the methodology can, in principle, be applied to nonlinear, distributed parameter systems and (2) the simulation models, such as the U.S. Geological Survey's SUTRA flow and mass transport model (Voss, 1984), can be validated, calibrated, and tested prior to the optimization. The Jacobian and derivative evaluation can also be evaluated using conventional finite-difference methods.

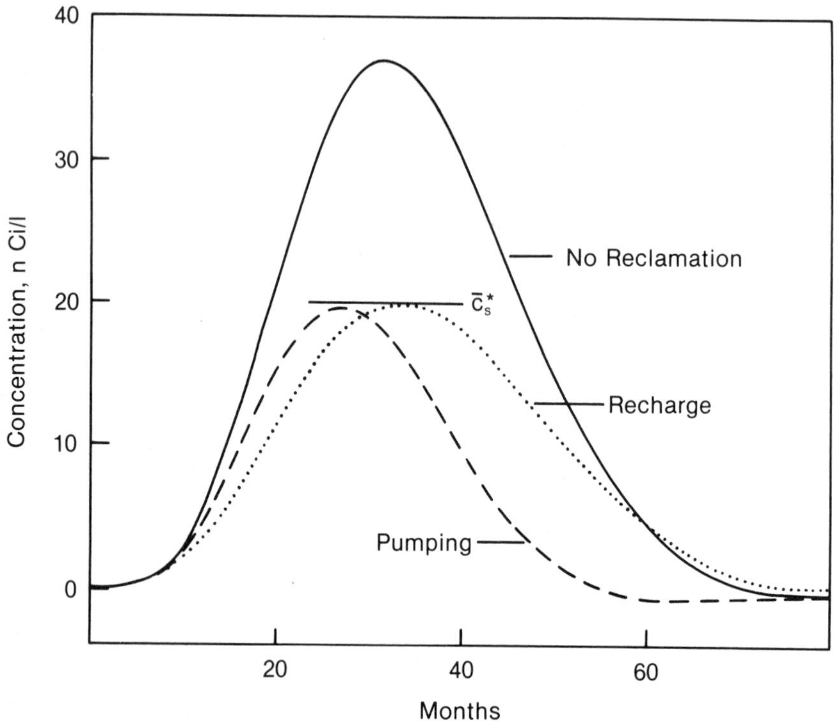

FIGURE 7.45 Concentration profiles at well 3—no reclamation, contaminant removal options, and in-ground dilution options (Adapted from S. Gorelick, C.I. Voss, P.E. Gill, W. Murray, M.A. Saunders, and M.H. Wright, "Aquifer Reclamation Design: The Use of Contaminant Transport Simulation Combined with Nonlinear Programming." *Water Resources Research*, 20(4): 415–427, 1984. Copyright © 1984 by the American Geographical Union. Reprinted with permission)

TABLE 7.16

	Optimal Pumping/Recharge Rates (in cfs) at Wells			
Strategy	A	B	C	D
Contaminant Removal	0.0	0.296	0.0	0.310
In-Ground Dilution	0.0	−0.163	0.0	−0.148

From S.M. Gorelick, O.I. Voss, P.E. Gill, W. Saunders, and M.H. Wright, "Aquifer Reclamation Design: The Use of Contaminant Transport Simulation Combined with Nonlinear Programming." *Water Resources Research*, 20(4):415–427, 1984. Copyright © 1984 by the American Geophysical Union. Reprinted with permission.

The time frame for solution of the groundwater quality optimization model begins at the inception of reclamation-well pumping and continues until the plume has passed all of the water supply wells. Since the recharge and pumping strategies are unknown, a priori, the travel time of the plume is also unknown. Rather than writing groundwater quality constraints for all times, the constraints are given for each numerical time step of the groundwater quality simulation model. As discussed by Gorelick, et al. (1984), a simulation period of 6.67 years gave an adequate representation of plume migration and redistribution. However, numerical experiments with the simulation model can reduce the time frame for the management model. Sensitivity analysis demonstrated that the critical time needed for the plume peak curve is between 1.75 and 5 years. The resulting optimization model has 120 nonlinear constraints.

The results of the transient design model are summarized in Table 7.16, which shows the optimal well selection and the corresponding pumping or injection rates. Concentration profiles at water supply well 3 are shown in Figure 7.45 for no reclamation, contaminant removal options, and in-ground dilution (recharge) options. Approximately 7 minutes of CPU time were required to solve the optimization model on an IBM 3081 computing system.

7.5 CLOSING COMMENTS

In this chapter, we have examined how simulation and optimization modeling can be used to develop, or to infer, optimal pumping and waste management policies. Although the simulation models have been extensively used in the United States, the optimization models are experimental in the sense that they have not been applied to actual field conditions, nor have the effects of the modeling assumptions been thoroughly validated. The limitations of these models are

1. The large size of the optimization models. The discretization of the response equations dramatically increases the size of the constraint matrix of the planning problem. There are, of course, the attendant problem of optimality of the planning policies and the stability problems associated with the solution of the mass transport problem.
2. The constraint set is also a nonconvex set. Solution algorithms cannot guarantee that global or even local solutions will be generated by the optimization methods.
3. The viability of the preference assumptions. Again, if the water supply and water quality resources are equally important, large-scale decomposition procedures are required for solution of the conjunctive use problem.

These questions are the basis of research studies currently under investigation at Humboldt State University and UCLA.

Two other approaches have been proposed to obtain useful solutions to the conjunctive use planning problem. In a series of papers, Gorelick (1982) and Gorelick and Remson (1982 a, b) have used the influence coefficient or response matrix method to generate the groundwater quality response equations. The approach is based on superposition of the individual well responses of the aquifer system. The method has been successful in generating water quality equations for relatively long planning horizons while minimizing the size of the resulting optimization problem. This is in contrast to the example problem that analyzed groundwater quality variations over a relatively short planning horizon and included, as constraints, the entire set of hydraulic and/or mass transport equations. However, the major limitation of the approach is the linearity assumption of the pumping and injection well response of the aquifer system.

Alternatively, screening models can also be used for optimal groundwater quality management. These types of models, which are commonly used in water resources systems optimization, involve both simulation and optimization modeling (Loucks, et al., 1981). Typically, a simplified optimization model of the groundwater system is initially used to screen the possible waste management alternatives. The optimal policies of the model are then simulated in a more detailed mathematical model of the groundwater quality system. The simulation model would more clearly define the probable hydrologic and environmental impacts and long-run costs and benefits of the management policies. If the policies are not feasible from the simulation studies, then additional constraints and/or objectives can be incorporated in the optimization model and the process can be repeated until acceptable management policies are identified. For the groundwater quality management problem, the screening approach would, for example, first involve a steady-state optimization model of the groundwater quality system. The steady-state model represents the ultimate response of the basin to waste disposal and, from this perspective, the model is indeed conservative (Willis, 1976). The optimal policies of the model can, in turn, be incorporated into a detailed simulation model of the groundwater system (see for, example, Section 7.2). The iterative process continues again until an acceptable set of policies is found. For highly nonconvex programming models, the screening approach may be the most computationally efficient approach to groundwater quality management.

7.6 REFERENCES

Baxter, G.P., & Wallace, C.C., "Changes in Volume Upon Solution in Water of the Halogen Salts of the Alkali Metals." *J. Amer. Chem. Soc.* 80:70–104, 1973.

Bredehoeft, J.D. "Comment On: Numerical Solution to the Convective-Diffusion Equation," by C.A. Oster, J.C. Sonnichsen, and R.T. Jaske. *Water Resources Research*, 7(3):755–756, 1971.

Bresler, E. "Simultaneous Transport of Solutes and Water Under Transient Unsaturated Flow Conditions." *Water Resources Research*, 9:975–986, 1973.

Cabrera, G., & Mariño, M.A. "A Finite Element Model of Contaminant Movement in Groundwater." *Wat. Res. Bulletin*, 12(2):317–325, 1976.

California Department of Water Resources. "Waste Water Reclamation: State of the Art." *Bulletin 189*, 1–43, 1973.

Chaudhari, N.M. "An Improved Numerical Technique for Solving Multidimensional Miscible Displacement Equations." *J. S. Pet. Engr.*, 277–284, 1971.

Cooley, R.L. "Finite Element Solutions for the Equations of Groundwater Flow." *Hydrol. Water Res. Publ. 18*, University of Nevada, Reno, 1974.

Danskin, J.M. "The Theory of Max-Min with Applications." *J. of SIAM on Applied Math.*, 14(4):641–664, 1966.

Environmental Protection Agency. "Wastewater Treatment and Reuse by Land Application," Vol. II, EPA-660/2-73-006B. Washington, D.C.: EPA, 1973.

Everett, H. "Generalized Lagrange Multiplier Method for Solving Problems of Optimum Allocation of Resources." *Operations Research*, 11:399–417, 1963.

Gardner, A.O., Jr., Peaceman, D.W., & Pozzi, A.L. "Numerical Calculation of Multidimensional Miscible Displacement by the Method of Characteristics." *J. Society of Petroleum Engineers*, 22–35, March 1964.

Geoffrion, A.M. "Generalized Benders Decomposition." *J. of Optimization Theory and Applications*, 10(4):237–244, 1972.

Giddings, T. "The Utilization of A Groundwater Dam for Leachate Containment at a Landfill Site." *Proc. of the Second National Symposium on Aquifer Restoration and Groundwater Monitoring*. Columbus, OH: National Water Well Association, 1982.

Gorelick, S.M., & Remson, I. "Optimal Dynamic Management of Groundwater Pollutant Sources." *Water Resources Research*, 18(1):71–76, 1982.

Gorelick, S.M. "A Model for Managing Sources of Groundwater Pollution." *Wat. Res. Research*, 18(4):773–781, 1982.

Gorelick, S.M., & Remson, I. "Optimal Location and Management of Waste Disposal Facilities Affecting Groundwater Quality." *Water Resources Research*, 18(1):43–51, 1982.

Gorelick, S.M., Voss, O.I., Gill, P.E., Murray, W., Saunders, M.A., & Wright, M.H. "Aquifer Reclamation Design: The Use of Contaminant Transport Simulation Combined with Nonlinear Programming." *Water Resources Research*, 20(4):415–427, 1984.

Haji-Djafari, S., & Wiggert, D.C. "Two-Dimensional Analysis of Tracer Movement and Transient Flow in a Phreatic Aquifer." Second Int. Symposium on Finite Element Methods, Rapallo, Italy, 1976.

Heinrich, J.C., Huyakorn, P.S., Zienkiewics, O.C., & Mitchell, A.R. "An Upwind Finite Element Scheme for Two-Dimensional Convective-Transport Problems." *Int. J. Numer. Metho. Engr.*, 11:131–143, 1977.

Henry, H.R. "Transitory Movements of the Saltwater Front in an Extensive Artesian Aquifer." *U.S. Geol. Surv. Prof. Pap. 450-B*, 1962.

Herbst, B.M. "Collocation Method and the Solution of Conduction-Convection Problems." *Int. J. Numer. Method. Engr.*, 17:1093–1102, 1981.

Huyakorn, P.S. "An Upwind Finite Element Scheme for Improved Solution of the Con vective-Diffusion Equations." Res. Rep. No. 76-WR-2, Water Resources Center, Princeton University, 1976.

Huyakorn, P.S., & Nilkuha, K. "Solution of Transient Transport Equation Using Upstream Finite Element Scheme." *Appl. Math. Model,* 3:7–17, 1979.

Konikow, L.F., & Bredehoeft, J.D. "Modeling Flow and Chemical Quality Changes in an Irrigated Stream-Aquifer System." *Water Resources Research,* 10(3):546–562, 1974.

Konikow, L.F. "Modeling Chloride Movement in the Alluvial Aquifer at the Rocky Mountain Arsenal, Colorado." *U.S. Geological Survey Water Supply Paper 2044,* 1977.

Landon, R.A., & Sylvester, K.A. "Delineation of Subsurface Oil Contamination and Recommended Containment/Cleanup." *Proceedings of the Second National Symposium on Aquifer Restoration and Groundwater Monitoring.* Columbus OH: National Water Well Association, 1982.

Lantz, R.B. "Quantitative Evaluation of Numerical Diffusion (Truncation Error)." *J. Society of Petroleum Engr.,* 315–320, September 1971.

Lapidus, L., & Amundson, N.R. "Mathematics of Adsorption in Beds: VI. The Effect of Longitudinal Diffusion in Ion Exchange and Chromatographic Columns." *J. Phys. Chem.,* 56:984–993, 1952.

Lapidus, L., & Pinder, G.F. *Numerical Solution of Partial Differential Equations in Science and Engineering.* New York: Academic Press, 1982.

Lasdon, L. *Optimization Theory for Large Systems.* New York: Macmillan, 1970.

Lehr, J.H., & Nielsen, D.M. "Aquifer Restoration and Groundwater Rehabilitation—A Light at the End of the Tunnel." *Groundwater,* 20(6):650–656, 1982.

Linaweaver, F.D., & Clark, C.S. "Cost of Water Transmission." *J. Amer. Water Works Ass.,* 50(12):1549–1560, 1964.

Loucks, D.P., Stedinger, J.R., & Haith, D.A. *Water Resources Systems Planning and Analysis.* Englewood Cliffs N.J.: Prentice-Hall, 1981.

Mangasarian, O.L. *Nonlinear Programming.* New York: McGraw-Hill, 1969.

Mariño, M.A. "Solute Transport in Saturated-Unsaturated Porous Media." In *Modeling, Identification and Control in Environmental Systems,* G. C. Vansteenkiste, Ed. Amsterdam: North-Holland, 1978.

Mariño, M.A. "Analysis of the Transient Movement of Water and Solutes in Stream-Aquifer Systems." *J. of Hydrology,* 49:1–17, 1981.

McBride, K.K. "Decontamination of Groundwater for Volatile Organic Chemicals: Select Studies in New Jersey." *Proc. of the Second National Symposium on Aquifer Restoration and Groundwater Monitoring.* Columbus OH: National Water Well Association, 1982.

McKee, J.E., & McMichael, F.C. *Final Report on Wastewater Reclamation at Whittier Narrows.* Pasadena: Calif. Inst. of Technology, 1966.

Molsather, L.R., & Barr, K.D. "Retrofit Leachate Collection System for an Existing Landfill: A Case History." *Proc. of the Second National Symposium on Aquifer Restoration and Groundwater Monitoring.* Columbus OH: National Water Well Association, 1982.

Neuman, S.P., & Narasimhan, T.N. "Mixed Explicit-Implicit Iterative Finite Element Scheme for Diffusion-Type Problems, 1, Theory." *Int. J. Numerical Method. Eng.,* 11:309–323, 1977.

Ohneck, R.J., & Gardner, G.L. "Restoration of an Aquifer Contaminated by an Accidental Spill of Organic Chemicals." *Proc. of the Second National Symposium on Aquifer Restoration and Groundwater Monitoring.* Columbus OH: National Water Well Association, 1982.

Pinder, G.F. "A Galerkin Finite Element Simulation of Groundwater Contamination on Long Island, New York." *Water Resources Research*, 9(6):1657–1669, 1973.

Pinder, G.F., & Shapiro, A. "A New Collocation Method for the Solution of the Convection-Dominated Transport Equation." *Water Resources Research*, 15(5), 1979.

Pinder, G.F., & Cooper, H.H., Jr. "A Numerical Technique for Calculating the Transient Position of the Saltwater Front." *Water Resources Research*, 6(3):875–888, 1970.

Rockafellar, R.T. "Duality and Stability in Extremum Problems Involving Convex Function." *Pacific J. Math.* 21:167–187, 1967.

Rossman, L.A., & Vanecek, R.T. "A Partitioning Procedure for Water Quality Management Models." *Water Resources Bulletin*, 14(4):842–855, 1978.

Segol, G., Pinder, G.F., & Gray, W.G. "A Galerkin Finite Element Technique for Calculating the Transient Position of the Saltwater Front." *Water Resources Research*, 11(2):343–347, 1975.

Stover, E.L. "Removal of Volatile Organics from Contaminated Groundwater." *Proceedings of the Second National Symposium on Aquifer Restoration and Groundwater Monitoring*. Columbus OH: National Water Well Association, 1982.

Sun, Ne-Zheng. *Mathematical Modeling and Numerical Methods in Groundwater Flow* (in Chinese). Beijing, China: Geological Publishing House, 1981.

Sun, Ne-Zheng, & Yeh, W. W-G. "A Proposed Upstream Weight Numerical Method for Simulating the Pollutant Transport in Groundwater." *Water Resources Research*, 19(6):1489–1500, 1983.

Tui, H. "Concave Programming Order Linear Constraints." *Dokl. Akad. Navl. SSR.*, 159:32–35, 1964.

Van Genuchten, M.T., & Gray, W.C. "Analysis of Some Dispersion Corrected Numerical Schemes for Solution of the Transport Equation." *Int. J. of Numerical Methods in Engr.*, 12:387–404, 1978.

Van Genuchten, M.T. "On the Accuracy and Efficiency of Several Numerical Schemes for Solving the Convective Dispersive Equation. In *Finite Element in Water Resources*, W.G. Gray, et al., Eds. London: Pentech Press, 1977.

Voss, C.I. "SUTRA: A Finite Element Simulation Model for Saturated-Unsaturated Fluid Density Groundwater Flow with Energy Transport or Chemically Reactive Single Species Solute Transport." *U.S. Geol. Surv. Water Resources Invest.*, 1984.

Willis, R., & Dracup, J.A. "Optimization of the Assimilative Waste Capacity of the Unsaturated and Saturated Zones of an Unconfined Aquifer System." *UCLA ENG-7394*, University of California, Los Angeles, 1973.

Willis, R. "A Planning Model for the Management of Groundwater Quality." *Water Resources Research*, 15(6):1305–1312, 1979.

Willis, R. "Optimal Groundwater Quality Management: Well Injection of Waste Water." *Water Resources Research*, 12(1):47–53, 1976.

Wolfe, P. "A Duality Theorem for Non-Linear Programming." *Quarterly Journal of Applied Math.*, 19(3):239–244, 1961.

Yaniga, P.M. "Alternatives in the Decontamination for Hydrocarbon Contaminated Aquifers." *Proceedings of the Second National Symposium on Aquifer Restoration and Groundwater Monitoring*. Columbus OH: National Water Well Association, 1982.

8. THE INVERSE PROBLEM IN GROUNDWATER SYSTEMS

8.1 INTRODUCTION

The simulation and optimization models presented in Chapters 6 and 7 have been used to predict and optimize the hydraulic and water quality response of groundwater systems. These physically based mathematical models are distributed parameter models since the response of the aquifer system is governed by partial differential equation(s) and the parameters imbedded in the equations are spatially dependent. The hydraulic or water quality parameters of the models are not directly physically measurable and a major element in the modeling process is the estimation, from historical observations of the hydraulic head or mass concentrations, of the unknown system parameters. Historically, aquifer parameters have been determined using trial and error and graphical matching techniques. These methods are based on analytical or closed-form solutions of the governing equations of the aquifer system (Theis, 1935). The techniques are generally not applicable to distributed parameter, inhomogeneous, anisotropic groundwater systems.

The problem of parameter identification in distributed parameter systems has been studied extensively during the last two decades. The inverse problem of parameter identification concerns the optimal determination of the system's hydraulic and mass transport parameters from observations collected in the spatial and time domains. The number of observations is finite and limited, whereas the spatial domain of the parameter field is continuous. For an inhomogeneous aquifer, the dimension of the parameter is theoretically infinite. In field applications, however, the spatial variables are approximated by a finite-difference or finite-element methods and the aquifer system is subdivided into several subregions. Each subregion is characterized by a constant parameter value. The reduction of the number of parameters from the infinite dimension to a finite dimensional form is called *parameterization*.

There are two types of errors associated with the inverse problem: (1) the system modeling error, as represented by a performance criterion, and (2) the error associated with parameter uncertainty (Yeh, 1986). An increase in parameter dimension (the number of unknown parameters associated with parameterization) will generally improve the system modeling error, but will increase the parameter uncertainly and conversely. The optimum level of parameterization depends on the quantity and quality of the observational data.

In the following sections, we will explore the ill-posedness of the inverse problem and present solution methods for the parameter estimation problem in groundwater systems.

8.2 ILL-POSEDNESS

The inverse problem is ill-posed. The ill-posedness is generally characterized by the nonuniqueness and instability of the identified parameters. The instability of parameter estimation algorithms is caused by small errors in the observed head or mass concentrations that cause serious errors in the identified parameters.

Chavent (1974) examined the uniqueness problem in connection with parameter identification in distributed parameter systems. The results demonstrated that the inverse problem is always nonunique in distributed parameter systems where only point measurements are available (measurements made at a limited number of locations in the spatial domain). And, as a result, parameter estimates will differ according to the initial estimates of the parameters. There are also no guarantees that the identified parameters are close to the true, unknown parameter values. Simulation or optimization models based on the parameter estimates may distort future groundwater predictions or optimal groundwater development alternatives.

The uniqueness problem in parameter identification is intimately related to identifiability. *Parameter identifiability* addresses the question of whether it is possible to obtain unique solutions of the inverse problem for the unknown hydraulic or water quality parameters, from data collected in the spatial and time domains. Kitamura and Nakagiri (1977) formulated the parameter identification problem as the one-to-one property of the inverse problem, that is, the one-to-one property of mapping from the space of system outputs to the space of parameters. However, the uniqueness of such a mapping is extremely difficult to establish and often nonexistent. *Identifiability* was defined as the capability to determine uniquely the system parameters in all points of its domain by using the input-output relation of the system and the input-output data. Kitamura and Nakagiri (1977) also obtained some results for parameter indentifiability or nonidentifiability for a system characterized by a linear, one-dimensional, parabolic partial differential equation.

Another definition of *identifiability*, given by Chavent (1979b), is based on the Output Least Square Error criterion. If the criterion is used for solving the inverse problem of parameter identification, the parameter is said to be Output Least Square identifiable if and only if a unique solution of the optimization problem exists and the solution depends continuously on observations. Chavent (1983) presented a weaker sufficient condition for Output Least Square identification.

Identifiability is usually not achievable in the case of point measurements where data is only available at a limited number of locations in the spatial domain. In view of the various uncertainties involved in groundwater modeling, a groundwater model can only be used to approximate the behavior of an aquifer system. If a small, prescribed error is allowed in prediction, Yeh and Sun (1984) developed an extended identifiability criterion which can be used for designing an optimum pumping test to assist aquifer parameter identification. The extended identifiability criterion is called the *δ-identifiability* and is based upon the concept of weak uniqueness.

8.3 THE PARAMETER ESTIMATION PROBLEM

To illustrate the techniques that have been used to solve the inverse problem, consider the unsteady flow equation for an inhomogenous, isotropic confined aquifer system. The governing equation (Eq. 2.55) can be expressed as

$$\frac{\partial}{\partial x}\left(T\frac{\partial h}{\partial x}\right) + \frac{\partial}{\partial y}\left(T\frac{\partial h}{\partial y}\right) = Q + S\frac{\partial h}{\partial t} \tag{8.1}$$

The boundary and initial conditions of the problem can be expressed as

$$\begin{aligned} h(x, y, 0) &= h_0(x, y) \qquad (x, y) \in \Gamma_0 \\ h(x, y, t) &= h_1(x, y, t) \qquad (x, y) \in \Gamma_1 \\ T\frac{\partial h}{\partial n} &= h_2(x, y, t) \qquad (x, y) \in \Gamma_2 \end{aligned} \tag{8.2}$$

where

$$\begin{aligned} h(x, y, t) &= \text{head} \\ T(x, y) &= \text{transmissivity} \\ S &= \text{storage coefficient} \end{aligned}$$

$$
\begin{aligned}
Q(x, y) &= \text{source/sink term (known)} \\
x, y &= \text{space variables} \\
t &= \text{time} \\
\Gamma_0 &= \text{flow region} \\
\Gamma &= \text{boundary of the aquifer} \\
&\quad (\Gamma_1 \cup \Gamma_2 = \Gamma) \\
\partial/\partial n &= \text{normal derivative} \\
h_0, h_1, h_2 &= \text{specified functions}
\end{aligned}
$$

In the parameter estimation problem, we will assume that the storage coefficient is known throughout the solution domain. The parameter chosen for identification is the transmissivity function, $T(x, y)$, which is assumed to be time invariant.

In general, a numerical scheme is required to obtain solutions of Equation 8.1, subject to conditions (Eq. 8.2), provided that parameter values are properly estimated. Various finite-difference or finite-element methods have been developed for numerical simulation studies. In solving the inverse problem, it is essential to have an efficient forward solution scheme, particularly when using an iterative nonlinear least square estimation. An example is the following classical Crank-Nicolson scheme

$$
\begin{aligned}
&\tfrac{1}{2}[T_{i+1/2,j}(h^{k+1}_{i+1,j} - h^{k+1}_{i,j})/(\Delta x)^2 \\
&- T_{i-1/2,j}(h^{k+1}_{i,j} - h^{k+1}_{i-1,j})/(\Delta x)^2 \\
&+ T_{i+1/2,j}(h^{k}_{i+1,j} - h^{k}_{i,j})/(\Delta x)^2 \\
&- T_{i-1/2,j}(h^{k}_{i,j} - h^{k}_{i-1,j})/(\Delta x)^2] + \tfrac{1}{2}[T_{i,j+1/2}(h^{k+1}_{i,j+1} \\
&- h^{k+1}_{i,j})/(\Delta y)^2 - T_{i,j-1/2}(h^{k+1}_{i,j} - h^{k+1}_{i,j-1})/(\Delta y)^2 + T_{i,j+1/2}(h^{k}_{i,j+1} - h^{k}_{i,j}) \\
&- T_{i,j-1/2}(h^{k}_{i,j} - h^{k}_{i,j-1})/(\Delta y)^2] \\
&= Q_{i.j} + S(h^{k+1}_{i,j} - h^{k}_{i,j})/\Delta t
\end{aligned}
\tag{8.3}
$$

where i, j denotes the x, y spatial location, k is an index of the time step, and the arithmetic mean is used to describe the intermediate transmissivities, i.e., $T_{i+1/2,j} = \frac{1}{2}\{T_{i+1,j} + T_{i,j}\}$.

The finite-difference equations can be solved by an alternating direction method (Douglas, 1962) which is locally second-order correct in space and time.

8.4 PARAMETER DIMENSION AND PARAMETERIZATION

Hydraulic or water quality parameters, such as dispersion or transmissivity coefficients, are continuous functions of the spatial domain of the groundwater system. In the parameter estimation problem, these continuous functions are replaced and approximated by finite dimensional functions. The reduction of parameter dimension is accomplished by parameterization using either the zonation or the interpolation method.

In the zonation approach, the flow region is divided into a number of subregions or zones and a constant parameter value is used to characterize each zone. The unknown parameter function is then represented by a number of constants that reflect the number of zones used in the zonation process. The dimension of parameterization (or parameter dimension) is represented by the total number of zones. The approach has been used by Coats, et al. (1970), Emsellem and de Marsily (1971), Yeh and Yoon (1976), and Cooley (1977, 1979).

In the interpolation method, finite elements are generally used to discretize the flow or transport domain. The unknown hydraulic (or water quality) parameters are then approximated by a linear combination of the basis functions,

$$T(x, y) = \sum_i T_i N_i \tag{8.4}$$

where N_i is the ith basis function and T_i is the nodal value of parameter T. The parameter dimension corresponds to the number of unknown nodal parameter values (DiStefano & Rath, 1975; Yoon & Yeh, 1976; Yeh & Yoon, 1981). Other interpolation methods that have also been used to approximate the transmissivity distribution in aquifer systems include splines (Sagar, et al., 1975; Yakowitz & Noren, 1976), the polynomial method (Garay, et al., 1976) and Kriging (Clifton & Newman, 1982). Groundwater hydraulic parameters have also been represented by geostatistical models as suggested by Kitanidis and Vomvoris (1983).

The optimal determination of the shape of the zones in the zonation approach or the location of nodes and the associated basis function in the interpolation method, are unresolved research problems in parameter estimation. Most of the published literature relies on a trial-and-error approach or hydrological mapping to determine the zones or basis functions used in the parameterization problem. However, a recent paper, by Sun and Yeh (1985) suggest a systematic way to identify the parameter structure.

8.5 PARAMETER IDENTIFICATION METHODS

Solution algorithms for the parameter estimation problem in groundwater systems are classified as either equation error criterion or output error criterion methods (Yeh, 1986). It should be noted that the equation error and output error classifications of the inverse solution methods are intrinsically consistent with Neuman's classifications of direct and indirect methods (Newman, 1973). The equation error criterion method treats the model parameters as dependent variables in a formal inverse boundary value problem. Algorithms that have been used for solution of the inverse problem include the energy dissipation method (Nelson, 1968), linear programming (Kleinecke, 1971), optimization with a flatness criterion (Emsellem & de Marsily, 1971), the multiple objective decision process (Neuman, 1973), the Galerkin method (Frind & Pinder, 1973), the algebraic approach (Sagar, et al., 1975), the inductive method (Nutbrown, 1975), linear programming and quadratic programming (Hefez, 1975), penalty function methods (Navarro, 1977), and the matrix inversion method allied with Kriging (Yeh, et al., 1983). Table 8.1 presents a summary of the equation error criterion methods that have been used for solution of the inverse problem.

The output error criterion approach to parameter estimation is based on an optimization or control model of the inverse problem. The objective of the optimization model minimizes an output error criterion. The parameter estimates are required to satisfy the aquifer system's hydraulic or water quality equations and possible upper or lower bounds. Algorithms for solution of the model have been designed to iteratively update parameter values until the model's response is sufficiently close to the historically observed conditions in the groundwater system.

Control-oriented techniques, based on quasilinearization (Bellman & Kalaba, 1965), have been developed for the solution of the aquifer parameter problem. The algorithms include quasilinearization (Yeh & Tauxe, 1971; DiStefano & Rath, 1975), minimax and linear programming (Yeh & Becker, 1973), maximum principle (Lin & Yeh, 1974; Yakowitz & Noren, 1976). Vermuri and Karplus (1969) formulated the inverse problem in terms of optimal control and a gradient method was used to solve the problem. Chen, et al. (1974) and Chavent (1975) also treated the problem in an optimal control approach and solved it using steepest descent and conjugate gradient methods. Kalman filtering techniques have also been proposed in the literature for parameter identification (Chen & Steinfeld, 1972; McLaughlin, 1975; Wilson, et al., 1978). Kitanidis and Vomvoris (1983) used the technique of maximum likelihood estimation and Kriging.

Mathematical programming techniques developed in the field of operations research have been used to solve the inverse problem of parameter identification in groundwater hydrology and in the field of petroleum engineering. The algorithms include gradient search procedures (Jacquard & Jain, 1965; Thomas, et al., 1972), decomposition and multilevel optimization (Haimes, et al., 1968), linear programming (Coats, et al., 1970; Slater & Durrer, 1971; Yeh, 1975), quadratic programming (Yeh, 1975; Chang & Yeh, 1976), the Gauss-Newton method (Jahns, 1966; McLaughlin, 1975), the modified Gauss-Newton method (Yoon & Yeh, 1976; Yeh & Yoon, 1976, Cooley, 1977; Cooley, 1982), the Newton-Raphson method (Neuman & Yakowitz, 1979), and the conjugate gradient method (Neuman, 1980). Some typical parameter identification models, based on the output error criterion method, are presented in Table 8.2.

TABLE 8.1 Parameter Identification Models, Equation Error Criteria (After Yeh, 1986)

Applicable Conditions	*Numerical Method*	*Parameters to be Identified*	*Data Processing*	*Prior Information or Constraints*	*Inverse Solution Procedure*	*Special Features and Comments*	*Reference*
Two or three dimensional, confined/unconfined, steady state	finite difference	K, T	none	none			Nelson (1960, 1961)
Two or three dimensional, confined/unconfined, unsteady state		K, T	generalized orthogonal regression	boundary condition in permeability	energy dissipation method		Nelson (1968)
Two dimensional, confined unsteady state		T, S	none	none	linear programming		Kleinecke (1971)
Two dimensional, steady-unsteady state		T, S, Q	none	flatness of parameters	minimizing norm of error flow	gradually increasing the number of zones	Emsellem and de Marsily (1971)
Two dimensional, confined, steady state	finite element	T	none	T known along a line crossed by all streamlines	matrix inverse	$T(x, y)$ is represented by finite element	Frind & Pinder (1973)
Two dimensional, confined, steady state, anisotropic	finite element	T_x, T_y	none	lower upper bound on parameters	parametric linear programming	penalty term used to control instability	Neuman (1973)
Two dimensional, unsteady state, isotropic/anisotropic		T_x, T_y S, Q	spline or Lagrange interpolation	none	algebraic approach	inverse problems are reduced locally to algebraic equations of small dimensions	Sagar, et al. (1975)

TABLE 8.1 (continied)

Applicable Conditions	*Numerical Method*	*Parameters to be Identified*	*Data Processing*	*Prior Information or Constraints*	*Inverse Solution Procedure*	*Special Features and Comments*	*Reference*
Two dimensional, confined, unsteady state	finite difference	T, S, Q	none	none	linear programming or quadratic programming	compares five different optimization criteria	Hefez, et al. (1975)
Two dimensional, confined, unsteady state	finite difference	T, S	none	limitation on local variability of T	direct integration of P.D.E.		Nutbrown (1975)
Two dimensional, confined, unsteady state	finite difference	T, S	none	initial estimates	minimizing a quadratic objective function with penalty function		Navarro (1977)
Two dimensional, confined, unsteady state	finite difference	T	kriging	none	generalized matrix inversion	instability is controlled by parameterization; determines optimum parameter dimensions	Yeh, et al. (1983)

Table presents typical models in chronological order. K, hydraulic conductivity; T, transmissivity; S, storage coefficient; Q, sink/source.

TABLE 8.2 Parameter Identification Models, Output Error Criteria (After Yeh, 1986)

Applicable Conditions	*Numerical Method*	*Parameters to be Identified*	*Prior Information or Constraints*	*Inverse Solution Procedure*	*Special Features and Comments*	*Reference*
Two dimensional, confined, unsteady state	finite difference	T, S	none	Gauss-Newton	for oil reservoir	Jacquard & Jain (1965)
Two dimensional, confined, unsteady state	finite difference	T, S	none	Gauss-Newton	statistical measures of estimated parameters are provided; for oil reservoir	Jahns (1966)
Two dimensional, unconfined unsteady state	finite difference	K, S	none	maximum principle in conjunction with steepest descent method	computation carried out on a hybrid computer	Vemuri & Karplus (1969)
One dimensional, unconfined, unsteady state	finite difference	D	none	quazilinearlization		Yeh & Tauxe (1971)
Two dimensional, unsteady state	finite difference	K, ϕ	upper-lower bounds on parameters	Gauss-Newton, step size is determined by quadratic interpolation	box-type constraints are imposed on parameters; for oil reservoir	Thomas, et al. (1972)
One dimensional, leaky aquifer	finite difference	$T, S, K'/b'$	none	quasilinearization	radial flow	Marino & Yeh (1973)
Two dimensional, unsteady state	finite difference	K, ϕ	none	steepest descent and conjugate gradient	parameters are considered as continuous function of position; gradients obtained by optimal control theory	Chen, et al. (1974)
Two dimensional, unsteady state	finite difference	K, ϕ	upper-lower bounds on parameters	steepest descent	for oil reservoir; gradients are generated by solving the adjoint model	Chavent, et al. (1975)

TABLE 8.2 (continued)

Applicable Conditions	*Numerical Method*	*Parameters to be Identified*	*Prior Information or Constraints*	*Inverse Solution Procedure*	*Special Features and Comments*	*Reference*
One dimensional, unconfined, unsteady state	finite difference	D	none	quasilinearization; maximum principle; gradient; influence coefficient; linear programming	compares five different algorithms	Yeh (1975a)
One dimensional, confined, unsteady state	finite difference	D	upper and lower bounds; linear constraints	quadratic programming	radial flow	Yeh (1975b)
Two dimensional, confined, unsteady state	finite element	T	structure constraints	quasilinearization	transmissivity function is represented by finite element	DiStefano & Rath (1975)
One dimensional	finite difference	K, ϕ	mean and covariance matrix of parameters	conjugate gradient, Gauss-Newton, Marquardt	a Bayesian penalty term is added to the objective function	Gavalas, et al. (1976)
Two dimensional confined, unsteady state	finite element	K	upper-lower bounds on parameters	Gauss-Newton with Rosen's gradient projection	permeability function is represented by finite element	Yoon & Yeh (1976)
Two dimensional unconfined, unsteady state	finite difference	T	upper-lower bounds on parameters	Gauss-Newton with Rosen's gradient projection	stepwise zoning procedure using statistical measures of parameters; covariance matrix of estimated parameters is provided	Yeh & Yoon (1976)
Two dimensional, steady state	finite element	K, Q, flux	none	modified Gauss-Newton (nonlinear regression by linearization)	statistical measures of model and parameters are provided	Cooley (1977)

One dimensional		ϕ, k	mean and covariance matrix of parameters	Gauss-Newton	for oil reservoir; covariance matrix of estimated parameter is provided; determines optimum level of parameterization	Shah, et al. (1978)
Two dimensional steady state	finite element	T	prior estimation of parameters added to objective	Newton-Raphson	covariance matrix of parameter estimates is provided	Neuman & Yakowitz (1979)
Two dimensional steady state	finite element	T	prior estimation of parameters added to objective	conjugate gradient	variational theory is used; use log transmissivities	Neuman (1980)
Two dimensional confined, unsteady state	finite difference	T	upper-lower bounds on parameters	Gauss-Newton with Rosen's gradient projection	finite element is used to represent $T(x, y)$; determines optimum parameter dimension; considers parameter uncertainty	Yeh & Yoon (1981)
Two dimensional steady state	finite element	K, Q, flux	prior estimates of parameters with or without reliability added to objective	modified Gauss-Newton (nonlinear regression by linearization)	two types of prior information are included in the analysis	Cooley (1982)
Steady state		K	point measurement of permeability and hydraulic head	maximum likelihood and kriging	parameter is represented as a "random field"	Kitanidis & Vomvoris (1983)
Two dimensional confined, unsteady state	finite element	K	none	Gauss-Newton with Rosen's gradient projection	generalized least squares; considers correlated errors	Sadeghipour & Yeh (1984)

TABLE 8.2 (continued)

Applicable Conditions	*Numerical Method*	*Parameters to be Identified*	*Prior Information or Constraints*	*Inverse Solution Procedure*	*Special Features and Comments*	*Reference*
Two dimensional steady state	finite difference	T	point measurements of transmissivity and head	cokriging	parameter is represented as a random field	Hoeksema & Kitanidis (1984)
Two dimensional confined, unsteady state	finite element	T	none	Gauss-Newton	identification of parameter structure	Sun & Yeh (1985)
Two dimensional, steady state	analytical solution	T	point measurements of transmissivity and head	Gaussian conditioned mean	parameter is represented as a random field	Dagan (1985)
Two dimensional steady state, leakage included	finite difference	T	point measurements of transmissivity and head		comparison of Gaussian conditional mean and kriging estimation	Hoeksema & Kitanidis (1985)

Table presents typical models in chronological order, K'/b', leakance; D, diffusivity; ϕ, porosity. See Table 8.1 for additional definitions.

8.6 EQUATION ERROR CRITERION PARAMETER ESTIMATION MODEL (DIRECT METHOD AS CLASSIFIED BY NEUMAN)

The equation error criterion method for parameter estimation requires an explicit formulation of the unknown parameters. Assuming head observations are available at each grid point in the aquifer system and substituting these observations into Equation 8.3, the Crank-Nicolson scheme can be expressed as,

$$
\begin{aligned}
&(h_{i+1,j}^{k+1/2} - h_{i,j}^{k+1/2})T_{i+1,j} - (h_{i,j}^{k+1/2} \\
&- h_{i-1,j}^{k+1/2})T_{i-1,j} + (h_{i,j+1}^{k+1/2} - h_{i,j}^{k+1/2})T_{i,j+1} \\
&- (h_{i,j}^{k+1/2} - h_{i,j-1}^{k+1/2})T_{i,j-1} + (h_{i+1,j}^{k+1/2} \\
&+ h_{i-1,j}^{k+1/2} + h_{i,j+1}^{k+1/2} + h_{i,j-1}^{k+1/2} - 4h_{i,j}^{k+1/2})T_{i,j} \\
&= \frac{2(\Delta x)^2}{\Delta t} S(h_{i,j}^{k+1} - h_{i,j}^{k}) + 2(\Delta x)^2 Q + \epsilon_{i,j}^{k+1/2}
\end{aligned}
\tag{8.5}
$$

where Δy is assumed to be equal to Δx and

$$h_{i,j}^{k+1/2} = \tfrac{1}{2}(h_{i,j}^{k} + h_{i,j}^{k+1}) \qquad \text{and} \qquad T_{i+1/2,j} = \frac{1}{2}(T_{i,j} + T_{i+1,j})$$

To account for the lack of equality in the equations, due to errors in observations, an unknown error term $\epsilon_{i,j}^{k+1/2}$ is added to Equation 8.5. In practice, only a limited number of field observations is available and interpolation schemes, such as cubic splines (Yakowitz & Noren, 1976) and Kriging (Yeh, et al., 1983) have been used to obtain head values at every computational grid associated with the numerical scheme that is based upon either finite-difference or finite-element approximations. The error term consists of interpolation errors as well as noise in the observations. Equation 8.5 can be simplified to

$$A_t T_g = b_t + \epsilon_t, \qquad t = 1, 2, \ldots, N \tag{8.6}$$

where

A_t = coefficient matrix, a function of h

T_g = transmissivity vector containing transmissivity values at all grid points

N = total number of time steps

b_t = column vector, a function of h

In a more compact matrix form, this becomes

$$\mathbf{A}\mathbf{T}_g = \mathbf{b} + \boldsymbol{\epsilon} \tag{8.7}$$

where

$$\mathbf{A} = [\mathbf{A}_1^T, \mathbf{A}_2^T, \cdots, \mathbf{A}_N^T]^T$$

$$\mathbf{b} = [\mathbf{b}_1^T, \mathbf{b}_2^T, \cdots, \mathbf{b}_N^T]^T$$

$$\boldsymbol{\epsilon} = [\boldsymbol{\epsilon}_1^T, \boldsymbol{\epsilon}_2^T, \cdots, \boldsymbol{\epsilon}_N^T]^T$$

T is a transpose operator when used as a superscript. It should be noted that whether finite-difference or finite-element is used as the forward solution method, the resulting equation error will always have the form of Equation 8.7. However, we have used a typical finite-difference method to demonstrate how to formulate the inverse problem by the equation error criterion. The advantage of this formulation is that Equation 8.7 is linear and $\mathbf{T}_g$ can be determined by minimizing the equation error $\boldsymbol{\epsilon}$.

From Equation 8.7, the least squares error (or residual sum of squares) can be expressed by

$$\boldsymbol{\epsilon}^T\boldsymbol{\epsilon} = (\mathbf{A}\mathbf{T}_g - \mathbf{b})^T (\mathbf{A}\mathbf{T}_g - \mathbf{b}) \tag{8.8}$$

Minimizing the least squares error, the transmissivity vector can be estimated as

$$\hat{\mathbf{T}}_g = (\mathbf{A}^T\mathbf{A})^{-1} \mathbf{A}^T\mathbf{b} \tag{8.9}$$

where $\hat{\mathbf{T}}_g$ is the estimated transmissivity vector of $\mathbf{T}_g$. Note that Equation 8.9 implicitly assumes homoscedasticity and lack of correlation among residuals, i.e., ($E(\boldsymbol{\epsilon}) = 0$ and $E(\boldsymbol{\epsilon}\boldsymbol{\epsilon}^T) = \alpha^2 I$). The solution is also highly dependent on the level of discretization used in the numerical solution of the governing equation. The solution of Equation 8.9 is generally unstable in the presence of noise. Another disadvantage is that the matrix $\mathbf{A}$ may be rank deficient due to the multicollinearity of its rows and columns. If $\mathbf{A}^T\mathbf{A}$ is not fully ranked, difficulty will be encountered in matrix inversion. The technique of ridge regression is a possible way to resolve the problem of collinearity.

8.7 OUTPUT ERROR CRITERION PARAMETER ESTIMATION MODEL (INDIRECT METHOD AS CLASSIFIED BY NEUMAN)

Parameter estimation models that are based on the output error criterion generally involve the minimization of a "norm" of the difference between observed and calculated heads at specified observation points. The optimization is constrained by the governing equations and possible regularity conditions for parameters. In contrast to the equation error criterion method, the output error criterion approach is applicable to situations where the number of observations is limited, and the method does not require differentiation of the observational data. However, the optimization model of the inverse problem is usually nonlinear and nonconvex which guarantees the possibility of locally optimal solutions (parameter estimates).

The objective function of the parameter estimation model can be represented by the least squares error as

$$\min_{T(x,y)} J = [\mathbf{h}_D - \mathbf{h}_D^*]^T [\mathbf{h}_D - \mathbf{h}_D^*] \tag{8.10}$$

where $\mathbf{h}_D$ is the vector of calculated heads at the observation wells, based upon estimated values of parameters, and $\mathbf{h}_D^*$ is the vector of observed heads.

In the parameter identification model, the transmissivity, $T(x,y)$, can be parameterized by either a zonation or interpolation method as mentioned earlier. The parameters and heads are also constrained by the hydraulic response equations (Eqs. 8.1 and 8.2).

The example problems demonstrate how quasilinearization and quadratic programming can be used to solve the output error criterion, parameter estimation problem.

Example Problem 8.1 Quasilinearization (After Yeh, 1975a)

The diffusivity of the unconfined aquifer system shown in Figure 8.1 is to be estimated from hydraulic head data. Using quasilinearization, develop a parameter estimation algorithm for identification of the aquifer's diffusivity.

The governing equation of the aquifer system can be expressed as

$$\frac{\partial h}{\partial t} = D \frac{\partial}{\partial x}\left[h \frac{\partial h}{\partial x}\right] \tag{8.11}$$

subject to the following initial and boundary conditions:

$$h = h(x), \quad 0 \leqslant x \leqslant L,\ t = 0; \quad h = h_0(t), \quad x = 0, \quad t > 0;$$

$$\frac{\partial h}{\partial x} = 0, \quad x = L, \quad t > 0 \tag{8.12}$$

in which h is the head in the aquifer, x is the space variable, t is the time variable, and D is diffusivity of the aquifer. The terms $h_0(t)$ and $\partial h/\partial x$ correspond to the known boundary conditions.

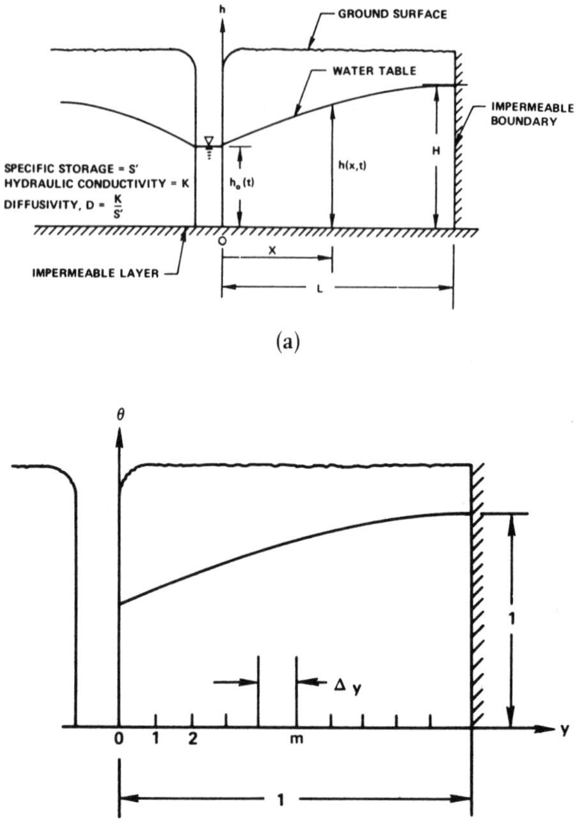

(a)

(b)

FIGURE 8.1 (a) System configuration; (b) Discretized configuration and normalization (From W. W-G. Yeh, "Aquifer Parameter Identification." *J. Hydraulic Div. Am. Soc. Civ. Eng.*, 101(HY9): 1975, p. 1198. Reprinted with permission from the American Society of Civil Engineers)

Introducing the variables,

$$\theta = \frac{h}{H}, \qquad y = \frac{x}{L}, \qquad \text{and} \qquad \tau = \frac{Ht}{L^2} \tag{8.13}$$

where H is the maximum height of the water table above the impermeable layer, a known constant, and L is the distance from the river to the water divide, also a known constant. Equation 8.11 may be expressed as

$$\frac{\partial \theta}{\partial \tau} = D \frac{\partial}{\partial y}\left[\theta \frac{\partial \theta}{\partial y}\right] \tag{8.14}$$

subject to

$$\theta = \frac{h(y)}{H}, \qquad 0 \leq y \leq 1, \qquad \tau = 0;$$

$$\theta = \frac{h_0(t)}{H}, \qquad y = 0, \qquad \tau > 0; \qquad \frac{\partial \theta}{\partial y} = 0, \qquad y = 1, \qquad \tau > 0 \tag{8.15}$$

The parameter chosen for identification is the diffusivity, D. It is assumed that observations on θ are available at an observation well within the system. The objective is to uncover this unknown parameter, based on some specified criterion function, along with these given observations and appropriate initial and boundary conditions. The space variable, y, of Equation 8.14 is discretized using the central difference scheme, while the time variable, τ, is being kept continuous. After simplification, the finite-difference model may be expressed as

$$\dot{\theta}_i = \frac{D}{2(\Delta y)^2}\{\theta_{i+1}^2 - 2\theta_i^2 + \theta_{i-1}^2\} \tag{8.16}$$

$$\theta = 1.0, \qquad 0 \leq y \leq 1, \qquad \tau = 0; \qquad \theta = 0.5, \qquad y = 0, \qquad \tau > 0;$$

$$\theta_n = \theta_{n-1}, \qquad y = 1, \qquad \tau > 0; \qquad \dot{\theta}_i = d\theta i/d\tau; \qquad i = 1,2, \ldots, (n-1)$$

The criterion function is the minimization of the following expression

$$S_0 = \sum_{j=1}^{T} [\theta_m(\tau_j) - \theta_m^*(\tau_j)]^2 \tag{8.17}$$

in which $\theta_m^*(\tau_j)$ are the observations made from time 0 to T at the mth discretized point corresponding to time τ_j, $j = 1, 2, \ldots T$.

In the parameter estimation model, the response equations are nonlinear functions of the unknown diffusivity and the hydraulic head distribution. The hydraulic equations can be linearized, using a generalized Taylor series, about initial estimates of the diffusivity, D^0, and the hydraulic head, h^0. The linearized equations may be expressed as

$$\dot{\theta}_i^1 = D^0 \frac{1}{2(\Delta y)^2}(\theta_{i+1}^{02} - 2\theta_i^{02} + \theta_{i-1}^{02}) + (\theta_i^1 - \theta_i^0)\left\{\frac{D^0}{2(\Delta y)^2}\right\}(-4\theta_i^0) + \quad (8.18)$$

$$(\theta_{i+1}^1 - \theta_{i+1}^0)\left\{\frac{D^0}{2(\Delta y)^2}\right\}(2\theta_{i+1}^0) + (\theta_{i-1}^1 - \theta_{i-1}^0)\left\{\frac{D^0}{2(\Delta y)^2}(2\theta_{i-1}^0)\right\}\delta +$$

$$(D^1 - D^0)\left\{\frac{1}{2(\Delta y)^2}(\theta_{i+1}^{02} - 2\theta_i^{02} - \theta_{i-1}^{02})\right\}$$

where $i = 1, 2, \ldots (n - 1)$, and $\delta = 0$ for $i = 1$ and $\delta = 1$ for $i \neq 1$. The superscript 1 represents the new approximation to the head and parameter values.

The general solution of the linearized equations can be obtained using the method of complementary functions. The solution is a linear combination of the particular solutions (p and q) of Equation 8.18, or

$$\theta_i^1 = D^1 p_i + q_i, \qquad i = 1, 2, \ldots, (n - 1) \quad (8.19a)$$

where

$$\dot{p}_i = p_i\left\{\frac{D^0}{2(\Delta y)^2}(-4\theta_i^0)\right\} + p_{i+1}\left\{\frac{D^0}{2(\Delta y)^2}(2\theta_{i+1}^0)\right\} +$$

$$(8.19b)$$

$$P_{i-1}\left\{\frac{D^0}{2(\Delta y)^2}(2\theta_{i-1}^0)\right\}\delta + \frac{1}{2(\Delta y)^2}(\theta_{i+1}^{02} - 2\theta_i^{02} + \theta_{i-1}^{02})$$

where $i = 1, 2, \ldots, (n - 1)$, $\delta = 0$ for $i = 1$, and $\delta = 1$ for $i \neq 1$. The initial and boundary conditions are given as

$$p = 0,\ 0 \leq y \leq 1,\ \tau = 0;\ p = 0,\ y = 0,\ \tau > 0;\ p_n = p_{n-1},\ y = 1, \tau > 0 \quad (8.19c)$$

The remaining particular solution is given from the solution of the linear differential equation

$$\dot{q}_i = (q_i - \theta_i^0)\left(\frac{D^0}{2(\Delta y)^2}\right)(-4\theta_i^0) +$$

$$(8.20a)$$

$$(q_{i+1} - \theta_{i+1}^0)\left(\frac{D^0}{2(\Delta y)^2}\right)(2\theta_{i+1}^0) + (q_{i-1} - \theta_{i-1}^0)\left\{\frac{D^0}{2(\Delta y)^2}\right\}(2\theta_{i-1}^0)\delta$$

where $i = 1, 2, \ldots, (n - 1)$, $\delta = 0$ for $i = 1$, and $\delta = 1$ for $i \neq 1$. The initial and boundary conditions for the solution are

$$q = \frac{h(y)}{H}, \; 0 \leq y \leq 1, \; \tau = 0;$$

$$q = \frac{h_0(t)}{H}, \; y = 0, \; \tau > 0; \quad q_n = q_{n-1}, \; y = 1, \; \tau > 0 \tag{8.20b}$$

The solution of the linearized model at the mth discretized (observation) point can then be expressed as

$$\theta_m^1 = D^1 p_m + q_m \tag{8.21}$$

The new estimate of the diffusivity, D^1, is obtained by minimizing the least squares objective,

$$S_1 = \sum_{j=1}^{T} [(D^1 p_m + q_m) - \theta_m^*(\tau_j)]^2 \tag{8.22}$$

The optimal parameter estimate is given from the first-order optimality conditions as

$$D^1 = \frac{\sum_{j=1}^{T} [\theta_m^*(\tau_j) p_m - q_m p_m]}{\sum_{j=1}^{T} p_m^2} \tag{8.23}$$

Equations 8.18 through 8.23 describe a complete cycle of the quasilinearization algorithm. Convergence of the algorithm occurs if at any iteration,

$$S_1 \leq \eta \tag{8.24}$$

where η is a small tolerance or there is no improvement in S_1.

Table 8.3 contains the generated observations for the fifth discretized point in the aquifer system, assuming a diffusivity value of 1. These observations were obtained by direct numerical integration of the Boussinesq equation with the given boundary and initial conditions using the Runge-Kutta method described in Section 4.4.2. The spatial and time increments for the numerical integration were $\Delta y = 0.1$ and $\Delta\tau = 0.002$. The results of applying the quasilinearization algorithm are shown in Table 8.4.

TABLE 8.3 Observed Values of Dimensionless Head at Fifth Discretized Point, D = 1.0

j (1)	τ_j (2)	$\theta_5^*(\tau_j)$ (3)
1	0	1.000
2	0.1	0.905
3	0.2	0.828
4	0.3	0.772
5	0.4	0.728
6	0.5	0.692
7	0.6	0.663
8	0.7	0.639
9	0.8	0.618
10	0.9	0.601
11	1.0	0.587

The generated observations and the corresponding parameters are referred to as the *true value*. The initial value of D was assumed to be 0.1; the value is actually 10 times smaller than the true value. The algorithm converged within five iterations.

Example Problem 8.2 Quadratic Programming (After Yeh, 1975b)

Develop a parameter identification model for the estimation of the storage coefficient and transmissivity of a confined aquifer, radial flow system. Formulate the optimization model using quadratic programming and quasilinearization. Determine the optimal parameter estimates from the data presented in Table 8.5.

The governing equation of the aquifer system, in radial coordinates, may be expressed as

$$\frac{1}{r}\frac{\partial}{\partial r}\left[rT(r)\frac{\partial h}{\partial r}\right] = S\frac{\partial h}{\partial t} + Q \tag{8.25a}$$

with the initial and boundary conditions,

$$t = 0, \qquad h = h^0, \qquad 0 \leqslant r < r_e$$

$$r = 0, \qquad h = h^*(t), \qquad t \geqslant 0 \tag{8.25b}$$

$$r = r_e, \qquad \frac{\partial h}{\partial r} = 0, \qquad t \geqslant 0 \tag{8.25c}$$

where r_e is the radius of influence and $h^*(t)$ is the observed head.

TABLE 8.4 Results for D of Successive Approximations by Quasilinearization

Zero	First	Second	Third	Fourth	True D
0.1	0.482	0.923	0.977	1.001	1.000

Introducing the dimensionless variables,

$$r_n = \frac{r}{r_e}, \qquad h_n = \frac{h}{h^0}, \qquad \tau = \frac{t}{r_e^2}$$

the radial flow equation may be expressed as

$$\frac{1}{r_n}\left[D(r_n)\frac{\partial}{\partial r_n}\left(r_n\frac{\partial h_n}{\partial r_n}\right) + r_n\frac{\partial h_n}{\partial r_n}\frac{\partial D(r_n)}{\partial r_n}\right] = \frac{\partial h_n}{\partial \tau} + V \tag{8.26a}$$

and,

$$\tau = 0, \qquad h_n = 1, \qquad 0 \leqslant r_n \leqslant 1 \tag{8.26b}$$

$$\begin{aligned} r_n = 0, &\qquad h_n = \frac{h(\tau)^*}{h^0}, \qquad \tau \geqslant 0 \\ r_n = 1, &\qquad \frac{\partial h_n}{\partial r_n} = 0, \qquad \tau \geqslant 0 \end{aligned} \tag{8.26c}$$

where $D = T/S$, $V = r_e^2Q/(h^0S)$, and h^0 is the initial head, a constant.

TABLE 8.5 Observed Values of the Dimensionless Head

			h_n				
r_{n_i}	D_i	$\tau =$	0.15	0.30	0.45	0.50	0.65
1	0.600		0.861	0.741	0.638	0.549	0.472
2	0.580		0.942	0.880	0.820	0.764	0.712
3	0.560		0.966	0.923	0.879	0.835	0.792
4	0.520		0.978	0.947	0.912	0.826	0.839
5	0.480		0.986	0.962	0.934	0.903	0.871
6	0.420		0.990	0.972	0.950	0.923	0.895
7	0.360		0.993	0.979	0.961	0.938	0.913
8	0.280		0.995	0.984	0.969	0.950	0.927
9	0.200		0.996	0.988	0.975	0.959	0.938
10	0.100		0.997	0.990	0.980	0.965	0.946
11	0.000		0.997	0.990	0.980	0.965	0.946

$V = 0.010$

The flow equation of the aquifer system is transformed using an implicit-central difference scheme to minimize truncation error and stability problems. The discretized equations may be expressed as

$$\left(\frac{D_i}{r_{n_i}}\right)\frac{1}{2(\Delta r_n)^2}[(r_{n_i}+r_{n_{i+1}})(h_{n_{i+1,j}}-h_{n_{i,j}})-(r_{n_i}+r_{n_{i-1,j}})(h_{n_{i,j}}-h_{n_{i-1,j}})]$$
$$+\frac{1}{4(\Delta r_n)^2}[(h_{n_{i+1,j}}-h_{n_{i-1,j}})(D_{i+1}-D_{i-1})]-V=\frac{h_{n_{i,j}}-h_{n_{i,j-1}}}{\Delta\tau} \tag{8.27}$$

where $i = 2, 3, \ldots, 10$ and $j = 2, 3, \ldots, t_t$. The transformed initial and boundary conditions are

$$h_{n_{i,1}} = 1, \qquad \forall i$$
$$h_{n10,j} = h_{n11,j}, \qquad \forall i \tag{8.28}$$
$$h_{n1,j} = \text{given observed dimensionless head}$$

where t_t is the total observed time, Δr_n is the spatial step size ($= 0.1$), and $\Delta\tau$ is the time step. It is assumed that,

$$h_{n1,j} = e^{-\tau} = e^{-(j-1)\Delta\tau}$$

The objective function of the model is described by the least squares objective

$$J = \sum_{j=2}^{t_t}\sum_{i=2}^{9}[h_{n_{i,j}} - h^*_{n_{i,j}}]^2 \tag{8.29}$$

where $h^*_{n_{i,j}}$ denotes the observed values of $h_{n_{i,j}}$.

The parameter estimation model is described by the finite-difference response equations (Eq. 8.27), the least squares objective (Eq. 8.29), and possible upper and lower bounds on the parameter values. The optimization model is a nonconvex programming problem.

The quasilinearization-quadratic programming algorithm is conceptually similar to the quasilinearization algorithm presented in Example Problem 8.1. The basic cycle begins at iteration 0. The required information for the algorithm includes a set of observations, $h^*_{n_{i,j}}$, and the initial estimates of the parameters D_i^0 and V^0. The hydraulic response equations are linearized about these initial estimates and the parameter estimation model is transformed into a quadratic programming (QP) problem. The solution of the QP problem generates new parameter estimates which are then the basis for a further linearization of the hydraulic response equations. The algorithm continues updating the parameter estimates until the least squares objective reaches an acceptable level or the differences in the least squares objective over successive iterations satisfy a given tolerance or error level. The algorithm is summarized by the following steps:

Step 1. Linearize the hydraulic response equations (Eq. 8.27) about the initial parameter estimates using a generalized Taylor series expansion. After retaining only the first-order terms, the equations may be expressed as

$$-\frac{h^0_{ni,j} - h^0_{ni,j-1}}{\Delta\tau} - V^1 + (D^1_{i+1} - D^1_{i-1})\left[\frac{1}{4(\Delta r_n)^2}(h^0_{ni+1,j} - h^0_{ni,j})\right] + \tag{8.30a}$$

$$D^1_i\left[\left(\frac{1}{r_{ni}}\right)\frac{1}{2(\Delta r_n)^2}\{(r_{ni} + r_{ni+1})(h^0_{i+1,j} - h^0_{ni,j}) - (r_{ni} + r_{ni-1})(h^0_{ni,j} - h_{ni-1,j})\}\right] +$$

$$(h^1_{ni,j} - h^0_{ni,j})\left\{\left(\frac{D^0_i}{r_{ni}}\right)\frac{1}{2(\Delta r_n)^2}[(r_{ni} + r_{ni+1}) + (r_{ni} + r_{ni-1})] - \frac{1}{\Delta\tau}\right\} +$$

$$(h^1_{ni+1,j} - h^0_{ni+1,j})\left[\left(\frac{D^0_i}{r_{ni}}\right)\frac{1}{2(\Delta r_n)^2}(r_{ni} + r_{ni+1}) + \frac{D^0_{i+1} - D^0_{i-1}}{4(\Delta r_n)^2}\right] +$$

$$(h^1_{ni-1,j} - h^0_{ni-1,j})\left[\frac{D^0_i}{r_{ni}}\frac{1}{2(\Delta r_n)^2}(r_{ni} + r_{ni-1}) + \frac{-(D^0_{i+1} - D^0_{i-1})}{4(\Delta r_n)^2}\right](1 - \delta(i,2)) +$$

$$(h^1_{ni,j-1} - h^0_{ni,j-1})\frac{1}{\Delta\tau}(1 - \delta(i,2)) = 0$$

where $i = 2, 3, \ldots, 10$; $j = 2, 3, \ldots, t_t$, and $\delta(k, l) = 1$ if $k = l$, and $\delta(k, l) = 0$ if $k \neq l$. The linearized response equations represent a system of $(t_t - 1) \times 9$ linear algebraic equations. The corresponding initial and boundary conditions are

$$\begin{aligned} h^1_{ni,1} &= 1, \quad \forall i \\ h^1_{n10,j} &= h^1_{n11,j}, \quad \forall j \\ h^1_{n1,j} &= e^{-(j-1)\Delta\tau} \end{aligned} \tag{8.30b}$$

Step 2. Formulate the quadratic programming model as

$$\min_{V^1, D^1_i} J = \sum_{j=2}^{t_t}\sum_{i=2}^{9}[h^1_{ni,j} - h^*_{ni,j}]^2 \tag{8.31}$$

subject to the linearized response equations (Eq. 8.30a), the boundary and initial conditions (Eq. 8.30b), and the parameter bounds

$$0 \leq V^1 \leq V_{max} \tag{8.32a}$$

$$D_{i_{min}} \leq D_i^1 \leq D_{i_{max}}, \qquad \forall i \tag{8.32b}$$

$$h_{ni,j}^1 \geq 0, \quad \forall i \text{ and } j \tag{8.32c}$$

where V_{max} is the upper bound of V and $D_{i_{max}}$ and $D_{i_{min}}$ are the upper and lower bounds of the diffusivity, D_i.

Equations 8.30, 8.31, and 8.32 are a standard quadratic programming problem. Solution algorithms for the model are presented in Chapter 5.

Step 3. Update the solutions of the original partial differential equations (Eq. 8.27) using the parameter estimates identified in Step 2 of the algorithm. Check the original least squares criterion (Eq. 8.29). If the error is less than the stopping criterion, the algorithm has terminated. Otherwise, go to Step 4.

Step 4. The algorithm returns to Step 1. The second iteration begins by solving the system of linearized response equations to obtain $h_{ni,j}^1$ using the updated parameter values, V^1 and D_i^1, from Step 2, along with V^0, D_i^0, and $h_{ni,j}^0$. The algorithm continues by replacing the 0 approximation with the update parameter estimates

TABLE 8.6 Results of Successive Approximations

	Zero	*First*	*Second*	*Third*	*Fourth*	*Fifth*	*True*
D_1	0.05	0.217	0.411	0.514	0.527	0.600	0.600
D_2	0.05	0.251	0.386	0.496	0.507	0.580	0.580
D_3	0.05	0.285	0.360	0.478	0.487	0.560	0.560
D_4	0.05	0.360	0.321	0.427	0.444	0.520	0.520
D_5	0.05	0.435	0.281	0.375	0.400	0.479	0.480
D_6	0.05	0.426	0.217	0.344	0.347	0.420	0.420
D_7	0.05	0.417	0.152	0.312	0.294	0.361	0.360
D_8	0.05	0.209	0.141	0.179	0.201	0.278	0.280
D_9	0.05	0.000	0.129	0.046	0.108	0.195	0.200
D_{10}	0.05	0.000	0.293	0.363	0.083	0.098	0.100
D_{11}	0.05	0.000	0.456	0.679	0.058	0.000	0.000
V	0.005	0.065	0.039	0.020	0.020	0.010	0.010
J	1.89×10^{-1}	3.21×10^{-2}	8.10×10^{-4}	8.73×10^{-3}	1.26×10^{-4}	5.91×10^{-7}	—

TABLE 8.7 Final Results of the Fifth Iteration

			h_n				
r_{n_i}	D_i	$\tau =$	0.15	0.30	0.45	0.50	0.65
1	0.600		0.861	0.741	0.638	0.549	0.472
2	0.580		0.942	0.880	0.820	0.764	0.712
3	0.560		0.966	0.924	0.879	0.835	0.792
4	0.520		0.978	0.947	0.912	0.876	0.834
5	0.479		0.986	0.962	0.934	0.903	0.871
6	0.420		0.990	0.972	0.950	0.923	0.895
7	0.361		0.991	0.979	0.961	0.938	0.913
8	0.278		0.995	0.984	0.969	0.950	0.927
9	0.195		0.996	0.988	0.976	0.959	0.939
10	0.096		0.997	0.990	0.980	0.965	0.947
11	0.000		0.997	0.990	0.960	0.965	0.947

until convergence is achieved or no further improvement can be made in the least squares criterion.

The generated observations for each of the discretized points in the system are shown in Table 8.5. Step sizes of $\Delta r_n = 0.1$ and $\Delta\tau = 0.15$ were used to generate the data. The results were obtained by directly solving Equation 8.27 with the boundary and initial conditions given by Equation 8.28.

The initial estimates of the parameters in the identification model are $D_i^0 = 0.05$ and $V^0 = 0.005$. Table 8.6 shows the results of successive approximations of the algorithm. The head distribution after five iterations is shown in Table 8.7. The algorithm terminated when the value of the least squares objective was reduced to 5.91×10^{-7}. Total computing time on an IBM 360/91 was approximately 25 seconds.

8.8 *GAUSS-NEWTON ALGORITHM FOR THE PARAMETER ESTIMATION PROBLEM*

The Gauss-Newton algorithm has been extensively used for the solution of equation error criterion, parameter estimation models (e.g., Jacquard & Jain, 1965; Jahns, 1966; Thomas, et al., 1972; Gavalas, et al., 1976; Yoon & Yeh, 1976; Cooley, 1977; and Cooley, 1982). The algorithm is attractive because it does not require the calculation of the Hessian matrix and the rate of convergence is faster than classical gradient searching procedures. The algorithm is basically for unconstrained minimization problems, but constraints, such as upper and lower bounds, can be incorporated in the algorithm with minor modifications. The algorithm starts with a set of initial estimates of parameters and converges to a local optimum. If the objective function is convex, the local optimum is the global optimum. However, because of noise in the observations, the inverse problem is usually nonconvex and, hence, only a local optimum can be assured in the minimization.

To illustrate the development of the algorithm, consider the problem of estimating the transmissivity distribution in a regional aquifer system. Define **T** as the vector of parameters that contains the nodal values of the transmissivity, $[T_1, T_2, \cdots, T_L]$. The algorithm generates the following parameter sequence for an unconstrained minimization problem,

$$\mathbf{T}^{k+1} = \mathbf{T}^k - \rho^k \mathbf{d}^k \tag{8.33}$$

with

$$A^k \mathbf{d}^k = \mathbf{g}^k \tag{8.34}$$

where

$$A^k = [J_D(\mathbf{T}^k)]^T \cdot [J_D(\mathbf{T}^k)], \qquad (L \times L)$$

$$\mathbf{g}^k = [J_D(\mathbf{T}^k)]^T \cdot [h_D(\mathbf{T}^k) - h_D^*], \qquad (L \times 1)$$

J_D = Jacobian matrix of head with respect to **T**, $(M \times L)$

ρ^k = step size (scalar)

$\mathbf{d}^k$ = Gauss-Newton direction vector, $(L \times 1)$

M = number of observations

L = parameter dimension

The step size ρ^k, a scalar, can be determined by a quadratic interpolation scheme such that $J(\mathbf{T}^{k+1}) < J(\mathbf{T}^k)$, or simply by a trial-and-error procedure. Occasionally, the direction matrix $[J_D^T J_D]$ may become ill-conditioned. Corrective methods introduced by Levenberg (1944) and Marquardt (1963) alter the Gauss-Newton direction vector and permit continued optimization of the parameter estimation problem.

The elements of the Jacobian matrix are represented by the sensitivity coefficients,

$$J_D = \begin{bmatrix} \dfrac{\partial h_1}{\partial T_1} & \dfrac{\partial h_1}{\partial T_2} & \cdots & \dfrac{\partial h_1}{\partial T_L} \\ \dfrac{\partial h_2}{\partial T_1} & \dfrac{\partial h_2}{\partial T_2} & \cdots & \dfrac{\partial h_2}{\partial T_L} \\ \cdot & & & \\ \cdot & & & \\ \cdot & & & \\ \dfrac{\partial h_M}{\partial T_1} & \dfrac{\partial h_M}{\partial T_2} & \cdots & \dfrac{\partial h_M}{\partial T_L} \end{bmatrix} \tag{8.35}$$

where M is the total number of observations and L is the total number of parameters. The transpose of the Jacobian matrix is

$$J_D^T = \begin{bmatrix} \dfrac{\partial h_1}{\partial T_1} & \dfrac{\partial h_2}{\partial T_1} & \cdots & \dfrac{\partial h_M}{\partial T_1} \\ \dfrac{\partial h_1}{\partial T_2} & \dfrac{\partial h_2}{\partial T_2} & \cdots & \dfrac{\partial h_M}{\partial T_2} \\ \cdot & & & \\ \cdot & & & \\ \cdot & & & \\ \dfrac{\partial h_1}{\partial T_L} & \dfrac{\partial h_2}{\partial T_L} & \cdots & \dfrac{{}^{\tau}h_M}{\partial T_L} \end{bmatrix} \tag{8.36}$$

The evaluation of the sensitivity coefficients will be discussed in Section 8.9.

Example Problem 8.3 Two-Dimensional Parameter Identification Model

Develop a parameter estimation model for the confined groundwater system shown in Figure 8.2. The boundary of the groundwater system is impermeable with the exception of the upper section of the aquifer. In this region, the boundary head varies from 120 feet at node (0,1) to 170 feet at node (0,10). Represent the trans-

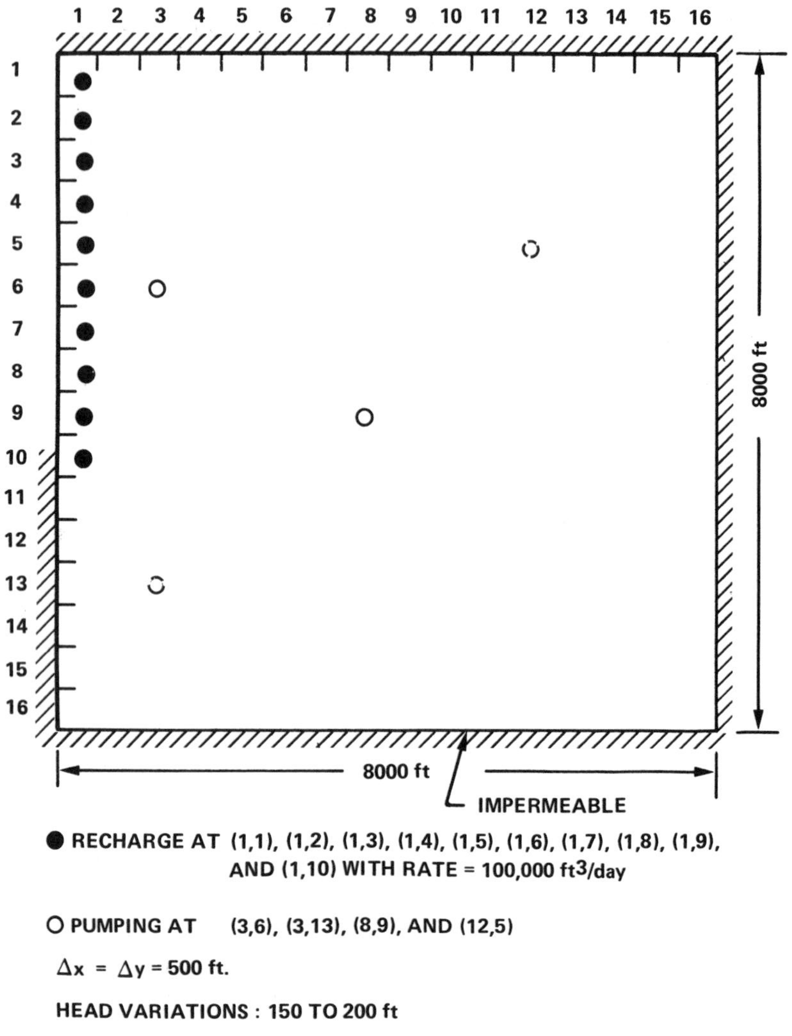

FIGURE 8.2 Aquifer configuration, case 1 (From W. W-G. Yeh and Y. S. Yoon "Aquifer Parameter Identification with Optimum Dimension in Parameterization." *Water Resources Research*, 17(3): 1981, p. 665. Copyright © 1981 by the American Goephysical Union)

missivity distribution using the finite-element interpolation method and evaluate the gradients of the Gauss-Newton algorithm.

The governing equation of the aquifer system is described by Equation 8.1. The Galerkin finite-element transformation of the confined flow equation generates the dynamic response equations, or

$$A\dot{\mathbf{h}} + B(T)\mathbf{h} + \mathbf{g} = 0 \tag{8.37}$$

where A is a diagonal matrix containing the storage coefficient, $B(T)$ is a positive-definite banded symmetric matrix of order n (the number of nodes used in the discretization of the spatial variables) and $\mathbf{g}$ contains the boundary condition data and source and sinks.

The Crank-Nicolson method is used to approximate the head distribution in time. The resulting set of algebraic equations may be expressed as

$$A\left[\frac{\mathbf{h}^{k+1} - \mathbf{h}^{k}}{\Delta t}\right] + B(T)\left[\frac{\mathbf{h}^{k+1} + \mathbf{h}^{k}}{2}\right] + \mathbf{g}^{k+1/2} = 0 \qquad (8.38)$$

where Δt = time increment, k is an index of the time step, $\mathbf{h}^k = \mathbf{h}(k \cdot \Delta t)$ and $\mathbf{g}^{k+1/2} = \mathbf{g}[(k + 1/2) \cdot \Delta t]$.

The equations may be solved by line or block-successive relaxation methods.

The gradients of the head with respect to the parameters can be obtained by differentiating Equation 8.38. These gradients, which are required by the Gauss-Newton algorithm, may be expressed as

$$\left[I + A^{-1}\frac{\Delta t}{2} B(T)\right]\mathbf{h}_{T_i}^{k+1} = \qquad (8.39)$$

$$\left[I - A^{-1}\frac{\Delta t}{2} B(T)\right]\mathbf{h}_{T_i}^{k} - A^{-1}\frac{\Delta t}{2} B(T_{T_i})(\mathbf{h}^k + \mathbf{h}^{k+1}) - A^{-1}\Delta t \mathbf{g}_{T_i}^{k+1/2},$$

$$i = 1, \ldots L$$

where $\mathbf{h}_{T_i} = \dfrac{\partial h}{\partial T_i}$, $T_{T_i} = \dfrac{\partial T}{\partial T_i}$, and $\mathbf{g}_{T_i} = \dfrac{\partial g}{\partial T_i}$.

The transmissivity distribution of the aquifer system is represented by the finite-element, interpolation method as

$$T(x, y) = \sum_i T_i N_i(x, y) \qquad (8.40)$$

where N_i are the finite-element basis functions and T_i are the nodal values of the transmissivity. In this example, bilinear basis functions on quadrilateral elements are used to represent the transmissivity field.

To ensure that the parameters remain physically plausible during the optimization process, upper and lower bounds on the transmissivities are also incorporated in the identification model.

$$T_{i,\min} \leq T_i \leq T_{i,\max} \qquad (8.41)$$

where $T_{i,\min}$ and $T_{i,\max}$ represent the lower and upper bounds which may be obtained from geologic information of the aquifer system.

The least squares objective function of the estimation model is given by Equation 8.10.

The optimization model is described by Equations 8.10, 8.38, 8.40, and 8.41. Example Problem 8.6 explores how the problem may be solved with the Gauss-Newton algorithm and investigates the effects of parameter dimensioning on the reliability of the optimal parameter estimates.

8.9 COMPUTATION OF SENSITIVITY COEFFICIENTS

Sensitivity coefficients, the partial derivatives of head with respect to each of the parameters, play an important role in the solution of the inverse problem. In the Gauss-Newton algorithm, elements of the Jacobian matrix are represented by the sensitivity coefficients, $\partial h_i/\partial T_l$, $i = 1,\ldots, M; l = 1,\ldots,L$. The three most common methods for evaluating the sensitivity coefficients for confined (linear) groundwater systems are presented below.

8.9.1 *Influence Coefficient Method*

The influence coefficient method (Becker & Yeh, 1972) uses the concept of parameter perturbation. The lth row of J_D^T is approximated by

$$\frac{\partial h_i}{\partial T_l} \approx \frac{h_i(T + \Delta T_l \cdot e_l) - h_i(T)}{\Delta T_l}, \qquad i = 1, \ldots, M \tag{8.42}$$

where

ΔT_l = small increment of T_l

e_l = the lth unit vector

The values of $h(\mathbf{T})$ and $h(\mathbf{T} + \Delta T_l \cdot e_l)$ are obtained by solving the governing equation (by simulation), subject to the imposed initial and boundary conditions. The method requires perturbing each parameter one at a time. If there are L parameters to be identified, the confined flow equation has to be solved (simulated) $(L + 1)$ times for each iteration of the nonlinear least squares minimization. The numerical representation of J_D^T is called the *influence coefficient matrix* (Becker & Yeh, 1972). The elements of the influence coefficient matrix, represented by a_{il}, are numerical approximations of the sensitivity coefficients

$$
\begin{array}{c|cccc}
 & h_1 & h_2 & \cdots\cdots\cdots & h_M \\
\hline
T_1 & a_{11} & a_{12} & \cdots\cdots\cdots & a_{1M} \\
T_2 & a_{21} & a_{22} & \cdots\cdots\cdots & a_{2M} \\
 & & & & \\
T_L & a_{L1} & a_{L2} & \cdots\cdots\cdots & a_{LM}
\end{array}
\tag{8.43}
$$

Each element in the matrix represents the ratio of change in the head to the change in a particular parameter. The value of ΔT_l is a small increment of T_l by which parameter T_l is perturbed. The appropriate value of ΔT_l is usually determined on a trial-and-error basis. Bard (1974) has suggested some guidelines in choosing the value of ΔT_l.

8.9.2 Sensitivity Equation Method

In this approach, a set of sensitivity equations are obtained by taking the partial derivatives with respect to each parameter in the flow equation and initial and boundary conditions. After taking the partial derivatives of the two-dimensional, confined flow equation (Eq. 8.1), the set of sensitivity equations can be expressed as

$$
\frac{\partial}{\partial x}\left[T\frac{\partial\left(\dfrac{\partial h}{\partial T_l}\right)}{\partial x}\right] + \frac{\partial}{\partial y}\left[T\frac{\partial\left(\dfrac{\partial h}{\partial T_l}\right)}{\partial y}\right] = S\frac{\partial\left(\dfrac{\partial h}{\partial T_l}\right)}{\partial t} \tag{8.44}
$$

$$
+\left[-\frac{\partial}{\partial x}\left(\frac{\partial T}{\partial T_l}\frac{\partial h}{\partial x}\right) - \frac{\partial}{\partial y}\left(\frac{\partial T}{\partial T_l}\frac{\partial h}{\partial y}\right)\right], \qquad l = 1, \ldots, L \tag{8.45}
$$

The associated initial and boundary conditions are

$$
\frac{\partial h(x, y, 0)}{\partial T_l} = 0, \qquad l = 1, \ldots, L \tag{8.46}
$$

$$
\frac{\partial h(x, y, t)}{\partial T_l} = 0, \qquad l = 1, \ldots, L \tag{8.47}
$$

$$T\frac{\partial\left(\dfrac{\partial h}{\partial T_l}\right)}{\partial n} = -\frac{\partial T}{\partial T_l}\frac{\partial h}{\partial n}, \qquad l = 1, \ldots L \tag{8.48}$$

The numerical values of $\partial h/\partial x$ and $\partial h/\partial y$ are obtained from the solution of the governing equation. Replacing $(\partial h/\partial T_l)$ by h and defining

$$Q = \left[-\frac{\partial}{\partial x}\left(\frac{\partial T}{\partial T_l}\frac{\partial h}{\partial x}\right) - \frac{\partial}{\partial y}\left[\frac{\partial T}{\partial T_l}\frac{\partial h}{\partial y}\right)\right] \tag{8.49}$$

the set of sensitivity equations have the same form as the confined or semiconfined flow equation. As a result, algorithms that are developed for the solution of the governing equation can be used to evaluate the set of sensitivity equations. The number of simulation runs required to generate the sensitivity coefficients per iteration is $(L + 1)$, which is the same as that of the influence coefficient method.

8.9.3 Variational Method

The variational method was first used for solving the inverse problem of parameter identification by Jacquard and Jain (1965) and Carter, et al., (1974, 1982). The models are associated with finite-difference schemes. Sun and Yeh (1985) extended the method to the case of a finite-element scheme. Following Carter, et al., (1974), the sensitivity coefficients can be computed by the equation

$$\frac{\partial h^{(j)}}{\partial T_l^{(i)}} = \iint_{(\Gamma_i)} \int_0^t \nabla q'(x, y, t - \tau)\nabla h(x, y, \tau)\, d\tau\, dx\, dy \tag{8.50}$$

$$j = 1, 2, \ldots, N_0; \qquad i = 1, 2, \ldots, N_n$$

where (Γ_i) is the exclusive subdomain of node i as defined by Sun and Yeh (1985); ∇ is the gradient operator; $h(x, y, t)$ is the solution of the governing equation; N_0 is the number of observation wells; N_n is the total number of nodes used in the numerical solution; $q'(x, y, t)$ is the time derivative of $q(x, y, t)$ which is the solution of the following set of adjoint equations

$$\frac{\partial}{\partial x}\left[T\frac{\partial q}{\partial x}\right] + \frac{\partial}{\partial y}\left[T\frac{\partial q}{\partial y}\right] = S\frac{\partial q}{\partial t} + G_j(x, y,)H(t) \tag{8.51}$$

subject to the following initial and boundary conditions

$$q(x, y, 0) = 0, \qquad (x, y) \in \Gamma_0 \tag{8.52}$$

$$q(x, y, t) = 0, \qquad (x, y) \in \Gamma_1 \tag{8.53}$$

$$\frac{\partial q}{\partial n}(x, y, t) = 0, \qquad (x, y) \in \Gamma_2 \tag{8.54}$$

where

$$G_j(x, y) = \frac{1}{P_j}, \qquad (x, y) \in \Gamma_j$$

$$G_j(x, y) = 0, \qquad \text{otherwise}$$

$$H(t) = 0, \qquad t \leq 0$$

$$H(t) = 1, \qquad t > 0$$

Γ_0 is the flow region, Γ is the boundary of the aquifer ($\Gamma_1 \cup \Gamma_2 = \Gamma$), and P_j is the area of subdomain (Γ_j).

Note that the adjoint equation for $q(x, y, t)$ (Eq. 8.51) has the same form as the confined flow equation (Eq. 8.1) and the same numerical scheme can be used to solve h and q. All sensitivity coefficients, $\partial h^{(i)}/\partial T_l^{(i)}$ ($j = 1, 2, \ldots, N_0$; $i = 1, 2, \ldots, N_n$), can be produced by solving the governing equation once and the adjoint equations for each observation well. The number of simulation runs required to calculate the sensitivity coefficients per iteration is ($N_0 + 1$) as compared to ($L + 1$), which is required by either the influence coefficient method or the sensitivity equation method.

The computational efficiency of the three methods depends on the number of unknown parameters. The variational method is preferable if $L > N_0$, the number of parameters exceeds the number of observation wells. On the other hand, if $N_0 > L$, the influence coefficient and sensitivity equation methods are preferable. To avoid instability when data contain noise, the number of parameters to be identified is usually less than the number of observation wells. In using the sensitivity equation method, caution must be exercised in that $\partial h_i/\partial T_l$ varies much more rapidly with time than h. DiStefano and Rath (1975) pointed out that, in order to obtain a set of sensitivity coefficients with acceptable accuracy, much smaller time steps are required in the simulation runs. In the influence coefficient method, the perturbation vector $\mathbf{\Delta T}$ can be appropriately chosen to cause sufficient change in $h(\mathbf{T}, t)$ and yet small enough so that numerical approximations of sensitivity coefficients are valid. However, sensitivity equation and variational methods are intrinsically much more accurate. The need for an efficient method for calculating the sensitivity coefficients has been pointed out by Dogru and Seinfeld (1981), McElwee (1982), and Sykes, et al., (1985).

The following example problems illustrate the calculation of the sensitivity coefficients and the evaluation of the numerical accuracy of the three approaches.

Example Problem 8.4 Sensitivity Coefficient Determination (After Li, et al., 1985)

Evaluate the accuracy of the three methods for calculating the sensitivity coefficients for the aquifer system shown in Figure 8.3 (Li, et al., 1985). Assume that the confined aquifer is homogeneous and isotropic. The dimension of the aquifer is 1400 m by 1400 m, and it is surrounded by impervious boundaries *AB* and *CD* and constant head boundaries *BC* and *DA*, where h = 100m. The transmissivity and the storage coefficient of the aquifer are 100 m/day^2 and 0.001, respectively. The aquifer is initially at a steady-state condition with piezometric head equal to 100 m throughout the aquifer. A pumping well extracting 10000 m^3/day is located at the center of the square aquifer, node 104. Observation wells are located at nodes 14, 28, 42, 56, 70, 84, 93, and 95.

The governing equation of the aquifer system is described by Equation 8.1. Because of the simple geometry and boundary conditions of the aquifer system,

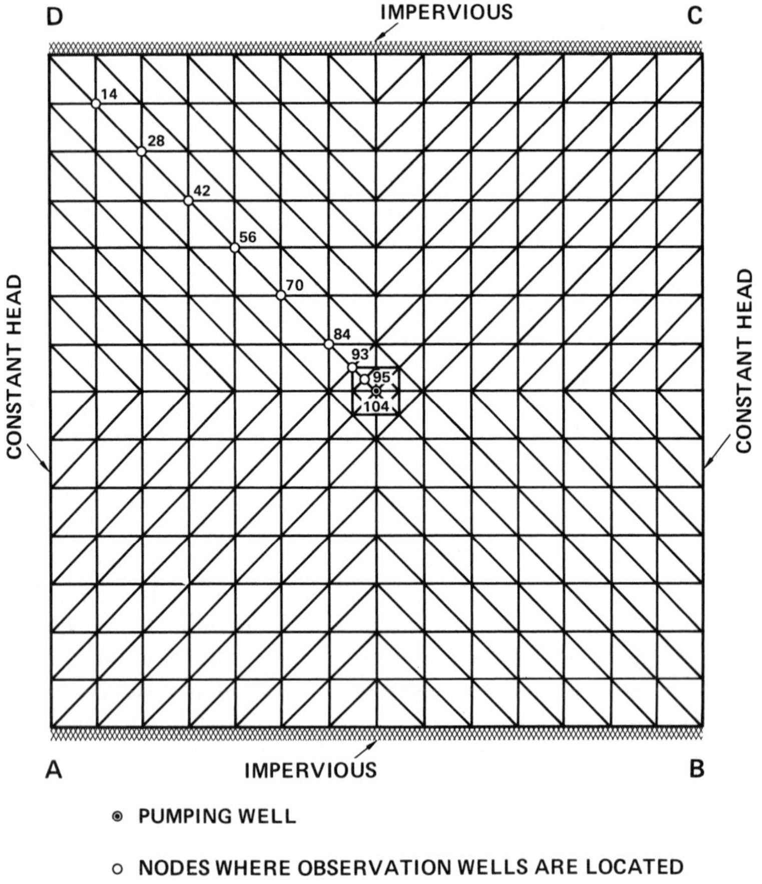

FIGURE 8.3 Aquifer configuration and finite-element discretization

the head distribution can be represented by the analytical solution (Chan, Mullineux, & Reed, 1976).

$$h(x,y,t) = 100 - s(x,y,\infty) + \frac{2Q}{a^2T} \sum_{m=1}^{\infty} \frac{\exp(-T\,\alpha_m^2 t/S)}{\alpha_m^2} \sigma(\alpha_m,x,\xi)$$
$$+ \frac{4Q}{a^2T} \sum_{m=1}^{\infty} \sum_{n=1}^{\infty} \frac{\exp(-T\,r_{m,n}^2 t/S)}{r_{m,n}^2} \sigma(\alpha_m,x,\xi)\, c(\beta_n,y,\eta) \tag{8.55}$$

where the steady-state solution $s(x,y,\infty)$ is given by

$$s(x,y,\infty) = \frac{Q}{aT} \sum_{m=1}^{\infty} \frac{\sigma(\alpha_m,x,\xi)}{\alpha_m \sinh(\alpha_m b)} \{\cosh\,[\alpha_m(a-|\eta - y|)]$$
$$+ \cosh\,[\alpha_m(a-\eta + y)]\} \tag{8.56}$$

where

$$s = \text{drawdown}$$
$$S = \text{storage coefficient}$$
$$T = \text{transmissivity}$$
$$Q = \text{rate of pumping well}$$
$$t = \text{time}$$
$$x,y = \text{Cartesian coordinates}$$
$$\xi,\eta = \text{coordinates of pumping well}$$
$$a = \text{dimension of the aquifer}$$
$$m,n = \text{integers}$$
$$\alpha_m = m\pi/a$$
$$\beta_n = n\pi/a$$
$$r_{m,n}^2 = \alpha_m^2 + \beta_n^2$$
$$\sigma(\alpha_m,x,\xi) = \sin(\alpha_m x)\sin(\alpha_m \xi)$$
$$c(\beta_n,y,\eta) = \cos(\beta_n y)\cos(\beta_n \eta).$$

The numerical solution of the confined flow equation in this example problem is obtained using the Galerkin finite-element method described in Chapter 4. The finite-element grid is shown in Figure 8.3. The flow region has been divided into 416 triangular elements; there are a total of 237 nodes.

Initially, the accuracy of the numerical model can be assessed by comparing the piezometric heads from the analytical solution with the numerical solution for 1.0, 2.0, 3.0, 5.0, 10.0, and ∞ (steady-state) days after pumping.

The piezometric heads, from the analytical expressions (Eqs. 8.55 and 8.56) are given in Table 8.8. The heads, from the numerical solution of Equation 8.1, are shown in Tables 8.9 and 8.10. The solutions of the unsteady-state flow regime are obtained with time step sizes of 0.2 and 1.0 days, respectively. Obviously, the values of the heads calculated with smaller time step sizes of 0.2 days are closer to the analytical solutions than those with larger time step sizes of 1.0 days. The numerical solutions compare favorably with analytical solutions, and the finite-element discretization results in acceptable truncation error.

The sensitivity coefficient matrix for the example problem can be reduced to

$$[J] = \left[\frac{\partial h_1}{\partial T}, \frac{\partial h_2}{\partial T}, \frac{\partial h_3}{\partial T}, \frac{\partial h_4}{\partial T}, \frac{\partial h_5}{\partial T}, \frac{\partial h_6}{\partial T}, \frac{\partial h_7}{\partial T}, \frac{\partial h_8}{\partial T}, \ldots, \frac{\partial h_L}{\partial T}\right]^T$$

where $\dfrac{\partial h_1}{\partial T} = \dfrac{\partial h^{14}}{\partial T}, \dfrac{\partial h_2}{\partial T} = \dfrac{\partial h^{28}}{\partial T}, \ldots, \dfrac{\partial h_8}{\partial T} = \dfrac{\partial h^{95}}{\partial T},$

and the superscripts denote the node numbers.

The sensitivity coefficients can be obtained by differentiating Equations 8.55 and 8.56. The coefficients for transient flow can be expressed as

$$\frac{\partial h}{\partial T}(x,y,t) =$$

$$-\frac{\partial s}{\partial T}(x,y,\infty) + \frac{2Q}{a^2T}\sum_{m=1}^{\infty}\left\{\left(-\frac{1}{T} - \alpha_m^2 t/S\right)\left[\frac{\exp(-T\alpha_n^2 t/S)}{\alpha_m^2}\sigma(\alpha_m, x, \xi)\right]\right\}$$

$$+ \frac{4Q}{a^2T}\sum_{m=1}^{\infty}\sum_{n=1}^{\infty}\left\{\left(-\frac{1}{T} - r_{m,n}^2 t/S\right)\left[\frac{\exp(-Tr_{m,n}^2 t/S)}{r_{m,n}^2}\sigma(\alpha_m, x, \xi)c(\beta_n, y, \eta)\right]\right\} \quad (8.57)$$

TABLE 8.8 Heads Calculated by Analytical Solution

Time After Pumping Day	*Nodal Number*							
	14	*18*	*42*	*56*	*70*	*84*	*93*	*95*
1	99.565	98.863	97.538	95.027	90.271	80.362	69.626	58.562
2	98.620	96.944	94.620	91.145	85.562	75.088	64.202	53.100
3	97.979	95.687	92.802	88.847	82.897	72.190	61.245	50.128
5	97.348	94.456	91.032	86.626	80.336	69.418	58.419	47.289
10	97.013	93.804	90.095	85.451	78.983	67.953	56.926	45.788
∞	96.984	93.746	90.013	85.349	78.864	67.825	56.795	45.657

TABLE 8.9 Heads Calculated by Finite-Element Method with Time Step of 0.2 Days

Time After Pumping Day	Nodal Number							
	14	28	42	56	70	84	93	95
1	99.538	98.853	97.627	95.316	90.844	80.996	71.239	60.149
2	98.681	97.080	94.847	91.477	86.016	75.430	65.502	54.371
3	98.049	95.832	93.023	89.147	83.286	72.439	62.454	51.309
5	97.396	94.555	91.182	86.832	80.610	69.536	59.502	48.345
10	96.993	93.768	90.051	85.413	78.974	67.762	57.700	46.536
∞	96.984	93.746	90.013	85.349	78.864	67.825	56.795	45.657

The steady-state, sensitivity coefficients are

$$\frac{\partial h}{\partial T}(x,y,\infty) = \frac{Q}{aT^2}\sum_{m=1}^{\infty}\frac{\sigma(\alpha_m,x,\xi)}{\alpha_m \sinh(\alpha_m a)}\{\cosh[\alpha_m(a-|\eta-y|)] + \cosh[\alpha_m(a-\eta+y)]\} \tag{8.58}$$

The sensitivity coefficients for the example problem are given in Table 8.11. These values can be considered as the exact solutions.

In the influence coefficient method, a forward finite-difference scheme is used to evaluate the sensitivity coefficients. Table 8.12 summarizes these sensitivity coefficients. The time step in the calculations is 0.2 days. An appropriate perturbation vector obtained experimentally is 1.0 m^2/day.

The sensitivity coefficients obtained from the sensitivity equation method are shown in Table 8.13. The time step is again 0.2 days.

TABLE 8.10 Heads Calculated by Finite-Element Method with Time Step of 1.0 Day

Time After Pumping Day	Nodal Number							
	14	28	42	56	70	84	93	95
1	99.495	98.833	97.785	95.933	92.325	83.762	74.570	63.701
2	98.835	97.431	95.471	92.458	87.401	77.176	67.344	56.238
3	98.272	96.286	93.718	90.082	84.435	73.742	63.793	52.657
5	97.573	94.901	91.685	87.468	81.349	70.340	60.320	49.168
10	97.069	93.917	90.265	85.681	79.284	68.098	58.041	46.879
∞	96.984	93.746	90.013	85.349	78.864	67.825	56.795	45.657

TABLE 8.11 Sensitivity Coefficients Calculated by Analytical Solution

Time After Pumping Day	*Nodal Number*							
	14	*28*	*42*	*56*	*70*	*84*	*93*	*95*
1	−0.006	−0.010	−0.011	−0.001	0.032	0.121	0.225	0.335
2	−0.002	−0.001	0.008	0.030	0.076	0.174	0.281	0.392
3	0.005	0.014	0.030	0.059	0.110	0.212	0.320	0.431
5	0.017	0.038	0.064	0.102	0.160	0.266	0.375	0.486
10	0.028	0.059	0.095	0.140	0.204	0.314	0.424	0.535
∞	0.030	0.063	0.100	0.147	0.211	0.322	0.432	0.543

Table 8.14 lists the sensitivity coefficients calculated by the variational method. The same time step sizes of 0.2 days are used in the computations of h and q. Since the aquifer system is homogenous, $\partial h_1/\partial T = \Sigma_{i=1}^{M} \partial h/\partial T_i$, where M is the total number of nodes.

Clearly, the results from the sensitivity equation and the variational methods are the same and very close to the analytical solutions. Although the results from the influence coefficient method are not as good as those from the other two methods, the accuracy of the influence coefficient method can be greatly improved if an optimal perturbation vector can be chosen. Therefore, in theory, all three methods should produce similar and acceptable accuracies.

The sensitivity coefficients from the influence coefficient method are listed in Table 8.15. It is evident that the results differ from the analytical solutions so that, in using the influence coefficient method, it is very important to choose the appropriate perturbation vector. It is relatively easy to determine the vector experimentally because accurate sensitivity coefficients can be calculated through the

TABLE 8.12 Sensitivity Coefficients Calculated by Influence Coefficient Method

Time After Pumping Day	*Nodal Number*							
	14	*28*	*42*	*56*	*70*	*84*	*93*	*95*
1	−0.004	−0.007	−0.007	−0.000	0.029	0.115	0.208	0.317
2	−0.002	−0.000	0.008	0.029	0.073	0.170	0.267	0.377
3	0.005	0.014	0.029	0.056	0.106	0.208	0.306	0.416
5	0.017	0.036	0.062	0.098	0.154	0.261	0.359	0.469
∞	0.030	0.062	0.099	0.145	0.209	0.320	0.418	0.530

Time step size (Δt) of 0.2 days was used.

TABLE 8.13 Sensitivity Coefficients Calculated by Sensitivity Equation Method (Δt = 0.2 days)

Time After Pumping Day	*Nodal Number*							
	14	28	42	56	70	84	93	95
1	−0.004	−0.007	−0.008	0.000	0.030	0.116	0.210	0.321
2	−0.002	−0.000	0.008	0.029	0.073	0.172	0.270	0.381
3	0.005	0.013	0.029	0.057	0.109	0.211	0.309	0.420
5	0.017	0.036	0.062	0.099	0.156	0.264	0.363	0.475
∞	0.030	0.062	0.100	0.146	0.210	0.322	0.423	0.535

TABLE 8.14 Sensitivity Coefficients Calculated by Variational Method (Δt = 0.2 days)

Time After Pumping Day	*Nodal Number*							
	14	28	42	56	70	84	93	95
1	−0.004	−0.007	−0.008	0.000	0.030	0.116	0.210	0.321
2	−0.002	−0.000	0.008	0.029	0.073	0.172	0.256	0.366
3	0.005	0.013	0.029	0.057	0.107	0.211	0.309	0.420
5	0.017	0.036	0.062	0.099	0.156	0.264	0.363	0.475
∞	0.030	0.062	0.100	0.146	0.210	0.322	0.423	0.535

TABLE 8.15 Sensitivity Coefficients Calculated by Influence Coefficient Method

Time After Pumping Day	*Nodal Number*							
	14	28	42	56	70	84	93	95
1	−0.006	−0.013	0.018	0.030	0.061	0.152	0.249	0.360
2	0.013	0.028	0.047	0.074	0.123	0.227	0.330	0.439
3	0.023	0.047	0.074	0.111	0.166	0.275	0.378	0.490
5	0.037	0.074	0.114	0.161	0.224	0.339	0.441	0.553
∞	0.058	0.115	0.171	0.232	0.307	0.425	0.529	0.641

Time step of 0.2 days and transmissivity increment of 0.1 m^2/day were used.

TABLE 8.16 Sensitivity Coefficients Calculated by Influence Coefficient Method

Time After Pumping Day	*Nodal Number*							
	14	*28*	*42*	*56*	*70*	*84*	*93*	*95*
1	−0.001	−0.002	−0.001	0.006	0.027	0.096	0.180	0.285
2	0.000	0.003	0.008	0.028	0.066	0.158	0.253	0.253
3	0.005	0.013	0.026	0.051	0.099	0.198	0.295	0.405
5	0.015	0.032	0.055	0.090	0.145	0.251	0.349	0.459
10	0.026	0.055	0.089	0.133	0.195	0.305	0.404	0.514
∞	0.030	0.062	0.099	0.145	0.209	0.320	0.419	0.530

Time step of 1.0 day and transmissivity increment of 1.0 m^2/day were used.

analytical solution. In field applications, however, the optimum perturbation vector is difficult to select and, as a result, the influence coefficient method may have a lower accuracy than the other two methods.

The sensitivity coefficients in Tables 8.16 through 8.18 are calculated by the influence coefficient method, the sensitivity equation method, and the variational method, respectively, with a 1.0 day time step. It can be seen that the results during the earlier pumping period differ from the analytical solutions, but during the later pumping period the results are close to the analytical solutions. Hence, the time step sizes have an effect on the accuracies of the sensitivity coefficients.

The accurate calculation of the sensitivity coefficients also depends on the accurate calculation of the piezometric heads. The factors that affect the calculation of the piezometric heads will also have an effect on the evaluation of the sensitivity coefficients. Numerical experiments have shown that the principal factors affecting

TABLE 8.17 Sensitivity Coefficients Calculated by Sensitivity Equation Method

Time After Pumping Day	*Nodal Number*							
	14	*28*	*42*	*56*	*70*	*84*	*93*	*95*
1	−0.002	−0.002	−0.001	0.006	0.028	0.097	0.182	0.288
2	0.000	0.003	0.010	0.028	0.067	0.160	0.256	0.366
3	0.005	0.012	0.026	0.052	0.100	0.200	0.298	0.410
5	0.015	0.032	0.056	0.091	0.147	0.253	0.353	0.464
10	0.027	0.056	0.090	0.134	0.197	0.308	0.408	0.520
∞	0.030	0.062	0.100	0.146	0.210	0.322	0.423	0.534

Time step of 1.0 day was used.

TABLE 8.18 Sensitivity Coefficients Calculated by Variational Method

Time After Pumping Day	Nodal Number							
	14	28	42	56	70	84	93	95
1	−0.002	−0.002	−0.001	0.006	0.028	0.097	0.182	0.288
2	0.000	0.003	0.010	0.028	0.067	0.160	0.256	0.366
3	0.005	0.012	0.026	0.052	0.100	0.200	0.298	0.410
5	0.015	0.032	0.056	0.091	0.147	0.253	0.353	0.464
10	0.027	0.056	0.090	0.134	0.197	0.308	0.408	0.520
∞	0.030	0.062	0.100	0.146	0.210	0.322	0.423	0.535

Time step of 1.0 day was used.

the accuracy of the sensitivity coefficients in all three methods are the finite-element or finite-difference discretization and the time step associated with the transient simulations.

Example Problem 8.5 Sensitivity Coefficient Evaluation (After Li, et al., 1985)

Compare the accuracy of the sensitivity equation and variational methods for the inhomogeneous, confined aquifer system shown in Figure 8.4 (Li, et al., 1985). The flow region is divided into two subregions, I and II, and each subregion has a constant transmissivity. The transmissivities of subregions I and II are 500 m^2/day and 2000 m^2/day, respectively. The storage coefficient of the aquifer is 0.0001. The south boundary (ABC) is a constant head boundary with a head of 100 m. The rest of the boundary is impermeable. Initially, the aquifer is at a steady-state condition with a constant head of 100 m throughout the aquifer system.

Equation 8.1 describes the groundwater hydraulics. Finite elements are to approximate the solution of the groundwater model. The discretized aquifer has 59 nodes and 92 elements. There is a pumping well at node 48 which extracts 10,000 m^3/day. Two observation wells, No. 1 and No. 2, are located at nodes 20 and 52, respectively, in subregions I and II.

Table 8.19 lists the sensitivity coefficients calculated by the sensitivity equation method, with a time step of 0.5 days. Using the same time step, the results calculated by the variational method are given in Table 8.20. Both methods give essentially the same results.

The sensitivity equation method and the variational method were also used again to calculate the sensitivity coefficients. The results, for a time step of 1.0 day, are listed in Tables 8.21 and 8.22, respectively. Again the results are consistent.

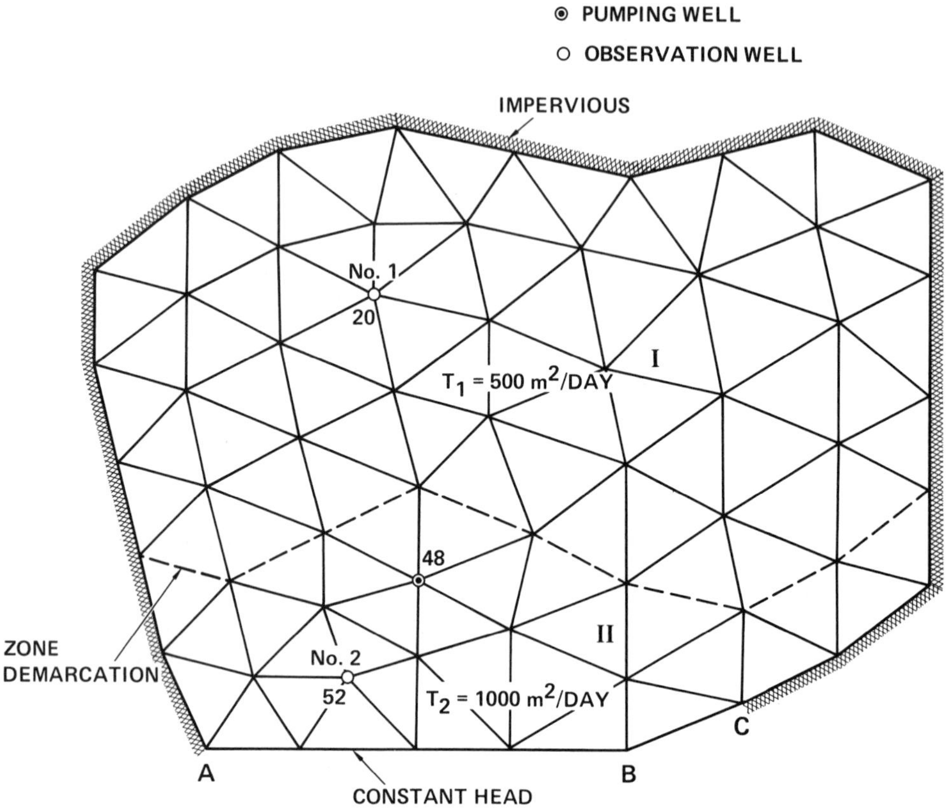

FIGURE 8.4 Example aquifer

TABLE 8.19 Sensitivity Coefficients by Sensitivity Equation Method with Time Step of 0.5 Days

Time After Pumping Day	*Number 1*		*Number 2*	
	$\frac{\partial h_1}{\partial T_1}$	$\frac{\partial h_1}{\partial T_2}$	$\frac{\partial h_2}{\partial T_1}$	$\frac{\partial h_2}{\partial T_2}$
0.5	−0.000391	0.001028	0.000191	0.000993
1.0	−0.000106	0.001729	0.000168	0.001239
1.5	0.000183	0.002111	0.000162	0.001331
2.0	0.000377	0.002309	0.000167	0.001373
2.5	0.000493	0.002410	0.000174	0.001394
3.0	0.000557	0.002461	0.000179	0.001403
3.5	0.000592	0.002486	0.000182	0.001408
4.0	0.000610	0.002498	0.000184	0.001410

TABLE 8.20 Sensitivity Coefficients by Variational Method with Time Step of 0.5 Days

Time After Pumping Day	Number 1		Number 2	
	$\frac{\partial h_1}{\partial T_1}$	$\frac{\partial h_1}{\partial T_2}$	$\frac{\partial h_2}{\partial T_1}$	$\frac{\partial h_2}{\partial T_2}$
0.5	−0.000391	0.001028	0.000191	0.000993
1.0	−0.000105	0.001729	0.000168	0.001239
1.5	0.000184	0.002111	0.000162	0.001331
2.0	0.000378	0.002309	0.000167	0.001373
2.5	0.000493	0.002410	0.000174	0.001394
3.0	0.000558	0.002461	0.000179	0.001403
3.5	0.000593	0.002486	0.000182	0.001408
4.0	0.000611	0.002498	0.000184	0.001410

8.10 PARAMETER UNCERTAINTY AND OPTIMUM PARAMETER DIMENSION

The identification of parameters in a distributed parameter system should, in principle, include the determination of both the parameter structure and its value. If zonation is used to parameterize the unknown parameters, parameter structure is

TABLE 8.21 Sensitivity Coefficients by Sensitivity Equation Method with Time Step of 1.0 Day

Time After Pumping Day	Number 1		Number 2	
	$\frac{\partial h_1}{\partial T_1}$	$\frac{\partial h_1}{\partial T_2}$	$\frac{\partial h_2}{\partial T_1}$	$\frac{\partial h_2}{\partial T_2}$
1.0	−0.000142	0.001507	0.000179	0.001146
2.0	−0.000261	0.002159	0.000170	0.001338
3.0	0.000483	0.002392	0.000175	0.001389
4.0	0.000575	0.002471	0.000181	0.001405
5.0	0.000610	0.002497	0.000184	0.001410
6.0	0.000622	0.002506	0.000185	0.001412
7.0	0.000627	0.002509	0.000186	0.001412
8.0	0.000628	0.002509	0.000186	0.001412

TABLE 8.22 Sensitivity Coefficients by Variational Method with Time Step of 1.0 Day

Time After Pumping Day	*Number 1*		*Number 2*	
	$\frac{\partial h_1}{\partial T_1}$	$\frac{\partial h_1}{\partial T_2}$	$\frac{\partial h_2}{\partial T_1}$	$\frac{\partial h_2}{\partial T_2}$
1.0	−0.000142	0.001507	0.000179	0.001146
2.0	−0.000262	0.002159	0.000170	0.001338
3.0	0.000483	0.002392	0.000176	0.001389
4.0	0.000576	0.002471	0.000181	0.001405
5.0	0.000611	0.002497	0.000184	0.001410

represented by the number and shape of zones. On the other hand, if finite-element is used for parameterization, parameter structure concerns the number and location of nodal values of parameters. Emsellem and de Marsily (1971) were the first to consider the problem of optimal zoning pattern. Yeh and Yoon (1976) suggested a systematic procedure based on a statistical criterion for the determination of an optimum zoning pattern. Shah, et al. (1978) also analyzed the relationship between the optimal dimension of parameterization and observations. The necessity to limit the dimension of parameterization has been further studied by Yeh and Yoon (1981), Yeh, et al. (1983), and Kitanidis and Vomvoris (1983).

The dimension of parameterization is directly related to the quantity and quality of data (observations). In field practice, the number of observations is limited and observations are corrupted with noise. Without controlling parameter dimension, instability often results (Yakowitz & Duckstein, 1980). If instability occurs in the inverse problem solution, parameters will become unreasonably small (sometimes negative, which is physically impossible) and/or large, if parameters are not constrained. In the constrained minimization, instability is characterized by the fact that, during the solution process, parameter values are bouncing back and forth between the upper and lower bounds. Reduction of parameter dimension can make the inverse solution stable. It has been generally recognized that as the number of zones (in the zonation case) is increased, the modeling error (least squares) decreases while the error in parameter uncertainty increases. The trade-off between the two types of errors can be evaluated parametrically and the optimum parameter dimension determined. This methodology is illustrated in Example Problem 8.6.

The error in parameter uncertainty is related to the covariance matrix of the estimated parameters (Yeh & Yoon, 1976; Shah, et al., 1978). The covariance matrix of the estimated parameters is defined by

$$\mathrm{Cov}(\hat{T}) = E\{(\overline{T} - \hat{T})(\overline{T} - \hat{T})^T\} \tag{8.59}$$

where

$\hat{T}$ = estimated parameters

$\overline{T}$ = true parameters

E = mathematical expectation

T = transpose of a vector when used as superscript

An approximation of the covariance matrix of the estimated parameters in nonlinear regression can be represented by the equation (Bard, 1974; Yeh & Yoon, 1976; Shah, et al., 1978; Yeh & Yoon, 1981),

$$\text{Cov}(\hat{T}) = \frac{J(\hat{T})}{M-L} [A(\hat{T})]^{-1} \tag{8.60}$$

where

$J(\hat{T})$ = least square error

M = number of observations

L = parameter dimension

$A = [J_D^T J_D]$

J_D = Jacobian matrix of h with respect to T

Equation 8.60 also assumes homoscedasticity and uncorrelated errors. Since these conditions are not generally satisfied, the actual covariance may be substantially larger than that given by Equation 8.60.

The covariance matrix of the estimated parameters also provides information regarding the reliability of each of the estimated parameters. A well-estimated parameter is generally characterized by a small variance as compared to an insensitive parameter that is associated with a large variance. By definition, the correlation matrix of the estimated parameters is

$$\overline{R} = \begin{bmatrix} \dfrac{c_{11}}{\sqrt{c_{11} \cdot c_{11}}} & \cdots & \dfrac{c_{1L}}{\sqrt{c_{11} \cdot c_{LL}}} \\ \\ \dfrac{c_{L1}}{\sqrt{c_{LL} \cdot c_{11}}} & \cdots & \dfrac{c_{LL}}{\sqrt{c_{LL} \cdot c_{LL}}} \end{bmatrix} \tag{8.61}$$

where c_{ij}'s are elements of the covariance matrix of the estimated parameter.

Correlation analysis of the estimated parameters indicates the degree of interdependence among the parameters with respect to the objective function of the identification model. Correlation of parameters is called the *collinearity problem*. Collinearity can cause slow rate of convergence in minimization algorithms and, in most cases, result in nonoptimal parameter estimates. Ridge regression and principal component analysis have been used for the analysis of the collinearity problem (Cooley, 1977).

Cooley (1977) treated the inverse problem as a problem in nonlinear regression. A finite-element scheme was used to solve the confined, steady-state groundwater flow equation. The parameters identified included transmissivity, hydraulic conductance, source/sink strength, and boundary flux. The nonlinear system of normal equations was solved using quasilinearization (Bellman & Kalaba, 1965) and a modified Gauss-Newton algorithm. Beale's nonlinearity measure was used to test the applicability of linear statistical analysis for the original nonlinear regression problem. The advantage of using a standard regression procedure for parameter identification is that it allows the application of established, formal statistical techniques for testing the validity of the modeling assumptions and model fit and the estimation of the reliability and significance of the model and its parameters. However, caution must be exercised, since statistics derived from linear statistical theory are not strictly applicable to the nonlinear case.

The following example problem examines the trade-offs between parameter dimensioning and the reliability of the parameter estimates.

Example Problem 8.6 Optimum Parameter Dimension (After Yeh and Yoon, 1981)

Determine the optimal parameter dimension for the aquifer system described in Example Problem 8.3. Evaluate, using the Gauss-Newton algorithm, the trade-off between the parameter reliability as represented by a norm of the covariance matrix and the parameter dimension. Assume the norm is defined as

$$\|\mathrm{Cov}(T_g)\| = \left[\sum_{i,j} C^2_{T_{gij}}\right]^{1/2}$$

where $T_g = G\,T_e$; $G = N_i(x,y)$; T_e = estimator of parameter $T(x,y)$.

The transmissivity distribution for the aquifer shown in Figure 8.2 is shown in Table 8.23 and plotted in Figure 8.5. This transmissivity map is used to generate observations along with the appropriate initial and boundary condition data. The generated observations are corrupted with noise of various degrees as represented by the corresponding standard deviations. The objective of the identification algorithm is to identify the unknown transmissivity map. The aquifer is initially at steady state with a head of 36.6 m (120 ft). A pumping test is conducted to produce

TABLE 8.23 Transmissivity Distribution, Case 1

Nodal Value in y	*Nodal Value in x*															
	1	*2*	*3*	*4*	*5*	*6*	*7*	*8*	*9*	*10*	*11*	*12*	*13*	*14*	*15*	*16*
1	290	260	232	206	182	160	140	122	106	92	80	70	72	56	52	50
2	290	260	232	206	182	160	140	122	106	92	80	70	72	56	52	50
3	290	260	232	206	182	160	140	122	106	92	80	70	72	56	52	50
4	290	260	232	206	182	160	140	122	106	92	80	70	72	56	52	50
5	290	260	232	206	182	160	140	122	106	92	80	70	72	56	52	50
6	290	260	232	206	182	160	140	122	106	92	80	70	72	56	52	50
7	290	260	232	206	182	160	140	122	106	92	80	70	72	56	52	50
8	290	260	232	206	182	160	140	122	106	92	80	70	72	56	52	50
9	290	260	232	206	182	160	140	122	106	92	80	70	72	56	52	50
10	290	260	232	206	182	160	140	122	106	92	80	70	72	56	52	50
11	290	260	232	206	182	160	140	122	106	92	80	70	72	56	52	50
12	290	260	232	206	182	160	140	122	106	92	80	70	72	56	52	50
13	290	260	232	206	182	160	140	122	106	92	80	70	72	56	52	50
14	290	260	232	206	182	160	140	122	106	92	80	70	72	56	52	50
15	290	260	232	206	182	160	140	122	106	92	80	70	72	56	52	50
16	290	260	232	206	182	160	140	122	106	92	80	70	72	56	52	50

Transmissivity distribution in 100 ft^2/d. Values correspond to nodal points used in the finite difference approximations which are defined at the centroid of each block. $S = 0.015$.

TABLE 8.24 Results of Successive Approximations, Case 1

*Parameter Dimension**	σ	J, $\times 10^3$	$\|C_{T_g}\|$, $\times 10^9$	$\Sigma_{i=1}^{256} (T_i - T_{i_{true}})^2$, $\times 10^9$
Homogeneous	1	1.61	0.009	15.8
	5	7.93	0.04	15.8
	10	28.1	0.16	15.9
2	1	0.34	0.005	1.02
	5	6.81	0.10	1.00
	10	27.2	0.52	1.19
3	1	0.272	0.008	0.11
	5	6.79	0.22	0.22
	10	27.2	0.72	0.71
4	1	0.272	0.018	0.03
	5	6.79	0.46	0.22
	10	27.2	1.93	1.07
5	1	0.271	0.028	0.06
	5	6.77	0.71	1.02
	10	27.1	2.58	3.31
6	1	0.270	0.83	0.38
	5	6.74	1.59	16.34
	10	27.0	6.98	15.90

*Parameter dimension is represented by the number of nodes that are to be identified.

head variations in the flow domain. Pumping wells are located at grid points (3,6), (3,13), (8,9), and (12,5). The respective pumping rates are 5660, 4242, 5660, and 5660 m^3/day. The generated head variations are between 36.6 and 51.85 m. Gaussian noises with standard deviations (σ of 1,5, and 10) were used to model the noise in the data. Observations are taken at 10 locations: (2,15), (3,3), (5,9), (6,11), (6,15), (7,7), (8,11), (10,3), (14,10), and (15,6). The time step for the solution of the hydraulic response equations is 10 days; the total identification period is 100 days.

To evaluate the trade-offs associated with parameter dimensioning, the parameter dimension was varied from 1, corresponding to a homogeneous aquifer, to 6. Table 8.24 shows the results of the successive approximations. Figures 8.5, 8.6, and 8.7 show the comparison of the transmissivity distributions between the true and identified values for various σ and parameter dimensions. The parameter dimension was gradually increased from 1 to 6 with the nodal arrangement shown in Figure 8.8. As can be seen from Figures 8.5 through 8.7, considerable improvement was obtained from an initial increase in parameter dimension (from 1

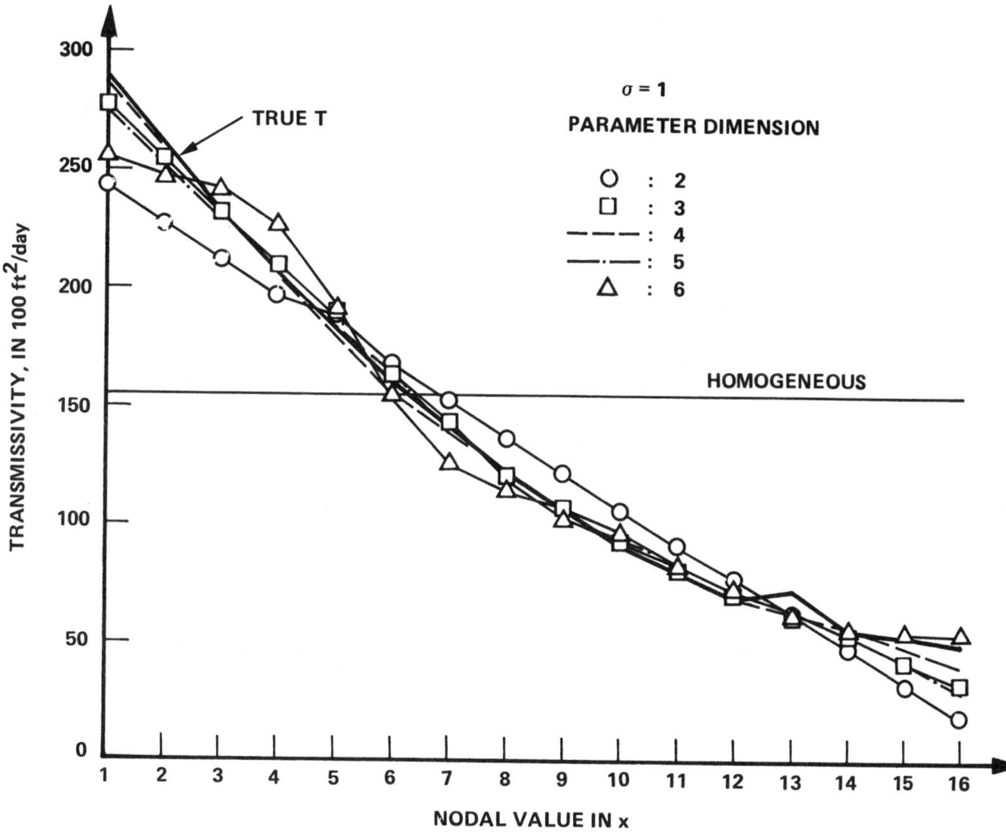

FIGURE 8.5 Comparison of true and identified transmissivity distributions, $\sigma = 1$. case 1 (From W. W-G. Yeh and Y. S. Yoon, "Aquifer Parameter Identification with Optimum Dimension in Parameterization." *Water Resources Research*, 17(3), 1981, p. 666. Copyright © 1981 by the American Geophysical Union)

to 4). However, any further increase in parameter dimension results in parameter instability. The instability phenomenon is typified by large σ. Figures 8.9 and 8.10 plot the results presented in Table 8.24 for $\sigma = 1$ and 5. Apparently, the least squares error decreases as the parameter dimension increases. However, the norm of the covariance matrix increases with increasing parameter dimension. Figure 8.11 shows the variation in parameter dimension with the least squares error (T_i are the identified nodal transmissivities and $T_{i\text{true}}$ are the true values used to generate the observations). A clearly defined minimum exists in Figure 8.11. In field applications, $T_{i\text{true}}$ would be unknown.

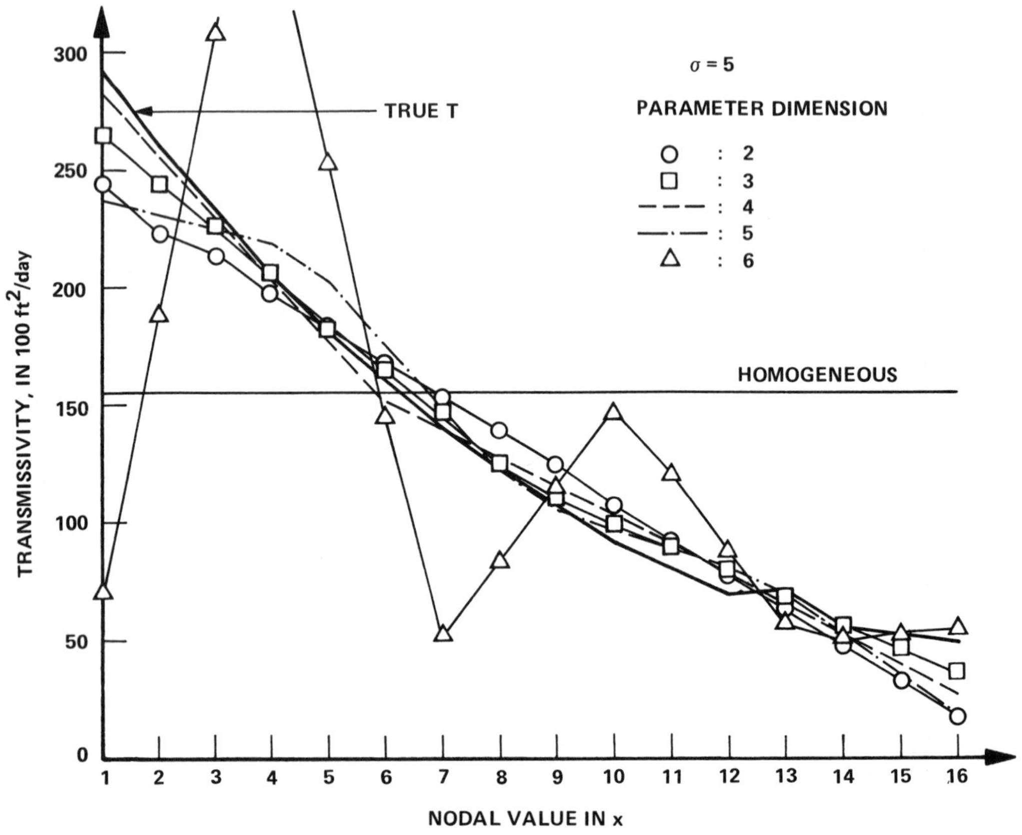

FIGURE 8.6 Comparison of true and identified transmissivity distributions, σ = 5, case 1 (From W. W-G. Yeh and Y. S. Yoon, "Aquifer parameter Identification with Optimum Dimension in Parameterization." *Water Resources Research*, 17(3), 1981, p. 667. Copyright © 1981 by the American Geophysical Union)

In Figure 8.11, minima occur in the neighborhood of a parameter dimension of 4. At this point, the slope of the least squares objective is asymptotically 0 and the slope of the norm of the covariance matrix starts to increase rapidly. Any further increase in parameter dimension will not significantly reduce the least squares error but will significantly increase the uncertainty of the estimated parameters. This implies that an optimum level of parameterization has been reached. It can also been seen from Figures 8.9 and 8.10 that the least squares objective and the covariance matrix norm are convex functions of the parameter dimension.

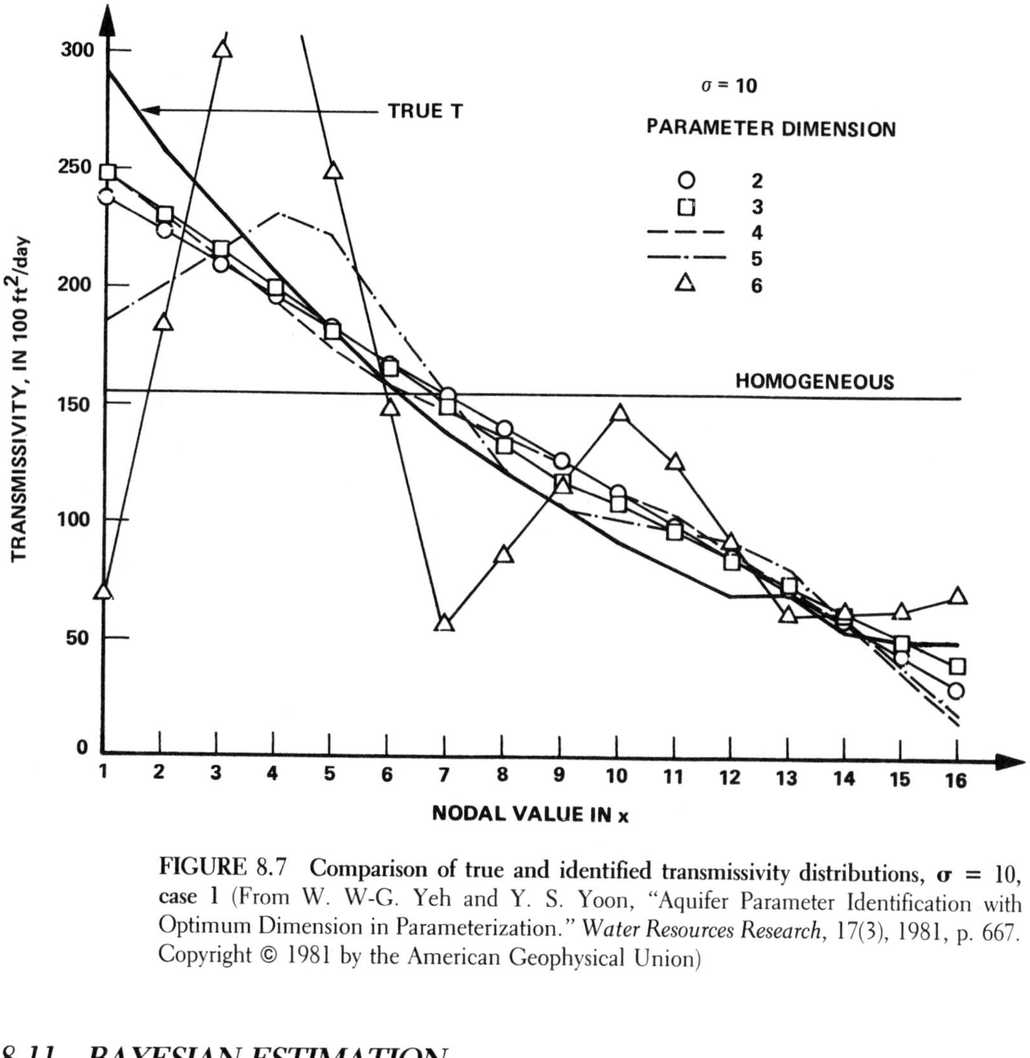

FIGURE 8.7 Comparison of true and identified transmissivity distributions, $\sigma = 10$, case 1 (From W. W-G. Yeh and Y. S. Yoon, "Aquifer Parameter Identification with Optimum Dimension in Parameterization." *Water Resources Research*, 17(3), 1981, p. 667. Copyright © 1981 by the American Geophysical Union)

8.11 BAYESIAN ESTIMATION

Bayesian estimation methods that incorporate prior information have also been applied to parameter identification (e.g., Gavalas, et al., 1976). The geological information required for Bayesian estimation includes the mean and covariance matrix of the parameters, which are

$$E\{\overline{T}\} = T_{\text{mean}} \tag{8.62}$$

$$E\{T_i - T_{\text{mean}})(T_j - T_{\text{mean}})\} = R_{ij} \tag{8.63}$$

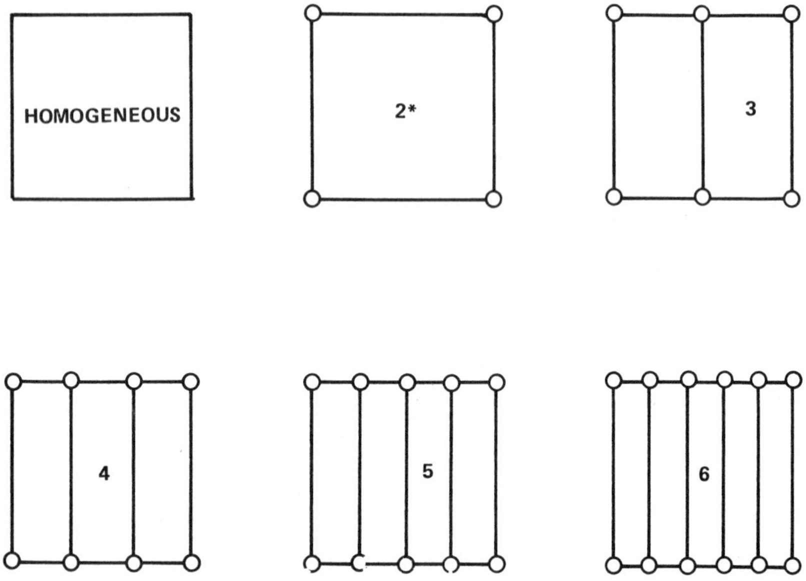

FIGURE 8.8 **Nodal arrangements, case 1** (From W. W-G. Yeh and Y. S. Yoon, "Aquifer Parameter Identification with Optimum Dimension in Parameterization." *Water Resources Research*, 17(3), 1981, p. 668. Copyright © 1981 by the American Geophysical Union)

The values of T_{mean} and R_{ij} are considered to be known and are the prior information which can be obtained from geological measurements in the field. Gavalas, et al. (1976) have shown that Bayesian estimation reduces to a quadratic minimization problem provided the parameters and the measurement errors are normally distributed and the model is linear in the parameters. When these conditions are not satisfied, a rigorous application of Bayesian estimation is impractical.

8.11.1 Composite Objective Function

Gavalas, et al. (1976) proposed the following practical approach which is similar to least squares minimization. The objective function of the model can be expressed as

$$J = \sum_{i=1}^{M} \frac{1}{\sigma_i^2} (h_i - h_i^*)^2 + \lambda (\overline{T} - T_{\mathrm{mean}})^T R^{-1} (\overline{T} - T_{\mathrm{mean}}) \qquad (8.64)$$

where λ is a weighting factor ($0 \leq \lambda \leq 1$) and σ_i^2, $i = 1,2, \ldots ,M$, is the variance

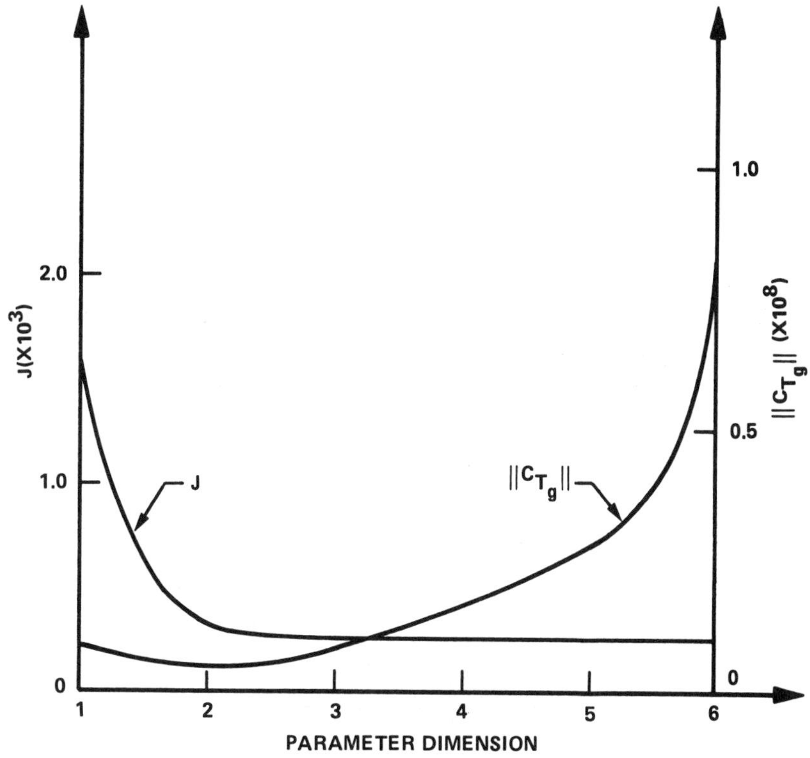

FIGURE 8.9 **Parameter dimension versus J and $\| C_{Tg}\|$, σ = 1.0, case 1** (From W. W-G. Yeh and Y. S. Yoon, "Aquifer Parameter Identification with Optimum Dimension in Parameterization." *Water Resources Research*, 17(3), 1981, p. 668. Copyright © 1981 by the American Geophysical Union)

of the measurement error which is assumed known. The second term in the objective function is the Bayesian term which penalizes the weighted deviation of the parameters from their mean value. It, in turn, requires the parameter to follow some preconceived pattern during the minimization process. Shah, et al. (1978) have demonstrated that, if reliable prior information is available, Bayesian estimation will lead to a smaller variance of the error of estimation.

8.11.2 Kalman Filter

The technique of Kalman filtering was originally developed in the field of optimal control (Kalman, 1960). It has been successfully applied in aerospace engineering for the problem of optimal estimation and control of vehicle trajectory. The ap-

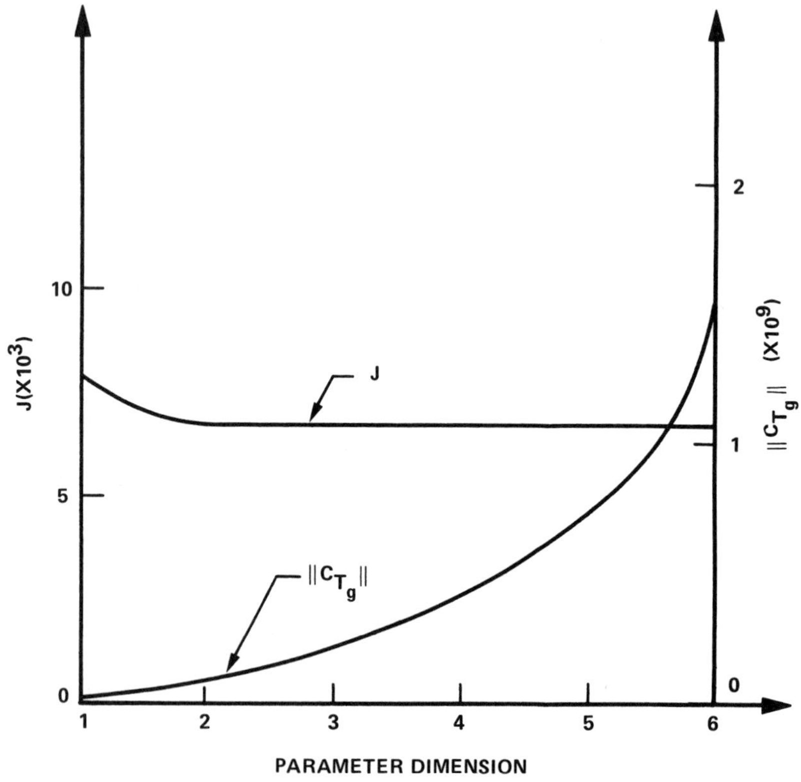

FIGURE 8.10 Parameter dimension versus J and $\| C_{Tg} \|$,σ = 5.0, case 1 (From W. W-G. Yeh and Y. S. Yoon, "Aquifer Parameter Identification with Optimum Dimension in Parameterization." *Water Resources Research*, 17(3), 1981, p. 669. Copyright © 1981 by the American Geophysical Union)

plication of Kalman filtering to parameter estimation in groundwater requires expressing the groundwater model in terms of a state-space formulation that consists of a vector state equation and a vector observation equation. For parameter estimation, the state vector is augmented to include the parameter vector as another state variable. If the errors in the state and observation equations have zero mean and are a white Gaussian process with known covariance matrices, Kalman filtering can be applied for simultaneous, recursive state and parameter estimation. Since prior information is generally required in the application of Kalman filtering, it can be classified in the Bayesian estimation category. Wilson, et al. (1978) used an extended Kalman filter for parameter estimation in groundwater. Their approach permits the utilization of prior information about the parameters and information taken from input-output measurements to improve estimates of parameters as well as the system state.

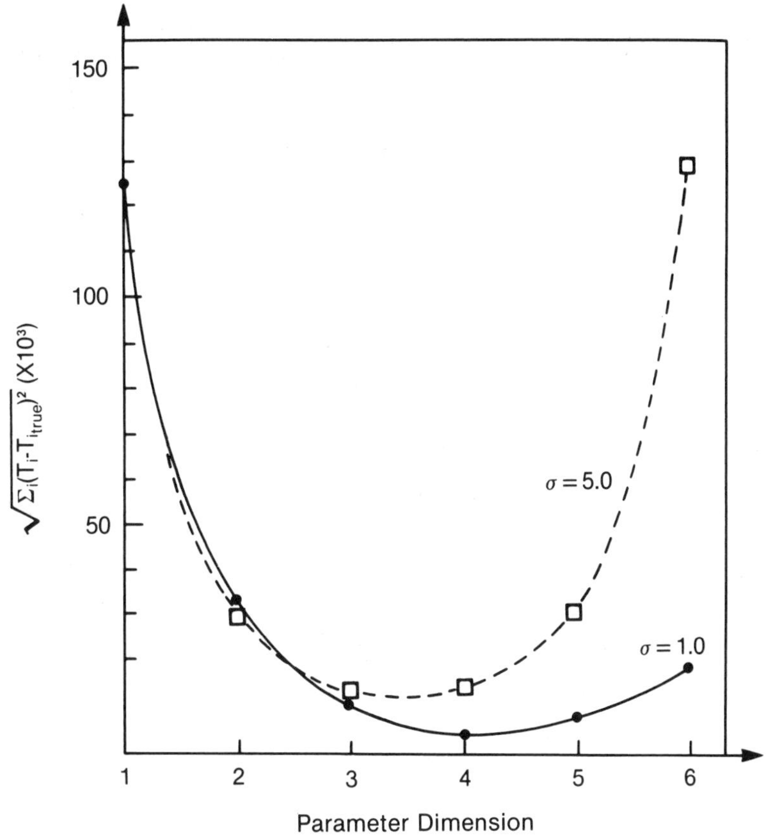

FIGURE 8.11 Parameter dimension versus $(\Sigma_{i=1}^{256}(T_i - T_{itrue})^2)^{1/2}$, σ = 1.0 and σ = 5.0, case 1 (From W. W-G. Yeh and Y. S. Yoon, "Aquifer Parameter Identification with Optimum Dimension in Parameterization." *Water Resources Research*, 17(3), 1981, p. 669. Copyright © 1981 by the American Geophysical Union)

8.12 OTHER STATISTICAL METHODS

In addition to the above-mentioned approaches for solving the inverse problems in groundwater, a few different types of statistical methods have been reported in the literature, which can be categorized by the following:

8.12.1 Statistical Methods that Incorporate Prior Information

Neuman and Yakowitz (1979) proposed a statistical approach to the inverse problem of parameter estimation. The approach differs from Bayesian estimation (Gavales, et al., 1976) since the prior information may include actual values of transmissivity determined from pumping tests or other measurements at specific locations in the aquifer; or, it may be based on statistical information about the spatial variability of transmissivities (not their actual values) in other aquifers consisting of similar materials. The composite least squares criterion proposed by Neuman and Yakowitz (1979) is similar to Equation 8.64 and can be expressed as

$$J = [h^* - f(\overline{T})]^T V_h^{-1} [h^* - f(\overline{T})] + \lambda(T^* - \overline{T})^T V_T^{-1} (T^* - \overline{T}) \qquad (8.65)$$

where

T^* = prior estimate of $\overline{T}$

V_T = known symmetric positive definite matrix

V_h = known matrix, symmetric and positive definite

λ = unknown positive parameter

$f(\overline{T})$ = model solution

h^* = observed head

The observed head (h^*) and the prior estimates of transmissivity (T^*) are related to true head (h) and true transmissivity ($\overline{T}$) by

$$h^* = h + \epsilon \qquad (8.66)$$
$$T^* = \overline{T} + v$$

and

$$E(\epsilon) = 0$$
$$\text{Var}(\epsilon) = \sigma_h^2 V_h \qquad (8.67)$$
$$E(v) = 0$$
$$\text{Var}(v) = \sigma_T^2 V_T \qquad (8.68)$$

It is assumed that V_h and V_T are known, but σ_h and σ_T do not enter the computations. The second term in the composite objective function provides a smoothing effect in the minimization. Neuman and Yakowitz (1979) proposed two methods, cross-validation and comparative residual analysis, to select the optimum value of λ. Neuman (1980) developed an efficient conjugate gradient algorithm for performing the minimization. He extended the variational method developed by Chavent (1975) for calculating the gradient with respect to the parameter in the case of generalized nonlinear least squares. The variational method presented by Chavent (1975) and Neuman (1980) is conceptually similar to Carter, et al. (1974), but differs in the objectives. Carter, et al. (1974) developed expressions that can be used to calculate the partial derivatives of the head with respect to the parameters, while Chavent (1975) and Neuman (1980) computed the partial derivatives of the least squares criterion with respect to the parameters.

Cooley (1982) proposed a method to incorporate prior information on the parameters into nonlinear regression models. The primary purpose of Cooley's work is the incorporation of prior information of unknown reliability; the approach is an extension of ridge regression. A secondary objective of the research is to incorporate Theil's (1963) approach into Cooley's (1977) nonlinear regression model where at least some prior information of known reliability is available. The approach is non-Bayesian in the sense that no prior distribution of parameter is assumed. The approach also differs from the method proposed by Neuman and Yakowitz (1979) and Neuman (1980) because the prior information consists of general nonlinear combinations of several types of parameters as opposed to direct estimates of a single type of parameter (transmissivity). Nonstochastic prior information, as represented by a set of approximate linearized equations, is incorporated in the ridge regression previously developed by Cooley (1977).

As presented in Cooley (1982), prior information having unknown reliability can be incorporated into the standard weighted least squares objective function by adding a penalty function. The resulting composite objective function consists of two terms. The first term is the weighted sum of squared errors in the hydraulic head, and the second term is the sum of the weighted errors in the parameters. Two scalars are used in the composite objective function to control the sum of the squared errors.

8.12.2 Geostatistical Approach

Geostatistical methods developed for parameter identification generally utilize the technique of Kriging. Kriging (Delhomme, 1976), a technique for interpolating nonstationary spatial phenomena, has been used by a number of researchers in the past (e.g., Delhomme, 1979) to estimate trransmissivity values at points where observations are not available. The technique of Kriging is based on the theory of regionalized variables. A *regionalized variable* is a variable that describes a phenomenon that is spread out in space (and/or in time) and shows a certain structure. Parameters describing groundwater flow such as transmissivities and the variation of head in an aquifer system can be considered as regionalized variables. A detailed Kriging algorithm can be found in Delfiner (1975). The question of data requirements in Kriging has been addressed by Hughes and Lettenmaier (1981).

The composite objective function presented by Aboufirassi and Marino (1984b) is also conceptually similar to Equation 8.64. Kriging has been used to estimate the missing values of head and the value of the error covariance matrix while Cokriging (Aboufirassi & Marino, 1984a) was used to estimate T^* and the associated error covariance matrix. *Cokriging* (Journel and Huijbregts, 1978), an extension of

Kriging to two or more variables, can be used to improve the accuracy of estimation of a variable that is not sufficiently sampled by considering its spatial correlation with other variables that are better sampled.

Kitanidis and Vomvoris (1983) proposed a geostatistical approach for solving the inverse problem. Their method consists of two main steps: (1) the structure of the parameter field is identified, i.e., mathematical representations of the variogram and the trend are selected and their parameters are established and (2) Kriging is applied to provide minimum variance and unbiased point estimates of hydrogeological parameters using all available information. In the approach, it is assumed that several point measurements of head and transmissivities (in logarithms) are available. In effect, parameterization is achieved by representing the hydrogeological parameters as a random field that can be characterized by the variogram and trend with a small number of parameters. The parameters estimated in the first step are the ones associated with the variogram and trend, thus, drastically reducing the parameter dimension. As demonstrated by Kitanidis and Vomvoris (1983), the reduction of parameter dimension has resulted in stable inverse problem solutions with the presence of errors. Hoeksema and Kitanidis (1984) have applied the geostatistical approach to the case of two-dimensional steady-state flow. A finite-difference numerical model of groundwater flow was used to relate the head and transmissivity variability and Cokriging was used to estimate the unknown transmissivity field. Dagan (1985) has also considered the geostatistical approach, but an analytical technique and Gaussian conditional mean are used in place of Kriging.

8.12.3 Generalized Least Squares

Correlation of error residuals in both time and spatial domain may occur for a number of reasons. Successive errors in time series tend to be positively correlated. Also, observations taken from adjacent pumping wells are affected by similar external conditions and may result in similar residuals. The presence of correlation in error terms suggests that there is additional information in the data that has to be included in the least squares minimization model. A well-established method that can be used to perform such minimization is the generalized least squares developed in the field of econometrics. In the case of an unsteady-state flow, the error vector for each time period can be approximated by a stationary first-order autoregression process (Judge, et al., 1980). A method suggested by Sadeghipour and Yeh (1984) requires the estimation of lag-one serial coefficient (ρ) and the common covariance matrix (W) for the error vector. Using the estimated values of ρ and W, the generalized least squares method as demonstrated by Sadeghipour and Yeh (1984) provides parameter estimates with minimum variance when errors are correlated. If ρ and W are fixed and errors are normally distributed, the generalized least squares criterion corresponds to the log likelihood function.

Sadeghipour and Yeh (1984) also presented a two-stage methodology for parameter estimation when ρ and W are unknown. Initially, it is assumed that $\rho = 0$ and $W = I$; the minimization produces the ordinary least squares parameter estimates. These estimates are used to estimate ρ and W. In the second step, the newly estimated ρ and W are used to perform the generalized least squares minimization. The process continues until convergence is reached.

8.13 *SUMMARY AND FUTURE RESEARCH DIRECTIONS*

The inverse problem of parameter identification in groundwater systems has been formulated based on the error criterion used. Typical solution methods have been presented for the inverse problem. The methods rely on either the equation error approach or the output error approach. The output error criterion appears to be widely used not only in groundwater but also in oil reservoir problems. Optimization methods originally developed in the fields of optimal control and operations research have been adopted to perform the minimization. Linear statistical methods have been used to establish the reliability of the estimated parameters. Bayesian estimation has also been used to incorporate prior information regarding the parameters to be estimated. The number of parameters (parameter dimension) that can be identified for a given situation depends on the quantity and quality of the data. In field applications, identifiability is relaxed, since measurements cannot be made at every spatial point as a function of time.

Parameter estimation models have utilized linear statistics for testing the validity and model fit, and estimating the reliability and significance of the model and its parameters. However, the inverse problem in groundwater is inherently nonlinear. Future research should be directed toward the development of nonlinear parameter estimation theories. But, caution must be exercised in that nonlinear methods are usually associated with high computational cost, and their practical applicability must be examined in view of the fact that linear estimation methods have served well in many instances.

The primary purpose of incorporating prior information into the inverse problem is to reduce the parameter uncertainty, not to improve the model fit. Prior information can only worsen the model fit. The incorporation of prior information as a penalty function in the composite, lease square objective function does not affect the feasible region of minimization. However, if prior information is used correctly, the inverse solution will produce stable and reliable parameter estimates that will be more useful in groundwater management and prediction. It is also intuitively obvious that inaccurate prior information will degrade the parameter estimates. A future area of research is the development of reliable prior parameter estimates that are compatible with sample information.

It has been made clear that parameterization must include the parameter structure and its values. The development of an efficient and systematic parameter structure identification procedure continues to be a future area of research.

The problem of identifiability has received little attention in the literature because of its intrinsic difficulty. But, some recently published results indicate that an extended identifiability can be used for groundwater modeling and management. A future area of research is to continue to study the problem of optimal pumping design in connection with aquifer parameter identification.

8.14 REFERENCES

Aboufirassi, M., & Mariño, M. A., "Kriging of Water Levels in the Souss Aquifer, Morocco." *Mathematical Geology*, 15(4): 537–551, 1983.

Aboufirassi, M., & Mariño M. A., "Cokriging of Aquifer Transmissivities from Field Measurements of Transmissivity and Specific Capacity." *Mathematical Geology*, 16(1): 19–35, 1984a.

Aboufirassi, M., & Mariño, M. A., "A Geostatistically-Based Approach to Identification of Aquifer Transmissivity in Yolo Basin, California." *Mathematical Geology*, 16(2): 125–137, 1984b.

Anger, G. (edited). "Inverse and Improperly Posed Problems in Differential Equations." *Proceedings of the Conference on Mathematical and Numerical Methods*. Halle/Saale, GDR, May 29 to June 2, 1979, Akademie-Verlag, Berlin, 1979.

Bard, Y. *Nonlinear Parameter Estimation*. New York: Academic Press, 1974.

Beale, E. M. L. "Confidence Regions in Non-Linear Estimation." *J. R. Stat. Soc. Ser. B*, 22: 41–76, 1960.

Bellman, R., & Kabala, R. *Quasilinearization and Nonlinear Boundary-Value Problems*. New York: Elsevier, 1965.

Beck, A. *Parameter Estimation in Engineering Science*. New York: Wiley, 1977.

Birtles, A. B., & Morel, E. H. "Calculation of Aquifer Parameters from Sparse Data." *Water Resources Research*, 15(4): 832–844, 1979.

Bruch, J. C., Jr., Lam, C. M., & Simundich, T. M. "Parameter Identification in Field Problems." *Water Resources Research*, 10(1): 73–79, 1974.

Carter, R. D., Kemp, L. F., Jr., Pierce, A. C., & Williams, D. L. "Performance Matching with Constraints." *Soc. Pet. Eng. J.*, 14(2): 187–196, 1974.

Carter, R. D., Kemp, L. F., Jr., & Pierce, A. C. "Discussion of Comparison of Sensitivity Coefficient Calculation Methods in Automatic History Matching." *Soc. Pet. Eng. J.*, 22(2): 205–208, 1982.

Chan, Y. K., Mullineaux, N., & Reed, J. R. "Analytic Solutions for Draw-downs in Rectangular Artesian Aquifers." *J. of Hydrology*, 31: 151–160, 1976.

Chang, S., & Yeh, W. W-G. "A Proposed Algorithm for the Solution of the Large-Scale Inverse Problem in Groundwater." *Water Resources Research*, 12(3): 365–374, 1976.

Chavent, G. "Identification of Functional Parameters in Partial Differential Equations." In Goodson, R. E. and M. Polis (Eds.), *Identification of Parameters in Distributed Systems*. ASME, United Engineering Center, New York, 1974, 31–48.

Chavent, G. "About the Stability of the Optimal Control Solution of Inverse Problems." In *Inverse and Improperly Posed Problems in Differential Equations* (Edited by G. Anger).

Proceedings of the Conference on Mathematical and Numerical Methods, Halle/Saale, GDR, May 29 to June 2, 1979, Akademie-Verlag, Berlin, 1979a, 45–58.

Chavent, G. "Identification of Distributed Parameter System: About the Output Least Square Method, Its Implementation, and Identifiability." In Isermann, R. (Ed.), *Identification and System Parameter Estimation*, Proceedings of the Fifth IFAC Symposium, Darmstadt, Federal Republic of Germany, 24–28 September 1979b, Pergamon Press, vol. 1, 85–97.

Chavent, G., Dupuy, M., & Lemonnier, P. "History Matching by Use of Optimal Control Theory." *Soc. Pet. Eng. J.*, 15(1): 74–86, 1975.

Chavent, G. "Local Stability of the Output Least Square Parameter Estimation Technique." *Mat. Applic. Comp.*, 2(1): 3–22, 1983.

Chen, W. H., Gavalas, G. R., Seinfeld, J. H., & Wasserman, M. L. "A New Algorithm for Automatic Historic Matching." *Soc. Pet. Eng. J.*, 14(6): 593–608, 1974.

Clifton, P. M., & Neuman, S. P. "Effects of Kriging and Inverse Modeling on Conditional Simulation of the Avra Valley Aquifer in Southern Arizona." *Water Resources Research*, 18(4): 1215–1234, 1982.

Coats, K. H., Dempsey, J. R., & Henderson, J. H. "A New Technique for Determining Reservoir Description from Field Performance Data." *Soc. Pet. Engr. J.*, 10(1): 66–74, 1970.

Cooley, R. L., & Sinclair, P. J. "Uniqueness of a Model of Steady-State Groundwater Flow." *J. Hydrology*, 31: 245–269, 1976.

Cooley, R. L. "A Method of Estimating Parameters and Assessing Reliability for Models of Steady State Ground Flow, 1, Theory and Numerical Properties." *Water Resources Research*, 13(2): 318–324, 1977.

Cooley, R. L. "A Method for Estimating Parameters and Assessing Reliability for Models of Steady State Groundwater Flow, 2, Application of Statistical Analysis." *Water Resources Research*, 15(3): 603–617, 1979.

Cooley, R. L. "Incorporation of Prior Information on Parameters into Nonlinear Regression Groundwater Flow Models, 1, Theory." *Water Resources Research*, 18(4): 965–976, 1982.

Cooley, R. L. "Incorporation of Prior Information on Parameter into Nonlinear Regression Groundwater Flow Models, 2, Applications." *Water Resources Research*, 19(3): 662–676, 1983.

Dagan, G. "Stochastic Modeling of Groundwater Flow by Unconditional and Conditional Probabilities: The Inverse Problem." *Water Resources Research*, 21(1): 65–72, 1985.

De Coursey, D. G., & Snyder, W. M. "Computer Oriented Method of Optimizing Hydrologic Model Parameters." *Journal of Hydrology*, 9: 34–53, 1969.

Delfiner, P. "Linear Estimation of Nonstationary Spatial Phenomena." In *Advanced Geostatistics in Mining Industry*. (M. Guarasico, Ed.) Dordrecht, Holland: D. Reidel, 1975, pp. 49–68.

Delhomme, J. P. Kriging in Hydrosciences, Ph.D. Thesis, University of Paris, France, 1976.

Delhomme, J. P. "Spatial Variability and Uncertainty in Groundwater Flow Parameters: A Geostatistical Approach." *Water Resources Research*, 15(2): 269–280, 1979.

DiStefano, N., & Rath, A. "An Identification Approach to Subsurface Hydrological Systems. *Water Resources Research*, 11(6): 1005–1012, 1975.

Dogru, A. H., Dixon, T. N., & Edgar, T. F. "Confidence Limits on the Parameters and Predictions of Slightly Compressible, Single Phase Reservoirs." *Soc. Pet. Eng. J.*, 17(1): 42–56, 1977.

Dogru, A. H., & Seinfeld, J. H. "Comparison of Sensitivity Coefficient Calculation Methods in Automatic History Matching." *Soc. Pet. Engr. J.*, 21(5): 551–557, 1981.

Douglas, J., Jr. "Alternating Direction Methods for Three Space Variables." *Number. Math.* 4: 41–63, 1962.

Emsellem, Y., & de Marsily, G. "An Automatic Solution for the Inverse Problem." *Water Resources Research*, 7(5): 1264–1283, 1971.

Frind, E. O., & Pinder, G. F. "Galerkin Solution of the Inverse Problem for Aquifer Transmissivity." *Water Resources Research*, 9(5): 1397–1410, 1973.

Garay, H. L., Haimes, Y. Y., & Das, P. "Distributed Parameter Identification of Groundwater Systems by Nonlinear Estimation." *J. Hydrology*, 30: 47–61, 1976.

Gavalas, G. R., Shah, P. C., & Seinfeld, J. H. "Reservoir History Matching by Bayesian Estimation." *Soc. Pet. Eng. J.*, 16(6): 337–350, 1976.

Gorelick, S. M., Evans, B., & Remson, I. "Identifying Sources of Groundwater Pollution: An Optimization Approach." *Water Resources Research*, 19(3): 779–790, 1983.

Guvanasen, V., & Volker, R. E. "Identification of Distributed Parameters in Groundwater Basins." *J. Hydrology*, 36: 279–293, 1978.

Haimes, Y. Y., Perrine, R. L., & Wismer, D. A. "Identification of Aquifer Parameters by Decomposition and Multilevel Optimization." Water Resources Center Contribution No. 123, University of California, Los Angeles, 1968.

Hefez, E., Shamir, V., & Bear, J. "Identifying the Parameters of an Aquifer Cell Model." *Water Resources Research*, 11(6): 993–1004, 1975.

Hoeksema, R. J., & Kitanidis, P. K. "An Application of the Geostatistical Approach to the Inverse Problem in Two-Dimensional Groundwater Modeling." *Water Resources Research*, 20(7): 1003–1020, 1984.

Hoeksema, R. J., & Kitanidis, P. K. "Comparison of Gaussian Conditional Mean and Kriging Estimation in the Geostatistical Solution of the Inverse Problem." *Water Resources Research*, 21(6): 825–836, 1985.

Hughes, J. P., & Lettenmaier, D. P. "Data Requirements for Kriging: Estimation and Network Design." *Water Resources Research*, 17(6): 1641–1650, 1981.

Hunt, B. W., & Wilson, D. D. "Graphical Calculation of Aquifer Transmissivity in Northern Canterbury, New Zealand." *Journal of Hydrology* (N.Z.), 13(2): 66–81, 1974.

Irmay, S. "Piezometric Determination of Inhomogeneous Hydraulic Conductivity." *Water Resources Research*, 16(4): 691–694, 1980.

Jackson, D. R., & Aron, G. "Parameter Estimation in Hydrology: The State of the Art." *Water Resources Bulletin*, 7(3): 457–471, 1971.

Jacquard, P., & Jain, C. "Permeability Distribution from Field Pressure Data." *Soc. Pet. Eng. J.*, 5(4): 281–294, 1965.

Jahns, H. O. "A Rapid Method for Obtaining a Two-Dimensional Reservoir Description from Well Pressure Response Data." *Soc. Pet. Eng. J.*, 6(4): 315–327, 1966.

Journel, A. G., & Huijbregts, J. C. *Mining Geostatistics*. New York: Academic Press, 1978.

Kalman, R. E. "A New Approach to Linear Filtering and Prediction Problems." *Trans. ASME J. Basic Eng.*, 82: 35–45, 1960.

Kashyap, D., & Chandra, S. "A Nonlinear Optimization Method for Aquifer Parameter Estimation." *J. Hydrology*, 57: 163–173, 1982.

Kitamura, S., & Nakagiri, S. "Identifiability of Spatially-Varying and Constant Parameters in Distributed Systems of Parabolic Type." *SIAM, J. Control and Optimization*, 15(5): 785–802, 1977.

Kitanidis, P. K., & Vomvoris, E. G. "A Geostatistical Approach to the Inverse Problem in Groundwater Modeling (Steady State) and One-Dimensional Simulations." *Water Resources Research*, 19(3): 677–690, 1983.

Kleinecke, D. "Use of Linear Programming for Estimating Geohydrologic Parameters of Groundwater Basins." *Water Resources Research*, 7(2): 367–375, 1971.

Kleinecke, D. "Comments on 'An Automatic Solution for the Inverse Problem' by Y. Emsellem and G. de Marsily." *Water Resources Research*, 8(4): 1128–1129, 1972.

Kubrusly, C. S. "Distributed Parameter System Identification, a Survey." *Int. Journal Control*, 26(4): 509–535, 1977.

Kruger, W. D. "Determining a Real Permeability Distribution by Calculations." *J. of Pet. Tech.*, 691–696, 1961.

Labadie, J. W. "Decomposition of a Large Scale Nonconvex Parameter Identification Problem in Geohydrology." Report No. ORC 72–73, Operations Research Center, University of California, Berkeley, 1972.

Levenberg, J. "A Method for the Solution of Certain Nonlinear Problems in Least Squares." *Quart. Appl. Math.*, 2: 164–168, 1944.

Li, J., Lu, A., Sun, N. Z., & Yeh, W. W-G. "A Comparative Study of Sensitivity Coefficient Calculation Methods in Groundwater Flow." Working paper, Dept. of Civil Engineering, University of California, Los Angeles, 1985.

Lin, A. C., & Yeh, W. W-G. "Identification of Parameters in an Inhomogeneous Aquifer by Use of the Maximum Principle of Optimal Control and Quasilinearization." *Water Resources Research*, 10(4): 829–838, 1974.

Lovell, R. E., Duckstein, L., & Kisiel, C. C. "Use of Subjective Information in Estimation of Aquifer Parameters." *Water Resources Research*, 8(3): 680–690, 1972.

Mariño, M. A., & Yeh, W. W-G. "Identification of Parameters in Finite Leaky Aquifer Systems." *J. of Hydraulic Division*, ASCE, 99(HY2): 319–336, 1973.

Marquardt, D. W. "An Algorithm for Least Squares Estimation of Nonlinear Parameters." *SIAM J.*, 11: 431–441, 1963.

Martensson, K. "Least Square Identifiability of Dynamic Systems." Technical Reports RB 7344, Dept. of Electr. Eng., U.S.C., Los Angeles, California, 1973.

McElwee, C. D. "Sensitivity Analysis and the Groundwater Inverse Problem." *Groundwater*, 20(6): 723–735, 1982.

McLaughlin, D. B. "Investigation of Alternative Procedures for Estimating Groundwater Basin Parameters." Report prepared for the Office of Water Research and Technology, U.S. Department of the Interior, Water Resources Engineering, Walnut Creek, California, 1975.

McLaughlin, D. B. *Hanford Groundwater Modeling—a Numerical Comparison of Bayesian and Fisher Parameter Estimation Techniques*. Richland WA: Rockwell International, Rockwell Hanford Operations, Energy Systems Group, 1979.

Navarro, A. "A Modified Optimization Method of Estimating Aquifer Parameters." *Water Resources Research*, 13(6): 935–939, 1977.

Nelson, R. W. "In-place Measurement of Permeability in Heterogeneous Media, 1, Theory of a Proposed Method." *J. of Geophysical Research*, 65(6): 1753–1758, 1960.

Nelson, R. W. "In-place Measurement of Permeability in Heterogeneous Media, 2, Experimental and Computational Considerations." *J. of Geophysical Research*, 66(8): 2469–2478, 1961.

Nelson, R. W. "Conditions for Determining a Real Permeability Distribution by Calculation." *Soc. Petrol. Eng. Jr.*, 2(3): 223–224, 1962.

Nelson, R. W. "In-place Determination of Permeability Distribution for Heterogeneous Porous Media Through Analysis of Energy Dissipation." *Soc. Pet. Eng. J.*, 8(1): 33–42, 1968.

Nelson, R. W., & McCollum, W. L. "Transient Energy Dissipation Methods of Measuring Permeability Distributions in Heterogeneous Porous Materials." Rep. CSC 691229, Water Resources Division, USGS, Washington, 1969.

Neuman, S. P. "Calibration of Distributed Parameter Groundwater Flow Models Viewed as a Multiple-Objective Decision Process Under Uncertainty." *Water Resources Research*, 9(4): 1006–1021, 1973.

Neuman, S. P. "Role of Subjective Value Judgement in Parameter Identification." In *Modeling and Simulation of Water Resources Systems*, G. C. Vansteenkiste, Ed. Amsterdam: North-Holland, 1975.

Neuman, S. P., & Yakowitz, S. "A Statistical Approach to the Inverse Problem of Aquifer Hydrology, 1, Theory." *Water Resources Research*, 15(4): 845–860, 1979.

Neuman, S. P., Fogg, G. E., & Jacobson, E. A. "A Statistical Approach to the Inverse Problem of Aquifer Hydrology, 2, Case study." *Water Resources Research*, 16(1): 33–58, 1980.

Neuman, S. P. "A Statistical Approach to the Inverse Problem of Aquifer Hydrology, 3, Improved Solution Method and Added Perspective." *Water Resources Research*, 16(2): 331–346, 1980.

Nutbrown, D. A. "Identification of Parameters in a Linear Equation of Groundwater Flow." *Water Resources Research*, 11(4): 581–588, 1975.

Pierce, A. "Unique Identification of Eigenvalues and Coefficients in a Parabolic Problem." *SIAM J. Control and Optimization*, 17(4): 494–499, 1979.

Pinder, G. F., Bredehoeft, J. D., & Cooper, H. H., Jr. "Determination of Aquifer Diffusivity from Aquifer Response to Fluctuation in River Stage." *Water Resources Research*, 5(4): 850–855, 1969.

Ponzini, G., & Lozej, A. "Identification of Aquifer Transmissivities: The Comparison Model Method." *Water Resources Research*, 18(3): 597–622, 1982.

Rowe, P. P. "An Equation for Estimating Transmissivity and Coefficient of Storage from River Level Fluctuation." *J. Geophysical Research*, 65(10): 3419–3424, 1960.

Sadeghipour, J., & Yeh, W. W-G. "Parameter Identification of Groundwater Aquifer Models: A Generalized Least Squares Approach." *Water Resources Research*, 20(7): 971–979, 1984.

Sagar, B., Yakowitz, S., & Duckstein, L. "A Direct Method for the Identification of the Parameters of Dynamic Nonhomogeneous Aquifers." *Water Resources Research*, 11(4): 563–570, 1975.

Shah, P. C., Gavalas, G. R., & Seinfeld, J. H. "Error Analysis in History Matching: The Optimum Level of Parameterization." *Soc. Pet. Eng. J.*, 18(3): 219–228, 1978.

Slater, G. E., & Durrer, E. J. "Adjustment of Reservoir Simulation Models to Match Field Performance." *Soc. Pet. Eng. J.*, 11(3): 295–305, 1971.

Smith, P. J., & Piper, B. S. "A Non-Linear Optimization Method for the Estimation of Aquifer Parameters." *J. Hydrology*, 39: 255–271, 1978.

Stallman, R. W. "Numerical Analysis of Regional Water Levels to Define Aquifer Hydrology." *Transactions, Am. Geophysical Union*, 37(4): 451–460, 1956.

Sun, N. Z., & Yeh, W. W-G. "Identification of Parameter Structure in Groundwater Inverse Problem." *Water Resources Research*, 21(6): 869–883, 1985.

Sykes, J. F., Wilson, J. L., & Andrews, R. W. "Sensitivity Analysis for Steady State Groundwater Flow Using Adjoint Operators." *Water Resources Research*, 21(3): 359–371, 1985.

Tang, D. H., & Pinder, G. F. "A Direct Solution to the Inverse Problem in Groundwater Flow." *Advances in Water Resources*, 2(2): 97–99, 1979.

Theil, H. "On the Use of Incomplete Prior Information in Regression Analysis." *Am. Stat. Assoc. J.*, 58(302): 401–414, 1963.

Theil, H. *Principles of Econometrics*. New York: John Wiley, 1971.

Theis, C. V. "The Relation Between the Lowering of the Piezometric Surface and the Rate and Duration of Discharge of a Well Using Groundwater Storage." *Trans. Amer. Geophys. Union*, 16: 519–524, 1935.

Thomas, L. K., Hellums, L. J., & Reheis, G. M. "A Nonlinear Automatic History Matching Technique for Reservoir Simulation Models." *Soc. Pet. Eng. J.*, 12(6): 508–514, 1972.

Vermuri, V., & Karplus, W. J. "Identification of Nonlinear Parameters of Groundwater Basin by Hybrid Computation." *Water Resources Research*, 5(1): 172–185, 1969.

Vermuri, V., Dracup, J. A., Erdmann, R. C., & Vermuri, N. "Sensitivity Analysis Method of System Identification and Its Potential in Hydrologic Research." *Water Resources Research*, 5(2): 341–349, 1969.

Wasserman, M. L., Emanuel, A. S., & Seinfeld, J. H. "Practical Applications of Optimal-Control Theory to History-Matching Multiphase Simulator Models." *Soc. Pet. Eng. J.*, 15(4): 347–355, 1975.

Wilson, J. L., & Dettinger, M. "State Versus Transient Parameter Estimation in Groundwater Systems." Paper presented at Specialty Conference on Verification of Mathematical and Physical Models in Hydraulic Engineering, Amer. Soc. of Civil Eng., Univ. of Md., College Park, Aug. 9–11, 1978.

Wilson, J., Kitanidis, P., & Dettinger, M. "State and Parameter Estimation in Groundwater Models." Paper presented at the Chapman Conference on Application of Kalman Filter to Hydrology, Hydraulics, and Water Resources, AGU, Pittsburgh PA, 1978.

Wismer, D. A., Perrine, R. L., & Haimes, Y. Y. "Modeling and Identification of Aquifer Systems of High Dimension." *Automatica*, 6: 77–86, 1970.

Yakowitz, S., & Duckstein, L. "Instability in Aquifer Identification: Theory and Case Studies." *Water Resources Research*, 16(6): 1045–1064, 1980.

Yakowitz, S., & Noren, P. "On the Identification of Inhomogeneous Parameters in Dynamic Linear Partial Differential Equations." *J. Math. Anal. Appl.*, 53: 521–538, 1976.

Yeh, W. W-G., & Tauxe, G. W. "A Proposed Technique for Identification of Unconfined Aquifer Parameters." *J. Hydrology*, 12: 117–128, 1971.

Yeh, W. W-G., & Tauxe, G. W. "Quasilinearization and the Identification of Aquifer Parameters." *Water Resources Research*, 7(2): 375–381, 1971.

Yeh, W. W-G., & Tauxe, G. W. "Optimal Identification of Aquifer Diffusivity Using Quasilinearization." *Water Resources Research*, 7(4): 955–962, 1971.

Yeh, W. W-G. "Aquifer Parameter Identification." *J. Hydraulic Div. Am. Soc. Civ. Eng.*, 101(HY9): 1197–1209, 1975a.

Yeh, W. W-G. "Optimal Identification of Parameters in an Inhomogeneous Medium with Quadratic Programming." *Soc. Pet. Eng. J.* 15(5): 371–375, 1975b.

Yeh, W. W-G., & Yoon, Y. S. "A Systematic Optimization Procedure for the Identification of Inhomogeneous Aquifer Parameters." *Advances in Groundwater Hydrology*, Z. A.

Saleen, Ed. Minneapolis MN: American Water Resources Association, 1976, 72–82.

Yeh, W. W-G., & Yoon, Y. S. "Parameter Identification with Optimum Dimension in Parameterization." *Water Resources Research*, 17(3): 664–672, 1981.

Yeh, W. W-G., Yoon, Y. S., & Lee, K. S. "Aquifer Parameter Identification with Kriging and Optimum Parameterization." *Water Resources Research*, 19(1): 225–233, 1983.

Yeh, W. W-G., & Sun, N. Z. "An Extended Identifiability in Aquifer Parameter Identification and Optimal Pumping Test Design." *Water Resources Research*, 20(12): 1837–1847, 1984.

Yeh, W. W-G. "Review of Parameter Identification Procedures in Groundwater Hydrology: The Inverse Problem." *Water Resources Research*, 22(1): 95–108, 1986.

Yoon, Y. S., & Yeh, W. W-G. "Parameter Identification in an Inhomogeneous Medium with the Finite-Element Method." *Soc. Pet. Eng. J.*, 217–226, 1976.

Yziquel, A., & Bernard, J. C. "Automatic Computing of a Transmissivity Distribution Using Only Piezometric Heads." Proc. Int. Conf. Finite Element in Water Resources, 2nd (C. A. Brebbia, W. G. Gray, & G. F. Pinder, Eds.). London: Pentech, 1978, 1.157–1.185.

INDEX